Lecture Notes in Mobility

Series Editor

Gereon Meyer, VDI/VDE Innovation und Technik GmbH, Berlin, Germany

More information about this series at http://www.springer.com/series/11573

Marcello Contestabile · Gil Tal ·
Thomas Turrentine
Editors

Who's Driving Electric Cars

Understanding Consumer Adoption and Use of Plug-in Electric Cars

Springer

Editors
Marcello Contestabile
Centre for Environmental Policy
Imperial College London
London, UK

King Abdullah Petroleum Studies
and Research Center
Riyadh, Saudi Arabia

Thomas Turrentine
Institute of Transportation Studies
University of California, Davis
Davis, CA, USA

Gil Tal
Institute of Transportation Studies
University of California, Davis
Davis, CA, USA

ISSN 2196-5544 ISSN 2196-5552 (electronic)
Lecture Notes in Mobility
ISBN 978-3-030-38381-7 ISBN 978-3-030-38382-4 (eBook)
https://doi.org/10.1007/978-3-030-38382-4

This Springer imprint is published by the registered company Springer Nature Switzerland AG
The registered company address is: Gewerbestrasse 11, 6330 Cham, Switzerland

Foreword

This book is timely—and needed. Technology will continue to improve, and battery costs will continue to drop. Engineers are making impressive progress. What remains most uncertain is consumer behavior and policy. Which consumers in which situations will buy which types of electric vehicles—battery and plug-in hybrid vehicles, as well as fuel cell electric vehicles? And what might or could policymakers do to accelerate sales—for small and large electric vehicles? The future of electric vehicles is firmly linked to policy and behavior.

As a researcher and policymaker, I closely follow electric vehicle happenings in China, Europe, and Japan and advise many car companies. For a variety of economic, environmental, and political reasons, it is crystal clear to me that electric vehicles will replace virtually all combustion engine cars and probably most trucks. It is a question of when, not if. EV market shares are increasing everywhere, automakers are rolling out an expanding array of electric models, supply chains are in place, and governments in almost all major car markets have adopted aggressive policies.

Within this sweeping transition are many hugely critical details, though. What types of vehicles at what prices will be most compelling to buyers? How important are range anxiety and public availability of electric chargers? How do consumers trade off less cost for greater range? We are all different, of course. How do these behaviors and attitudes vary across young and old, urban and rural, and male and female, as well as across different personality types? Should automakers emphasize pure battery EVs, or plug-in hybrid EVs—or even fuel cell EVs? And how should policymakers respond? What types of incentives might they offer and when? And how much should they invest in public charging? Where should chargers be placed and should they be fast chargers? None of us know exactly how all this will play out. But the more we learn and understand, the smarter our policy and investment decisions will be.

What we do know, as indicated by many authors in this book, is that the transition will be slow, unless the following happens.

Challenge #1: Battery (and hydrogen and fuel cell) costs must continue to fall. Today's EVs are still considerably more expensive to produce than gasoline cars, even with plummeting battery costs—especially so in the case of larger SUVs and pickup trucks. Automakers price today's EVs below cost to sell them, satisfy government mandates, and help meet aggressive fuel efficiency and greenhouse gas emission standards.

Challenge #2: Political will. Automakers are making massive investments in EVs. But apart from Tesla and some Chinese manufacturers, the large automakers are mostly producing just enough to satisfy government policies and build initial supply chains. Politicians and regulators need to boost a whole range of policies—fuel economy and greenhouse gas emission regulations, consumer purchase incentives, more vehicle chargers—to signal to both automakers and consumers that the nations are resolutely committed to electrifying most or all vehicles.

Challenge #3: Consumer embrace of EVs. Consumers are cautious about purchasing expensive, new products. Even when EV prices are lower than for comparable gasoline cars—as sometimes occurs because of regulatory demands—sales can be slow. The authors of this book provide state-of-the-art insights into consumer behavior and much useful guidance and insight about which policies are most effective.

How exactly the transition to the battery, plug-in hybrid, and fuel cell electric vehicles will unfold is unknown and unknowable. It will vary greatly from one region to another. Some strategies and paths are more expensive and complicated than others. But generally, the best policy approach is to provide substantial incentives to consumers initially, assure that a network of public chargers and hydrogen stations is prominently accessible, impose regulations and policies on automakers to motivate investments and sales, and be adaptive and agile as consumer preferences, technological innovations, and industry investments evolve.

Authors of this book provide a solid understanding of consumers needed to inform and design policies. While useful insights can be gained from previous experiences with other technologies such as renewable energy, it is essential to recognize that the EV market is highly heterogeneous, that early adoption is not necessarily a good guide for what will follow, and that market success will rely on policies that are responsive to the complexities of consumer behavior.

The authors of this book test their insights by scanning and evaluating experiences around the world. They lay the foundation for crafting well-designed measures that are adaptive and durable enough to bridge the transition to a sustainable market.

Davis, CA, USA Daniel Sperling
October 2019

Daniel Sperling is the author of *Three Revolutions: Steering Automated, Shared, and Electric Vehicles to a Better Future* (2018). He is also Founding Director of the Institute of Transportation Studies and Blue Planet Prize Professor of Engineering and Environmental Science and Policy at the University of California, Davis. He is a member of the California Air Resources Board.

Acknowledgements

We gratefully acknowledge the support of the King Abdullah Petroleum Studies and Research Center (KAPSARC) through its sponsorship of the Hybrid and Electric Vehicle Technology Collaboration Programme (HEV-TCP) of the International Energy Agency. As a sponsor of the HEV-TCP, KAPSARC initiated and led the Task 36 "Consumer Adoption and Use of EVs" of which this book is the main deliverable.

We also acknowledge the kind support of the HEV-TCP members and secretariat to Task 36 and this book.

Contents

Introduction: Understanding the Development of the Market for Electric Vehicles

Marcello Contestabile and Thomas Turrentine

Abstract Many books have been written about plug-in electric vehicle (PEV) technology. However passenger cars are consumer products and ultimately the market success of PEVs requires individual consumers to decide to adopt them. The decision processes behind buying and using PEVs are not yet well understood by the governments that support their adoption and the automotive industry that manufactures them. The aim of this book is therefore to provide stakeholders with an overview of the international evidence currently available as well as the necessary tools to navigate it. By comparing and contrasting the country-specific studies presented in the book, further insights can be derived, some questions answered and new questions asked. This introductory chapter discusses the motivation, intended audience, scope and contents of the book.

Keywords PEV adoption · PEV use · Consumer research · PEV markets

Why This Book

Many books have been written about plug-in electric vehicle (PEV) technology. However passenger cars are consumer products and ultimately the market success of PEVs requires individual consumers to decide to adopt them. Buying a new passenger car often involves a substantial financial commitment, typically only second to that of buying a home. At the same time, consumers choosing to buy a particular car model do not base their decision purely on the practical utility that they expect to derive from it, but crucially also on what that particular model symbolises to them. The decision to buy a PEV is even more complex as these vehicles combine a symbolic meaning

M. Contestabile (✉)
Centre for Environmental Policy, Imperial College London, London, UK
e-mail: marcello.contestabile@imperial.ac.uk

King Abdullah Petroleum Studies and Research Centre (KAPSARC), Riyadh, Saudi Arabia

T. Turrentine
Institute of Transportation Studies, University of California Davis Institute of Transportation Studies, Davis, CA, USA
e-mail: tturrentine@ucdavis.edu

M. Contestabile et al. (eds.), *Who's Driving Electric Cars*, Lecture Notes in Mobility,
https://doi.org/10.1007/978-3-030-38382-4_1

that is still emerging and evolving with the need for behavioural adaptations such as dealing with range limitations. This explains why at present the decision process of buying PEVs is not yet well understood by the very governments that support their adoption through sets of policy measures, and neither it is by the automotive industry that manufactures them. The aim of this book is to address this important gap and provide stakeholders with an overview of the international evidence currently available on consumer adoption and use of PEVs, as well as the necessary tools to navigate it.

Over the last few years the market for PEVs has truly taken off. Driven by strong government incentives, since 2010 the global PEV market has been growing at a rate of at least 40% per year. According to figures from the 2019 edition of the Global EV Outlook published by the International Energy Agency, in 2018 the global PEV fleet has exceeded 5.1 million cars, up two million from 2017 and with new PEV car registrations having almost doubled in a year. In addition to incentivizing adoption, governments continue to support PEV Research & Development (R&D) and increasingly also manufacturing, all of which has led to substantial technological learning and scale economies. In particular, lithium ion batteries for PEVs have become more reliable, their energy density has improved and their price fallen faster than many experts had anticipated. As a result, some experts are now predicting that PEVs will become cost-competitive in the coming decade. Moreover, industry is increasingly getting behind PEVs, with electricity providers becoming more interested in the flexible load that these vehicles can offer to manage intermittent renewable power and automotive original equipment manufacturers (OEMs) becoming committed to a transition from internal combustion engines to electric powertrains. This gives great hope that PEVs, together with greening of the grid, will provide a two-punch solution to reduce urban air pollution and meet radical carbon reduction goals.

However, the transition is still in its early phases and substantial obstacles lay ahead, overcoming which will require adopting suitable strategies. Strategies that are both effective and efficient, i.e. leading to the desired outcome rapidly and at an acceptable cost to society, need to be informed by the best available evidence. Consumers are, together with industry and government, one of the key actor groups in the transition to electric mobility, therefore understanding their response to technology and policy is essential. Knowledge in PEV technology is growing rapidly, thanks to the substantial collaborative R&D programmes currently in place that increasingly include support to manufacturing scale-up and the development of regional supply chains. Initiatives exist to share best practices in PEV public policies internationally. Consumers however have so far received comparatively little attention. The aim of this book is therefore clear: to gather the latest evidence on PEV consumers and their decisions to adopt and use PEVs internationally, and make it accessible to a wide range of stakeholders. The book supports the transition to electric mobility in the following ways: by helping policymakers and industry navigate the evidence available today on PEV consumers so that they can act on it now, and by identifying areas where future PEV consumer research will lead to more evidence being generated that both policymakers and industry can use to adapt and refine their strategies as the transition progresses.

The timing of this book is also not coincidental. The book, which was conceived as part of Task 36 "Consumer Adoption and Use of EVs" of the International Energy Agency's Hybrid and Electric Vehicle Technology Collaboration Programme (IEA HEV-TCP), exploits the great opportunity provided by the availability for the first time of a substantial amount of PEV consumer data. Until recently, the understanding of consumer decisions to adopt and use PEVs could only be based on the study of participants to small-scale PEV field trials and of buyers and users of conventional internal combustion engine vehicles. Now that PEVs have entered the everyday lives of millions of car users and that a growing number of PEV models are being offered on the market by automotive OEMs, this has radically changed. Important insights can now be generated directly from the study of actual PEV buyers and users, as well as rejecters, and this can inform governments seeking to continue to support the transition towards electric mobility as well as automotive OEMs developing new PEV models. However, this remains a specialist field and the evidence that is emerging from different markets, under different conditions and from studies that use different methods is difficult to navigate for the non-experts. This is particularly important as the stakeholders that need to make important decisions today are at risk of not taking into account the best available evidence or misinterpreting it. Of course, the body of evidence is already quite vast and is growing and being updated rapidly, therefore covering it all in one book would not be possible. What we aim to offer though is an accessible yet comprehensive picture of the state of the art in this important area of research that can benefit a broad range of stakeholders.

It is also important to stress that with this book we do not aim to arrive at final conclusions on the role of PEV consumers in the transition process and how their behaviour should inform policy and technology choices. PEV technology, policy and consumers are co-evolving and much will still change in future that we cannot foresee. Instead, we want to mark the start of a journey into PEV consumer behaviour for all those stakeholders who are actively engaged in the transition to electric mobility and need to develop a better understanding of all factors driving the market. After all, PEV technology has now firmly entered the common language of the non-experts, who can often be heard talking about new battery chemistries for example, while not long ago this was the domain of battery electrochemists and industry technologists only. However we find that, by comparison, the general understanding of PEV markets and consumers lags well behind that of its technology. Therefore, our aim is to promote the diffusion of this knowledge among relevant stakeholders, which we believe to be essential to developing the effective and efficient strategies that the transition to electric mobility requires. Undoubtedly this is just the beginning and we hope that this book will also help open the way to more continued efforts in this field.

Who It Is for

In the spirit of the IEA's TCPs, the book gathers contributions of leading experts studying PEV adoption and use from countries that are members of the HEV TCP and beyond, and aims to share the knowledge across countries and stakeholder groups outside of the expert community. Therefore, as part of Task 36 "Consumer Adoption and Use of EVs" of the HEV-TCP we have first built an international team of experts around this initiative, making sure that all main PEV markets were covered. We have then asked the experts to share their insights in a form that allowed non-experts with varying degrees of domain knowledge to navigate some of the complexity of this specialist field and use its growing body of evidence to inform their strategies both present and future. In order for the book to be useful and accessible to a non-expert audience, the main emphasis is on presenting the latest available evidence and its most significant policy implications. However an appreciation of the theory and methods used to generate such evidence is also essential for the reader to understand the strengths and limitations of the studies conducted and critically interpret their results. Hence the book also offers an overview, albeit non-systematic, of the main theories and methods that are used in this field of research. Here we have endeavoured to strike a difficult balance between keeping the subject accessible to the non-experts while being rigorous and providing the interested reader with the necessary concepts and references for further study. Lastly, the book also outlines directions in which the field is evolving and new insights that may be generated in future as a result, which we also deem of particular importance to our intended audience.

As previously mentioned the target audience of this book is broad and includes policymakers, technologists and researchers in the areas of energy and sustainable transport. In particular, the main stakeholders groups the book intends to cater for are the following:

- The International Energy Agency and its network of organisations and initiatives
- Policymakers in countries that are supporting PEV technology and market development, as well as in countries that are preparing to do so
- Governmental and non-governmental organizations that produce energy and technology outlooks and that support sustainable transport research and technology deployment
- Relevant industry players, particularly automotive OEMs, oil & gas companies, energy utilities, and PEV charging infrastructure providers
- The research community at large, also including early career researchers and postgraduate students with an interest in this field.

What It Covers

The content of this book reflects the fact that most PEV sales so far are concentrated in few "beachhead" markets in North America, Europe and Asia. North America is arguably where it all began, with the introduction in 1990 by the State of California of the Zero Emission Vehicle (ZEV) mandate, a ground-breaking piece of legislation that has played a key role in spurring PEV innovation and market development internationally. ZEV mandates have since been adopted by other States as well as across the Canadian border, and PEV markets in North America are developing accordingly. Europe is at the forefront of low emission transport policy also thanks to its very stringent fuel economy standard, which is becoming an increasingly strong driver to the electrification of passenger cars. Moreover, some European countries have long supported PEV adoption through strong incentives and this, combined with particularly favourable market conditions, has resulted in PEVs reaching very high shares of their passenger car markets. Norway in particular stands out as the world leader in terms of PEV market share by some distance, followed by Sweden and the Netherlands, and larger European passenger car markets such as the UK and Germany have more recently seen a strong acceleration in PEV sales too. Lastly, Asian PEV markets are also extremely important and China in particular deserves our attention as its government's aggressive industrial policy attempts to transform the world's biggest car market, clean up China's notorious dirty air, as well as make China a technological leader in batteries and electric cars. This has resulted in China having a PEV market share higher than those of the UK and Germany and, due to the size of the country, the single largest PEV stock in the world.

What all these "beachheads" have in common are policy programs designed to encourage sales and use of PEVs. However, each beachhead is an experiment, with distinct markets and sets of policies. This book is focused on the evidence emerging from the study of actual adopters from these markets and policy experiments. The authors are all transportation scientists, each studying the transition to PEVs in their own market, and regularly comparing notes with one another in order to derive more general lessons on market development, optimal policy, and infrastructure. In particular this set of scientists has been collecting detailed data during these early years of consumer purchase and use behaviour, and have analysed them using a suite of methods. There has been a revolution in transport data collection through GPS location and other powertrain data that has allowed the authors to log detailed driving and energy data from these early PEV owners, along with survey and other data collected by the state and energy companies. Meanwhile, analytical methods have also evolved and all of this has resulted in several policy-relevant insights which in this book are compared and contrasted in order to derive high-level messages for policymakers and other stakeholders.

PEV consumer research uses a range of qualitative and quantitative methods, and often mixed-method approaches combining both. Each individual method has its strengths, limitations and complementarities with other methods, all of which is not obvious to the non-expert and yet it is crucial for the correct design of the research and use of its results for policy-making purposes. While the emphasis of this book is on the evidence and not the methods—a systematic discussion of the latter is beyond its scope—contributions to the book have also been selected with a view to covering all main methods used in PEV consumer research, and the contributing authors have been asked to provide high-level critical discussions of the methods they have used as well as the results they have arrived at. Detailed discussions of individual methods can be found in more specialized publications, such as scientific journal articles and monographs, however this book allows the non-expert reader to appreciate the breadth of methods and data currently available to the researchers, see the main methods in action and understand how different approaches can be used for different problems. The reader will also be able to appreciate the future opportunities that access to more extensive data and the further evolution of methods promises to offer.

The book is divided into four parts. Part one, two, and three cover North America, Europe and China respectively, and constitute the core of the book. Each chapter takes a close look at the transition to electric mobility from the angle of a different national or state policy experiment and addresses different questions. In particular:

- Gil Tal and co-workers at the Plug-In Hybrid and Electric Vehicle Research Center begin their chapter by providing us with an overview of the historical development of PEV policy in California. This is a very fitting start to the book as California is where the journey of the modern PEV began. In the remainder of the chapter, they go on to provide us with selected insights from the many research projects on PEV adoption and use that they have recently conducted, particularly in relation to the changing demographics of PEV buyers, the impact of incentives on the development of the PEV market, and charging behaviour of different user segments and how this may change as a result of different electricity pricing strategies and PEV range
- Ken Kurani puts the spotlight on an important problem that is somewhat overlooked by policymakers, i.e.: the need for mainstream consumers to become aware and develop a positive attitude towards PEVs before they can adopt this technology. Ken measures changes in awareness and perception of PEVs in mainstream passenger car buyers in California between 2014 and 2017, a period of rapid market growth. He also studies the extent to which direct interactions between PEV owners and owners of non-PEVs can facilitate increased awareness and a more positive attitude towards PEV technology
- Tamara Sheldon and Rubal Dua explore the impact of PEV adoption on gasoline consumption in the US by modelling the additional conventional internal combustion engine vehicles that would have been on the road had the PEVs not been purchased. They challenge the general assumption on which counterfactual scenarios are based, i.e.: that PEVs replace conventional internal combustion engine vehicles (ICEVs) of the same segment in a like-for-like fashion. They develop an improved counterfactual scenario using a large commercial dataset of new vehicle purchases and a discrete choice model which allows them to arrive at more realistic results in terms of actual gasoline savings yielded by the adoption of PEVs
- Jonn Axsen and Zoe Long combine qualitative interviews, quantitative surveys and choice models in a mixed-method approach to the study of the Canadian PEV market. The results of this mixed-methods study are also fed into a technology adoption model that they have developed, which allows exploring the interplay between consumers, technology suppliers and policies. By running model simulations, the authors are able to explain the influence of constraints such as vehicle model offer and access to charging infrastructure on the currently limited market uptake of PEVs in Canada

- Erik Figenbaum discusses the evolution of the market in Norway, a 20-year experiment in policy. Norway has bragging rights to the highest market penetration of PEVs, which is well above 40% of total new passenger car sales, and hopes to eliminate all combustion vehicle sales by 2025. It owes this success to a web of policy incentives that were already in place when PEVs began an earnest technical comeback in 2007 and to a committed national government that provided strong and stable support to PEVs since 1990. Erik has been able to document almost every facet of this process, and relies on nation-wide survey data for the study of PEV adoption and use by Norwegian households, and how these are evolving over time
- Rick Wolbertus and Robert van den Hoed document the evolution of the PEV market in the Netherlands as a result of changing policy incentives. The Dutch PEV market was second only to Norway between 2012 and 2017, however a sharp reduction in government incentives caused sales to drop substantially. This particularly applies to Plug-in Hybrid Electric Vehicles (PHEVs) which had dominated early PEV sales in the Netherlands. Rick and Robert illustrate the strong correlation between the structure of government incentives and PHEV sales. They also address the hotly debated topic of the extent to which PHEVs are driven on electricity in the Netherlands; to do so they analyse a large dataset of PHEV charging sessions which they have collected
- Frances Sprei covers another very important PEV market in Europe: Sweden. Frances studies the suitability of Battery Electric Vehicles (BEVs) with their range limitations to the Swedish market by focusing on driving patterns as opposed to attitudes. In her study she relies on GPS data gathered as part of experiments in which the participants, samples of two-car households, trade in one of their petrol vehicles for a BEV for long periods of time. The data allows to observe how the two vehicle households adapt their driving patterns to the BEVs and also to estimate the total cost of ownership (TCO) of the vehicles based on real life utilization. One of the experiments also combines a survey in order to better understand perceptions and attitudes before and after the experiment
- Stephen Skippon and George Beard have been studying the PEV market in the UK since it started to develop. Having also tackled early adopters, here they focus on characterizing future mainstream adopters in the UK by studying current users of ICEVs. The study is mainly based on data from large-scale surveys of passenger car users in the UK, complemented by PEV use experiments on mainstream users. Stephen and George particularly discuss methodological approaches to controlling for biases when using data from current mainstream ICEV users who have no prior experience of PEVs
- Patrick Plötz and Elisabeth Dütschke discuss PEV adoption and use in Germany. The German market has only recently started to grow rapidly, also as a response to the diesel emissions scandal which has prompted the government to increase its support to PEV adoption. Patrick and Elisabeth have studied it extensively and present here an overview of their work as well as that of others. They analyse PEV adoption in Germany by user segment, relative sales of BEVs and PHEVs, distribution of adopters by city size and more. They also characterise electric

kilometres driven by BEVs and PHEVs in Germany and assess drivers and barriers to future adoption

- Shawn Ou, Zhenhong Lin and co-workers ask the question whether the Chinese PEV market will continue to grow as fast as it has done in recent years. They first provide a detailed overview of Chinese PEV policy and how it has evolved since year 2001, and illustrate how this has shaped the national PEV market and industry. Then they discuss a model that they have developed which simulates the evolution of the passenger car stock in China out to 2020. Using the best available market data as model input, Shawn and Zhenhong explore how changing policy support will influence the evolution of the Chinese market. Lastly, they take a closer look at Chinese PEV users and, drawing on existing survey studies, estimate the economic cost to Chinese PEV users of range anxiety as well as of limited availability of parking and charging infrastructure
- Yan Xing, Gil Tal and Yunshi Wang further deepen the study of PEV adopters in China. They use both qualitative and quantitative methods to perform a cross-sectional study of the PEV market in six Chinese cities, also including second and third tier cities which have so far received limited attention. They have conducted a large scale survey—the largest of its kind so far—and use the data so generated to explore the influence of policy incentives, individual factors including socio-demographics and attitudes, driving and charging behaviours on PEV adoption. They compare and contrast the differences across the six cities, with a view to understanding how PEV sales can be maintained at current levels in large cities, and substantially increased in smaller ones where limited adoption has occurred so far, while incentives are being phased out.

The book concludes with a reflection on the main lessons learned from the study of PEV adoption and use in markets so diverse, with different socio-economic and climate conditions, and different sets of policy support measures. Who has already adopted PEVs and what is keeping others from adopting? How long does the range of a BEV need to be? How important is public charging infrastructure? Are PHEVs important for achieving high levels of PEV penetration? To what extent are PHEVs really driven on electricity? How easily can leading PEV markets be replicated? And, how long do governments need to incentivise PEVs? The research conducted so far sheds some light on these questions and raises several others. The conclusions drawn from the studies presented in this book are encouraging. PEV owners are for the most part enthusiastic about their vehicles and electricity. They are responding positively to the continual improvements in the technology they have embraced. However, with the exception of Norway, it is still early in the other markets and there are many more consumers with fewer resources and with more conservative attitudes to change. As the price of batteries drops, governments and car companies will face greater challenges to keep up with infrastructure demands, less "innovative" car segments and the ups and downs of policy. Observers like to note that the transition to electric mobility is a marathon, not a sprint. This book is about the first few miles of that marathon.

North America

Electric Cars in California: Policy and Behavior Perspectives

Gil Tal, Ken Kurani, Alan Jenn, Debapriya Chakraborty, Scott Hardman and Dahlia Garas

Abstract This chapter is divided into two parts: in the first part we illustrate the evolution of PEV regulation and technology in California, which has both led and influenced PEV policy globally. In the second part we present selected results of recent research conducted at UC Davis, particularly looking at the profiles of current PEV adopters in California and how these have evolved in recent years, the impact of financial and non-financial incentives on adoption of PEVs, charging behavior as a function of dwelling and home charger type, travel patterns, electricity price and PEV range.

Keywords Zero emission vehicles (ZEV) · Plug-in electric vehicle (PEV) · Battery electric vehicles (BEV)

The Evolution of PEV Policy in California and the US

The Genesis of the ZEV Mandate

Developing a sustainable plug-in electric vehicle (PEV) industry is a complex, long term, multi-sectorial undertaking, involving numerous actors in industry and government. It is unlikely that a PEV industry would develop without a coordinated government and industry effort, given the overall maturity of the conventional internal combustion engine vehicle market and the many hurdles PEVs face in meeting costs, reliability, and market volume. This chapter will first provide an overview of the historic development of PEV policy and technology in California, and then will discuss PEV adoption and use research in California.

PEVs were a significant portion of the vehicle market early in the 20th century, prior to widespread availability of gasoline, the introduction of electric starter motors for gasoline powered vehicles, and good road systems that encouraged longer trips away from cities. PEVs used lead acid batteries, were mainly found in dense urban

G. Tal (✉) · K. Kurani · A. Jenn · D. Chakraborty · S. Hardman · D. Garas
University of California, Davis, USA
e-mail: gtal@ucdavis.edu

M. Contestabile et al. (eds.), *Who's Driving Electric Cars*, Lecture Notes in Mobility,
https://doi.org/10.1007/978-3-030-38382-4_2

areas, and were the preferred means of travel for wealthy women, who could have vehicles charged and delivered by PEV service companies. Gasoline vehicles were exceedingly dirty, noisy, and injuries were common when starting them with the engine crank. By the 1930s, however, PEVs disappeared from the market in the United States.

While a few automotive OEMs conducted limited research on PEVs, and particularly Battery Electric Vehicles (BEVs) in the 1950s and 60s, it was not until the oil crisis and worsening air quality, primarily in California, and particularly in the Los Angeles basin in the 1970s, that American car users and automotive OEMs began to reconsider PEVs. However, batteries had not progressed much, and energy densities were not sufficient for modern vehicles, which weighed over two thousand pounds on average and seated five people. Most battery powered vehicle experiments had to focus on small vehicles, with low speed DC drive systems. At this point, researchers in the Department of Energy in the US began to experiment in earnest in electric drives.

Los Angeles air quality in the 1960s and 1970s had deteriorated so much that smog alerts were becoming a regular feature of life and the snow-capped mountains and beautiful inland valleys were clouded in a semi-permanent gray-brown haze—or smog—only occasionally clearing. The problems with air quality and associated lung diseases prompted the development of the Clean Air Act, signed by President Richard Nixon, whose hometown of Wittier had some of the dirtiest air in Los Angeles. However, the problems of Los Angeles were comparably worse and intractable compared to most other cities, given the trapped air (inversions) in the region and its large volume of vehicles.

To deal with its special problems—as not only Los Angeles but some other parts of California have similar mountain bounded air basin issues—California was given special authority under the Clean Air Act to regulate air pollution, particularly from automobiles. Governor Ronald Reagan formed the California Air Resources Board (CARB) to measure, regulate and control emissions. In fact, over the next decade California would tighten car pollution regulations to the point it became embattled with car companies and the federal government. Even with great progress in emission controls, regulators and scientists at CARB determined in the 1980s that to finally clean up air in California's cities would require cars to have zero emissions. It was at this time that some maverick engineers would show regulators a very advanced electric vehicle that would encourage them to move ahead with a zero emission vehicle (ZEV) policy goal.

A central moment and event in the rebirth of electric mobility in the United States was the development of a lightweight, extremely aerodynamic, powerful two seat, AC motor driven vehicle called the Impact. It was developed at an advanced, largely autonomous division of GM in Los Angeles, under the guidance of Paul MacCready, an aeronautical inventor, and a team of electrical engineers. Paul MacCready was a visionary who was famous for building and flying a human powered aircraft over the English Channel. He assembled a team of electrical engineering experts who created a fast, sleek, extremely lightweight and aerodynamic battery powered vehicle. Additionally, this vehicle introduced the use of computers in controls of the battery systems and motors. A team of air quality regulators from CARB were introduced to the Impact in 1990 and Roger Smith, the CEO of General Motors declared the Impact "the future" on Earth Day 1990. The surprising power and performance of this vehicle encouraged California regulators at CARB to develop the ZEV mandate in 1990. The ZEV mandate required the six major vehicle manufacturers to sell 2% of their vehicles as ZEVs in 1998, increasing to 5% in 2001 (Collantes 2006). The six vehicle manufacturers were those with sales of over 75,000 per year, and included GM, Ford, Toyota, Honda, Chrysler and Nissan. This law stunned automakers, which were already embattled with regulators in California over the low emissions requirements.

The only powertrain with the possibility to meet the requirement of the ZEV mandate at the time was battery electric. However the regulation would also encourage the development of fuel cell, hybrid and eventually plug-in hybrid powertrains. This was done in a quest to find alternatives to the costly and range limited BEVs, which automakers thought would have very few buyers. The ZEV mandate had a multi-year build-up in which automakers would develop prototypes. CARB would hold hearings about whether such vehicles would sell, whether the technology was ready, and if cost goals (particularly for batteries) could be met. Initially much of the focus was on batteries, whether they could last the life of a vehicle, could store enough energy to power the vehicle for required speeds and distances, and could be cost effective. In 1996, CARB realized that the vehicle technology was not ready for market in 1998, and worked to develop a compromise with automakers.

As a result of the 1996 hearings and of discussions held in private meetings, the Air Resources Board postponed the ZEV mandate, and decided to conduct a market and vehicle demonstration, a suggestion Toyota had made in 1990. A Memorandum of Agreement was developed. With small numbers of pre-market vehicles, including an update of the GM Impact, called the EV1, and other vehicles from Chrysler, Toyota, Honda, Ford, and Nissan, automakers developed small pre-commercial vehicle programs. There were no charger standards, and unfortunately automakers split between two types of charger systems, Toyota and GM using inductive paddles and Ford, Chrysler and Honda using conductive plugs. This resulted in the "charger war", which was not resolved until the Society for Automotive Engineers (SAE) rules on J1772 in 2009.

During this period, a new battery technology, nickel metal hydride (NiMH), was scaled up for large format applications at the Ovonic Battery Company. This battery was partly the result of the Department of Energy Battery Research Program and was installed in a number of GM EV1s. NiMH would go on to be the battery chemistry used in the successful Prius hybrid platform as well as all other gasoline-electric hybrids that became available on the market. Lithium-ion chemistries were becoming popular in consumer electronics but at the time Nissan was the only maker to use a lithium-ion battery in a vehicle, the Altra.

Development of Plug-in Hybrid Vehicles

In the post ZEV mandate era, a notable happening was the development of the Nissan hyper-mini, which used lithium ion batteries. 17 of these vehicles were demonstrated in California at UC Davis. This period was mostly marked by a surge in interest and development of fuel cell hybrid electric vehicles by a number of auto companies and governments, in particular GM, Honda and Daimler.

The next most notable event in the USA for electric mobility was the development of plug-in hybrid electric vehicles (PHEVs). Several OEMs, including Volvo and Volkswagen, tinkered with PHEVs and promoted these designs in the 1990s. Since the early 1990s Dr. Andy Frank at UC Davis had been developing prototype PHEVs: vehicles with an integrated electric drive train, powered from a grid-charged battery, and a gasoline motor that can both proper the vehicle and charge the battery so as to overcome range limitations. These proof-of-concept vehicles were developed for US DOE-sponsored student competitions and were parallel, power-split vehicles employing an electric motor, very small gasoline engine, two clutches, and either a manual or continuously variable transmission (CVT). They were able to achieve fuel economies nearly double the conventional vehicle, and operate in an all-electric mode for 40–60 miles.

In 2004, enthusiasts of Frank's designs were able to put a larger battery in Toyota Prius hybrid vehicles and create a practical PHEV. This vehicle demonstrated the potential of PHEVs to organizations such as Southern California Edison, the Electric Power Research Institute, and Argonne National Labs. A coordinated research project among these partners encouraged a new round of interest in grid-connected vehicles, primarily in California. In 2007, GM, following interest in this type of technology, and discouraged with BEVs, began a program to develop a new PHEV with 40 miles of all electric range, the Chevy Volt.

Actors Supporting the PEV Market

In the US, the Department of Energy is the one of the primary actors in the Federal government for promoting new transportation technologies, carrying out and supporting research, providing loans to entrepreneurs, and in supporting market development. The Environmental Protection Agency is also an important federal actor: the role of the EPA is to enforce the Clean Air Act and other Environmental Laws, many of them originating in the 1970s. The EPA in effect regulates the tailpipe emissions of vehicles, power plants and industrial facilities. The Department of Commerce is the agency which distributes tax credits in support of the PEV market.

In California, the most important actor is the California Air Resources Board, which has regulatory power over automakers, and distributes cash rebates to electric vehicle buyers. Second is the California Energy Commission which awards research grants and importantly also implementation grants for electric vehicle planning activities and infrastructure development. Third most important public actor is the California Public Utilities Commission that regulates the power industry. This agency determines the price of electricity for private homes, businesses, and public charging companies. In recent years, the Governor's Office and the legislature have become active in this arena too; notably California Governor Jerry Brown has issued an executive order for Californian agencies to advance the PEV market.

Next are the power companies. The United States has over 2000 electric power companies, some working across big regional sectors and some much smaller, serving only local and rural communities, making any action in this sector exceedingly difficult to coordinate. Moreover, automotive OEMs are not accustomed to coordinating with power companies. To overcome this, the Electric Power Research Institute does coordinate a program of cooperation between GM and over 200 utilities, both private and public. Coordination is also needed because it is common for PEV buyers to live in an area served by one company and work in an area served by a different one. Electric utilities shape power rates and are important actors in both local and regional contexts. Electric utilities in the Western States already face challenges being forced to size their power demands around air conditioning loads at peak high temperature events in the dry desert summer afternoons and evenings. Therefore, the added load from PEV charging at these times can become a problem, however it can also be an asset if adequately managed.

It is also worth noting that the Clean Air Act allows States that are out of compliance with the Clean Air Act standards for air quality to choose the California compliance program. This has led other states to adopt the California Zero Emission Vehicle regulation. The governors of 8 states—California, Connecticut, Maryland, Massachusetts, New York, Oregon, Rhode Island and Vermont—have signed a memorandum of understanding (MOU) to take specific actions to put 3.3 million ZEVs on the road in their states by 2025, along with the recharging infrastructure required to support those vehicles. ZEVs include BEVs, PHEVs, and hydrogen fuel-cell-electric vehicles (FCEV); the technologies can be applied in passenger cars, trucks and transit buses.

In addition to this, some large cities and city regions in the United States have taken actions to promote PEVs, primarily through efforts of their mayors, in cooperation with regional partners. Some notable cities in this regard are Portland (Oregon), San Francisco, Los Angeles, San Jose, Austin (Texas) and Seattle (Washington).

Policy Evolution Post ZEV Mandate

In 1996, after a two-year demonstration period during which six automobile companies leased several hundred BEVs in California, the ZEV mandate was postponed by a decade because it was determined the technology was not mature enough, the price of vehicles was too high, and customers were not ready to adopt the initial design.

However, many first buyers disagreed with the arguments, saying they loved their vehicles, and wanted ARB to continue its mandate.

Despite the postponement, the ARB continued its interest in the ZEV goals and pursued FCEVs and PHEVs in the years after the BEV lease trials. Today, the ZEV program continues as a "technology development" program. While ZEVs were originally meant to meet local air quality goals, the State of California was getting serious about climate change too. California's electricity sector was already much more efficient than the rest of the nation thanks to the decoupling of electricity prices from demand: in the 1970s California electric regulators decided that electricity prices would go down when users conserved energy. As a result, California has enjoyed much more efficient use of energy than the rest of the United States which, combined with a greater share of hydroelectric power than many other parts of the nation, meant lower GHG emissions too.

While the electricity sector in California does not have as big a share of GHG emissions as in other parts of the US, its giant automobile culture (23 million light duty vehicles) accounts for more than 40% of CO_2 emissions in California. CARB was therefore given responsibility for CO_2, emissions too, and its programs began to focus the attention on the full lifecycle emissions of vehicles, including both in-use fuel economy and the carbon content of fuels such as hydrogen and electricity.

In 2009 the California Air Resources Board developed the Advanced Clean Cars (ACC) program, which combines the control of smog-causing pollutants and GHG emissions into a single coordinated package of requirements for model years 2015 through 2025 and assures the development of environmentally superior cars that will continue to deliver the performance, utility and safety car owners have come to expect.

The Zero Emission Vehicle regulation acts as the technology-forcing piece of the ACC program, pushing manufacturers to produce ZEVs and PHEVs in the 2018 through 2025 model years. In addition, the ACC program also includes amendments to the Clean Fuels Outlet (CFO) requirements that will assure that ultra-clean fuels such as hydrogen are available to meet vehicle demands brought on by amendments to the ZEV regulation. The ZEV regulation has so far set the following commercialization requirements for automotive OEMs:

- 2012–2014: 12,500 ZEVs—58,000 Plug-in hybrid EVs
- 2015–2017: 50,000 ZEVs—83,000 Plug-in hybrid EVs.

Moreover, in March 2012 Governor Brown issued an executive order directing the State government to help accelerate the market for ZEVs in California and setting a target of 1.5 million ZEVs by 2025. The Executive Order established several steps toward meeting its goals, including specific strategies and actions that State agencies should take. For the purposes of the executive order and its action plan, ZEVs include BEVs, PHEVs and FCEVs. The action plan was the result of an interagency working group led by the Governor's Office, including several state agencies and other entities, and builds upon existing work already underway at these agencies. In addition, the action plan benefitted from extensive input through workshops and comments from outside stakeholders. The action plan is continually adjusted to meet the needs of the changing market.

Beyond 2025 the driving force for lower emissions from vehicles will be climate change. In order to meet the 2050 GHG emission reduction goal, by 2035 the new vehicle fleet will need to be primarily composed of advanced technology vehicles such as PEVs and FCVs. Accordingly, the ACC program coordinates the goals of the LEV, ZEV, and CFO programs in order to lay the foundation for commercialization and support of ultra-clean vehicles.

Selected Studies of the PEV Market in California

Who Is Buying Electric Vehicles in California?

Prior research has taken a static view using one-time surveys to study who is buying electric vehicles (Nicholas et al. 2017; Dong et al. 2014). Here we present multi-year (2012–2017) questionnaire survey data on the socio-demographic profile of 11,037 PEV adopters in California using latent class cluster analysis. This analysis aims to explore the similarities among buyer groups and reveals that PEV buyers in California can be clustered into four heterogeneous groups: 49% are *High income families*, 26% *Mid/high income old families*, 20% *Mid/high income young families*, and about 5% are *Middle income renters*. A full detailed analysis and methods description can be seen in the publication by Lee et al. (2019).

Cluster size and descriptive statistics for the latent class cluster model can be seen in Table 1. The first cluster is the largest (48.9%). This cluster consists of high income, middle aged, mostly male, home owning, and highly educated households. We refer to this group as *High-income families*, due to their high income and large number of people in the household. The second and third clusters account for 25.9 and 19.6% of respondents, respectively. Although the households in these two groups were similar in terms of income, education, and number of drivers in households, they had some differences. Those in the second cluster were older home owning households. The third cluster consisted of younger households of which half rent and half own their home. Therefore, we refer to these two clusters as *Mid/high income old families*, and *Mid/high income young families*. Around 5% of PEV owners were classified into the fourth cluster. Persons in this cluster were middle aged, middle income, mostly male, with fewer people in the household and fewer cars, and mostly rent their home. We call this cluster: *Middle income renters*. Annual household income for this cluster is US $71,000 on average compared to the California state-wide median of US $77,359 in 2016 according to American community survey. Table 1 also shows the socio-demographic profile of 1650 Tesla owners for reference purposes. Their household characteristics are similar to *High-income families*, though they have higher household incomes and are slightly older. The proportion of females

Table 1 Socio-demographic characteristics of four clusters and Tesla owners (from Lee et al. 2019)

	High income families	*Mid/high income old families*	*Mid/high income young families*	*Middle income renters*	Tesla	ANOVA or Chi-square test results
N of PEV buyers	4676	2500	1786	425	1650	NA
Income (US $1000s)	252.2	127.5	127.3	71.1	311.1	F = 2416.4 (p = <0.001)
Age	43.5	53.5	30.7	47.2	46.5	F = 2864.8 (p = <0.001)
Proportion of females	0.24	0.26	0.33	0.48	0.22	Chi-sq. = 268.9 (p = <0.001)
Proportion of home owners	0.92	0.96	0.55	0.26	0.93	Chi-sq. = 3480.1 (p = <0.001)
N of vehicles	2.60	2.44	2.15	1.56	2.53	F = 963.9 (p = <0.001)
N of people in HH	3.23	2.54	2.79	1.74	2.89	F = 2415.2 (p = <0.001)
N of drivers	2.28	2.12	2.01	1.48	2.17	F = 580.1 (p = <0.001)
Education[a]	2.52	2.07	2.18	2.08	2.45	Chi-sq. = 1254.6 (p = <0.001)

[a]Education is measured on an ordinal scale from 1 = high school graduate, 2 = college graduate, 3 = postgraduate

in the Tesla owner group is smaller than the *High income families* group, but other characteristics are similar.

Year-to-year distributions of the socio-demographic clusters are shown in Fig. 1. This figure shows the distribution of the four clusters from 2012 to 2017. Although *High-income families* were the largest group across every year, their proportion has been gradually decreasing from about 55.6% in 2012 to about 40.5% in 2017. The total number of buyers in this cluster is still growing as the market for PEVs grows. The proportion of *Mid/high income old families* has not changed substantially except for the slight decrease in 2015; their proportion has so far remained between 22.7 and 30.5%. The proportion of *Mid/high income young families* and *Middle income renters* began to increase in 2013 and 2015, respectively. The market thus far mostly consists of *High income families*, with *Mid/high income old* and *Mid/high income young*

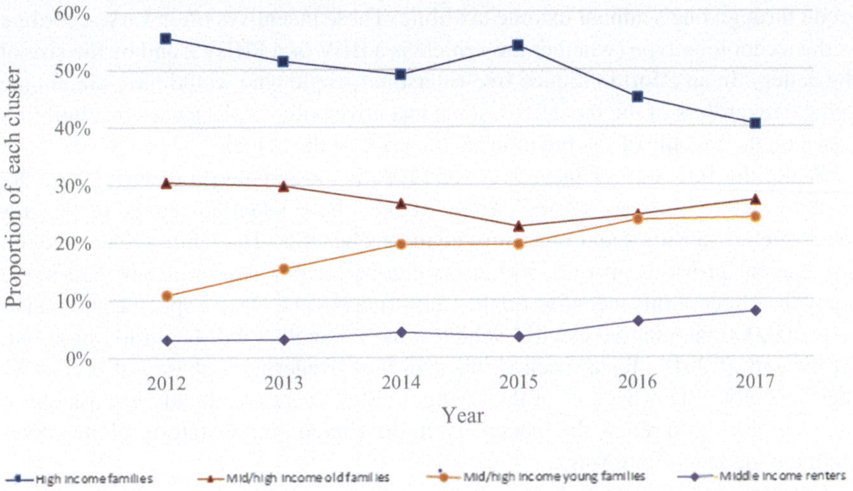

Fig. 1 Year-to-year changes of cluster proportions 2012–2017 (from Lee et al. 2019)

families also making up a sizeable share. The proportion *Middle income renters*, though growing, is still small at 7.9% in 2017.

The results of this analysis reveal that *High-income families* are the largest cluster of PEV buyers in California from 2012 to 2017. This group has the socio-demographic profile of typical early adopters; notably their high household incomes and high level of education. This cluster may not continue to be the largest group of PEV adopters. While they make up 49% of the PEV market 2012–2017, they only represent 3.6% of California households. For market growth to continue the mid/high- and middle-income clusters need to begin adopting PEVs in greater numbers than they are doing today. Policymakers will need to consider the different needs these consumers have for infrastructure and incentives compared to high income adopters to ensure these consumers can adopt PEVs. This may be especially true for those in the *Middle-income renter* cluster. These consumers are more likely to live in a rented home or multi-unit dwelling, this means that will have difficulties in accessing charging from home and may not be able to install a home location charger.

Understanding the Effect of Electric Vehicle Incentives

Incentives for electric vehicles are diverse: they include one-time monetary incentives that assist new car buyers, constant monetary benefits through special electricity rates, and non-monetary incentives such as carpool lane access or special parking privileges. Even within a category of incentives there can be a host of important differences. For example, purchase incentives can be offered at the "point-of-sale" (where the incentive discounts the sticker price), rebated at a later date, or as a tax credit through one's annual income tax filing. These incentives often vary according to the technology type (whether the vehicle is a BEV or a PHEV), and by the size of the battery. In an effort to reduce free-ridership (people who would have bought the vehicle regardless of the incentive), some incentive policies place caps on eligibility based on the income of the buyer or by the price of the vehicle.

While the full suite of incentives for PEVs is too lengthy to discuss, there are several notable "success stories" where markets have taken advantage of government offered subsidies and other miscellaneous benefits. The United States federal government provides up to $7500 in tax credits for purchasers of new PEVs; we study the effect of this incentive on the California buyers. Once a specific automaker sells 200,000 cumulative electric vehicles, the tax credits will begin to phase out. At the end of 2018, Tesla reached this cap, thus triggering a phase-out period for their incentives. However, as of the spring of 2019, there are already discussions at the federal level to renew the incentives. In the United States, state-level incentives augment the federal offerings.

California's Clean Vehicle Rebate Program has been more consistent in their offerings with $2500 and $1500 rebates provided to buyers of BEVs and PHEVs respectively. In 2015, California added an income cap that limited eligibility of the rebate for wealthy individuals. Other states, such as Connecticut, have limited the rebate based on a price cap on the PEV being purchased. The state of Washington exempts PEVs from sales taxes, because this reduces the price at the time of purchase; many practitioners argue this is a more effective subsidy than rebates or tax credits that consumers would receive in the future. The diversity of incentive programs and structures has led to many academic studies on their effects on market adoption.

The earliest studies examining these policies were either qualitative in nature (Bakker and Trip 2013) or were part of larger vehicle choice models (Hackbarth and Madlener 2013). Sierzchula et al. is the first study focused specifically on understanding the effect of financial incentives. The authors measure this econometrically and find a fairly low elasticity with a $1000 incentive corresponding to just 0.06% increase in market share. The use of more sophisticated data, methodologies and longer time frames of study has led to a confluence of research that find relative agreement in price elasticities: Vergis's econometric approach treating incentives holistically showed a 22% increase in market share attributed to the purchase incentive (Vergis and Chen 2015), DeShazo's focus on the California CVRP revealed an average $1838 in incentives led to approximately a 7% increase in PEV market share (DeShazo et al. 2017), Tal's survey-based study of federal incentives found a 32.5% increase corresponding to the $7500 incentive (Tal and Nicholas 2016), and Jenn's panel study of incentives coupled with knowledge of the technology and incentives revealed a 2.6% increase in sales corresponding to $1000 in incentives (Jenn et al. 2018). While these studies have some variation in their results, these can likely be attributed to differences in modeling approaches as well as discrepancies in region and the time period of examination. Hardman et al. provides a more comprehensive review of the purchase incentive literature identifying 35 studies of which 32 find incentives to be effective in promoting BEV sales (Hardman et al. 2017), but we find general agreement in the magnitude of the incentive effects. It should be noted that there has been some work to understand the differences in how a purchase incentive is structured and, for example, how a point-of-sale incentive can be more effective than a rebate credit (Matthews et al. 2017; DeShazo 2016).

Researchers have also investigated the impact of reoccurring incentives on the decision to adopt a PEV. Studies that investigate the impact of HOV lane access are of interest to this study. Early studies investigated the impact of HOV lane access on hybrid electric vehicles. These studies found that HOV lane access for hybrid vehicles was an effective policy measure in promoting the vehicles. More recent studies have begun to investigate the impact of HOV lane access for PEVs. These studies have used qualitative methods (Hardman and Tal 2016), questionnaire surveys (Krause et al. 2013; Tal and Xing 2017), and quantitative analysis. Consensus amongst these studies is less clear than the in the body of literature for financial incentives.

Demand Drivers for Plug-in Vehicle Charging Infrastructure

Utilities, government agencies, and automotive OEMs are increasingly investing in building vehicle charging infrastructure to encourage the adoption and maximize the use of PEVs. Undoubtedly, public infrastructure is required to encourage adoption and address issues like range limitations and access to charging opportunities for those who don't have home charging facilities (Nicholas et al. 2017; Dong et al. 2014; Plötz and Funke 2017; Morrissey et al. 2016; Tal et al. 2014; Zhang et al. 2016). But, as some prior experience has revealed, attempts to maximize the coverage of charging stations (e.g. under the federal EV project) with limited information regarding charging behavior can result in low or almost no utilization of publicly accessible charging points in some locations (Hagerty and Ramsey 2011). Moreover, building infrastructure is costly and like road congestion, when there is significant uptake of PEVs we may not be able to "build out of congestion".

Research so far has mostly focused on predicting the optimal supply of public infrastructure based on observed travel pattern of gasoline vehicles or by simulating possible driving scenarios of PEV owners (Guo et al. 2016; Nie and Ghamami 2013). To better estimate the need for infrastructure it is also essential to understand the characteristics of PEV drivers and their response to varied charging service propositions (Tal et al. 2014). Therefore, we have focused on understanding the importance of factors driving demand for infrastructure like the cost of charging, driver characteristics, access to charging infrastructure, and vehicle characteristics. It is observed that economic factors like differences in the cost of charging play an important role in the demand for charging location. PEV drivers tend to substitute toward workplace charging when they pay a higher electricity rate at home, more so when the latter is free. Additionally, socio-demographic factors like dwelling type and gender, as well as vehicle characteristics like range, have an influence on the choice of charging location.

Recent research conducted at UC Davis shows that heterogeneity in access to infrastructure can exist among PEV owners due to home ownership, dwelling type,

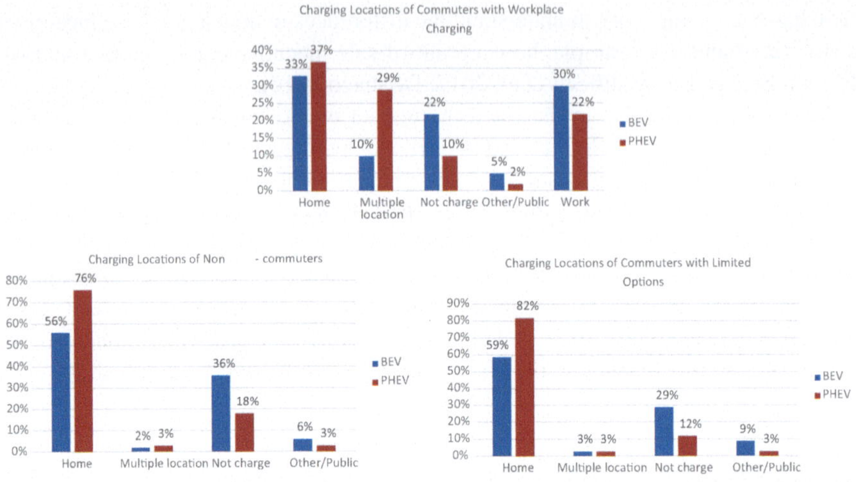

Fig. 2 Reported daily choice of charging location by PEV owners (not charge on this day, home only, workplace only, public only, more than one place)

whether the vehicle is used to commute or not, and if charging facility is available at the commute location, affecting their plug-in behavior. For example, PEV drivers who commute to work may have access to workplace charging infrastructure while non-commuters don't have the option. Also, renters and apartment dwellers may not have access to infrastructure at their residence. Choice of charging location as reported by PEV owners under different scenarios of accessibility is given in Fig. 2.

Since our objective is to identify and quantify the importance of the factors that influence the charging location choice of PEV owners, we focus on a sub-group of 3201 PEV drivers who use their vehicle for commuting to work and have access to charging infrastructure there. Analyzing their choice of charging location on weekdays we find that residential electricity rate plays a very important role in the choice decision. The probability of choice of home as a charging location falls as the electricity rate goes up (Fig. 3). Moreover, if workplace charging is free and the residential electricity rates goes up by 10%, the probability of home charging falls by 3.5 percentage points compared to 2.5 percentage points when the former is paid. In general, it is observed that, along with residential electricity rate the cost of workplace charging has a strong influence on the substitution behavior of PEV owners and their demand for infrastructure. When free, on average, the probability of choosing the workplace as the only charging location for that day goes up by 19 percentage points. Also, the probability of charging at multiple locations like home and work goes up marginally. Along with free workplace charging, the number of chargers at the workplace, time limits on charging, the possibility of congestion, and parking etiquette affect the choice of charging location.

Among other household characteristics, we find that dwelling type, access to Level 2 chargers at home, and ownership of solar photovoltaic panels affect the probability

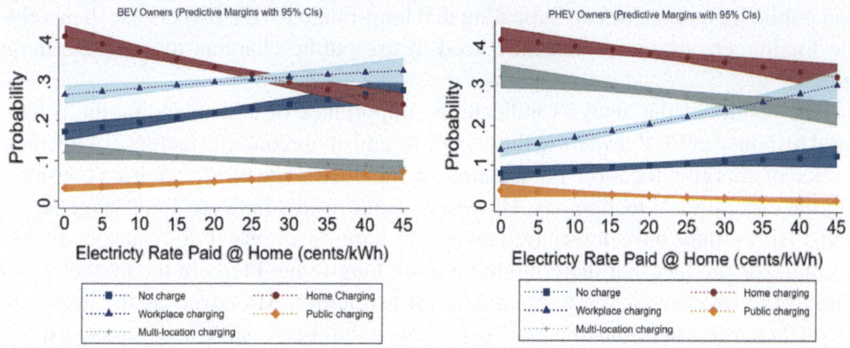

Fig. 3 Predicted probability of choice of charging locations (not charge on this day, home only, workplace only, public only, more than one place) by electricity rate paid at home

of home charging with the former two having a stronger influence. Dwelling type is often a determinant of access to home charging infrastructure. It is observed that, while the likelihood of charging at home is 34 percentage points for detached home-owners, it is only 23 percentage points for apartment residents. Also, keeping all other factors constant, having a Level 2 charger at home increases the probability of charging by 18 percentage points while decreasing the probability of workplace charging by 19 percentage points. Among the driver characteristics, we observe that female drivers have a lower probability of charging in public locations and older drivers are less likely to charge their vehicle on a weekday.

Finally, vehicle range plays an important role in driving charging needs and therefore the choice of charging location. In Fig. 4, we observe that as the range of BEVs goes up, the probability of both home and workplace charging on a particular day goes down. The probability of charging at multiple locations also goes down for higher range BEVs and PHEVs. However, the probability of choice of workplace

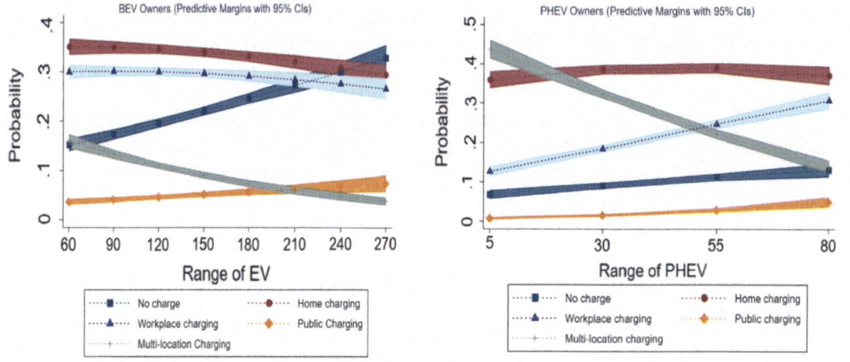

Fig. 4 Predicted probability of choice of charging location by vehicle range

and public locations goes up indicating that long-range PHEV owners use their vehicle for longer commutes and may need to use public chargers to complete these trips.

The results of the analysis indicate the importance of understanding the behavioral response of PEV owners to the economic and non-economic factors driving their choice of charging location. For example, infrastructure goals are often set considering the current PEV technology. However, as our results indicate, long-range BEVs and PHEVs may have lower frequency of charging events. Policymakers should account for the fact that there might be more long-range PEVs in the market when setting the investment needs for public infrastructure. Moreover, as the access to Level 2 chargers at home improves over time, it will be possible to complete a days' travel needs in a long-range PEV with only overnight charging. Pricing policies will also play a key role in determining the demand for charging infrastructure. Policymakers need to develop pricing schemes that will prevent an unnecessary shift of charging behavior from home to non-home locations and allow optimal use of the public infrastructure.

References

Bakker, S., & Trip, J. J. (2013). Policy options to support the adoption of electric vehicles in the urban environment. *Transportation Research Part D: Transport and Environment, 25,* 18–23.

Collantes, G. O. (2006). *The California zero-emission vehicle mandate: A study of the policy process, 1990–2004* (Research Report UCD-ITS-RR-06-09). Institute of Transportation Studies, University of California, Davis.

Daina, N., Sivakumar, A., & Polak, J. W. (2017). Electric vehicle charging choices: Modelling and implications for smart charging services. *Transportation Research Part C: Emerging Technologies, 81,* 36–56.

DeShazo, J. R. (2016). Improving incentives for clean vehicle purchases in the United States: Challenges and opportunities. *Review of Environmental Economics and Policy, 10*(1), 149–165.

DeShazo, J. R., Sheldon, T. L., & Carson, R. T. (2017). Designing policy incentives for cleaner technologies: Lessons from California's plug-in electric vehicle rebate program. *Journal of Environmental Economics and Management, 84*(2017), 18–43.

Dong, J., Liu, C., & Lin, Z. (2014). Charging infrastructure planning for promoting battery electric vehicles: An activity-based approach using multiday travel data. *Transportation Research Part C: Emerging Technologies, 38,* 44–55. https://doi.org/10.1016/j.trc.2013.11.001.

Guo, Z., Deride, J., & Fan, Y. (2016). Infrastructure planning for fast charging stations in a competitive market. *Transportation Research Part C: Emerging Technologies, 68,* 215–227.

Hackbarth, A., & Madlener, R. (2013). Consumer preferences for alternative fuel vehicles: A discrete choice analysis. *Transportation Research Part D: Transport and Environment, 25,* 5–17.

Hagerty, J. R., & Ramsey, M. (2011). Charging stations multiple but electric cars are few. *Wall Street Journal.*

Hardman, S., & Tal, G. (2016). Exploring the decision to adopt a high-end battery electric vehicle: Role of financial and nonfinancial motivations. *Transportation Research Record, 2572*(1), 20–27.

Hardman, S., Shiu, E., & Steinberger-Wilckens, R. (2016). Comparing high-end and low-end early adopters of battery electric vehicles. *Transportation Research Part A: Policy and Practice, 88.*

Hardman, S., Chandan, A., Tal, G., & Turrentine, T. (2017). The effectiveness of financial purchase incentives for battery electric vehicles—A review of the evidence. *Renewable and Sustainable Energy Reviews, 80,* 1100–1111.
Jenn, A., Springel, K., & Gopal, A. R. (2018). Effectiveness of electric vehicle incentives in the United States. *Energy Policy, 119,* 349–356.
Krause, R. M., Carley, S. R., Lane, B. W., & Graham, J. D. (2013). Perception and reality: Public knowledge of plug-in electric vehicles in 21 US cities. *Energy Policy, 63,* 433–440.
Lane, B., Sherman, C., Sperl, J., Krause, R., Carley, S., & Graham, J. (2014). Beyond early adopters of plug-in electric vehicles? Evidence from fleet and household users in Indianapolis. In *2014 Annual Meeting*. Transportation Research Board.
Lee, J. H., Hardman, S., & Tal, G. (2019). Who is buying electric vehicles in California? Characterising early adopter heterogeneity and forecasting market diffusion. *Energy Research & Social Science.*
Matthews, L., Lynes, J., Riemer, M., Del Matto, T., & Cloet, N. (2017). Do we have a car for you? Encouraging the uptake of electric vehicles at point of sale. *Energy Policy, 100,* 79–88.
Morrissey, P., Weldon, P., & O'Mahony, M. (2016). Future standard and fast charging infrastructure planning: An analysis of electric vehicle charging behaviour. *Energy Policy, 89,* 257–270.
Nicholas, M., Tal, G., & Ji, W. (2017). *Lessons from in-use fast charging data: Why are drivers staying close to home*. Institute of Transportation Studies.
Nie, Y. M., & Ghamami, M. (2013). A corridor-centric approach to planning electric vehicle charging infrastructure. *Transportation Research Part B: Methodological, 57,* 172–190.
Plötz, P., & Funke, S. A. (2017). *Mileage electrification potential of different electric vehicles in Germany* (8 pp.).
Russo, E. (2015). Public electric-car charging stations sit idle most of time. *The Seattle Times.*
Sierzchula, W., Bakker, S., Maat, K., & Van Wee, B. (2014). The influence of financial incentives and other socio-economic factors on electric vehicle adoption. *Energy Policy, 68,* 183–194.
Tal, G., & Nicholas, M. (2016). Exploring the impact of the federal tax credit on the plug-in vehicle market. *Transportation Research Record, 2572*(1), 95–102.
Tal, G., & Xing, Y. (2017). *Modeling the choice of plug-in electric vehicles in California: A nested logit approach* (No. 17-02193).
Tal, G., Nicholas, M. A., Davies, J., & Woodjack, J. (2014). Charging behavior impacts on electric vehicle miles traveled: Who is not plugging in? *Transportation Research Record, 2454*(1), 53–60.
Vergis, S., & Chen, B. (2015). Comparison of plug-in electric vehicle adoption in the United States: A state by state approach. *Research in Transportation Economics, 52,* 56–64.
Zhang, Y., Qian, Z. S., Sprei, F., & Li, B. (2016). The impact of car specifications, prices and incentives for battery electric vehicles in Norway: Choices of heterogeneous consumers. *Transportation Research Part C: Emerging Technologies, 69,* 386–401.

What Conversations Between PEV Owners and Owners of Non-PEVs in California Tell Us About Sustaining a Transition

Ken Kurani

Abstract The excitement of advocates, policymakers and automakers for the increasing number of makes and models of plug-in electric vehicles (PEVs) coming to market, increasing number of PEVs sold in successive years, and growing networks of PEV charging infrastructure is presently lost on the vast majority of the car-buying public—even in California, touted as being among the global PEV market leaders. Sustaining PEV market growth will become problematic if the number of car-owning households paying attention to PEVs is not growing, too. Survey data from California in 2014 and 2017 show no increase over time in the percentage of households that have already purchased, shopped for, or even started to gather information about PEVs. Conceptual frameworks regarding how innovations spread through relevant social groups stipulate communication is necessary between the few early actors and the many more potential later actors. With the goals of improving PEV policy and marketing, we explored what might be said between households who have purchased PEVs and some who have not. Workshops were convened in three regions throughout California representing high to low PEV sales and charging infrastructure development. In response to non-PEV owners' questions, PEV owners gave "account," i.e., told their stories of buying and driving a PEV which included informal tallies of costs, incentives, and benefits. This storytelling conveyed to non-PEV owners many of the signs of PEVs, i.e., how PEVs, charging infrastructure, and incentives are (to those who know the signs) symbols of transition to electric-drive vehicles. Routinely, but not universally, learning the symbol system of PEVs produced more positive evaluations of PEVs' symbolic, functional, and affective benefits among the non-PEV owners. The workshops reinforce the results from large sample survey research showing the lack of attention paid to PEVs by most car-owning households and suggest elements of a social narrative to promote a transition to electric-drive vehicles.

Keywords Multi-method research · Consumer · Symbol · Narrative · Communication · Market · Transition · Electric vehicle · Workshop · Survey

---

K. Kurani (✉)
University of California, Davis, USA
e-mail: knkurani@ucdavis.edu

M. Contestabile et al. (eds.), *Who's Driving Electric Cars*, Lecture Notes in Mobility,
https://doi.org/10.1007/978-3-030-38382-4_3

Plug-in Electric Vehicle Market Growth?

Creating and sustaining a transition from vehicles powered by internal combustion engines burning carbon-based fuels to electric vehicles supplied with renewable electricity is one of the cornerstones of sustainable transportation. In June 2017 a subset of the participating nations of the International Energy Agency's (IEA) Electric Vehicle Initiative announced "a collective aspirational goal to reach 30% sales share for electric vehicles by 2030" (IEA 2017).

Such aspirational statements mask disturbing prospects, even for less ambitious goals. On 6 July 2018, the IEA's "Tracking Clean Energy Progress" web page for electric vehicles (IEA 2018) opened with the headline "On track", then continued:

> 2017 witnessed record global sales of electric cars (1.1 million), leading to a global stock of over 3 million. Global sales increased 54% in 2017, helping EVs remain on track to reach the [Sustainable Development Strategy] SDS target.

Only further down the same page is this more cautious assessment offered:

> Despite rapid growth over the past decade of electric car sales, the penetration of electric cars is still limited to less than 1% of the global car fleet today. In order to be on track with the SDS, the global passenger car stock should consist of at least 14% electric cars by 2030, which requires an annual growth rate of about 40% between 2017 and 2030.

Plug-in electric vehicle (PEV) registrations from 2013 to 2018 for California and the US are illustrated in Fig. 1. Claims of rapid growth in the percent of PEVs sold year over year cannot mask that in eight years (starting in 2011 through 2018), cumulative registrations of new PEVs in California barely exceeded 500,000 units.

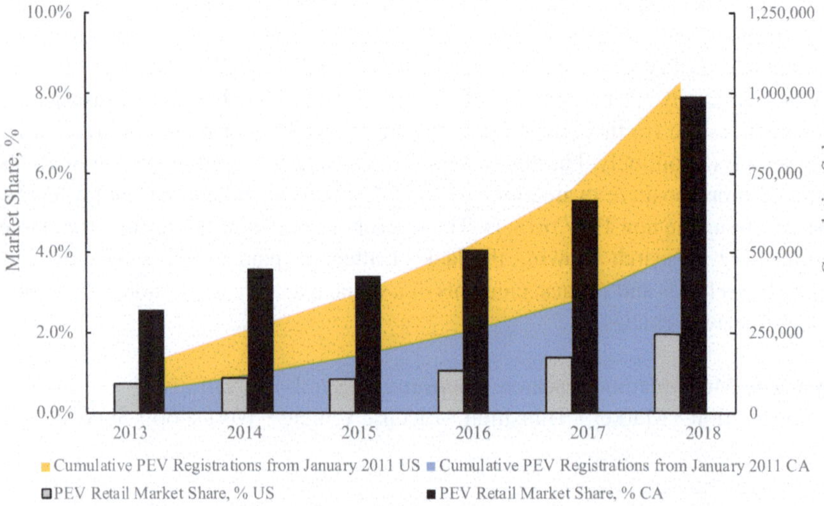

Fig. 1 Electric vehicle market in California and the US. *Data source* Alliance of Automobile Manufacturers (2018)

Over the same period, cumulative sales of all new light duty vehicles in the state exceeded 14.5 million.

Can PEV sales growth meet policy goals such as those stated by the IEA and California's for five million zero emission vehicles (ZEVs: PEVs plus fuel cell vehicles) sold by the year 2030? How might the experience of PEV buyers instigate and inform the growth in sales required to meet these goals? The rest of this chapter presents results from several studies of consumer awareness, knowledge, and evaluation of PEVs. In the next section, large sample survey data from 2014 and 2017 are compared to explore if and how consumers' consideration of PEVs changed over that period. The samples are repeated cross-sections and thus can speak to differences across the population at two points in time but not change over time of any given consumer. Then results are presented of workshops convened across California in 2014 with PEV owners and owners of non-PEVs. The workshops explored conversations between these two groups—early actors and the many more potential later actors who must buy a PEV if private vehicle owners are to be part of a successful transition to electric-drive vehicles. The discussion of these workshops draws on Kurani et al. (2018) which provides a more extensive description of the workshop design, conduct, and analysis as well as discussion of workshop topics not presented here.

The use of multiple research methods creates different viewpoints, allows different theoretical perspectives, and thus helps to address many shortcomings of single-methodology studies. The use of large sample surveys allows comment on the possible pervasiveness of results from small sample workshops. Small sample workshops allow, at least prospective, answers to why results from large sample surveys appear as they do. The use of repeated studies allows comment on older results whose present relevance might be questioned in a dynamic setting.

A Failure to Instigate Broader Consumer Engagement with PEVs Is a Barrier to Continued Market Growth

A barrier to PEV market growth in the US where the light-duty vehicle market is dominated by purchases by private households is the failure of even the idea of PEVs, much less the practice of buying them, to spread beyond a small core of households. Evidence for this includes the absence of difference in measures of consideration of PEVs between samples of car-owning households in California in 2014 and 2017; each sample contains about 1700 households.

Car-owners consideration of PEVs is defined as the extent to which a household has already (at the time they completed a survey) contemplated a vehicle that runs on electricity for their household. Respondents are asked to choose from these statements the one that most closely describes them:

- I have not—and would not—consider buying a vehicle that runs on electricity

- I have not considered buying a vehicle that runs on electricity—but maybe someday will
- The idea has occurred, but no real steps have been taken to shop for one
- Started to gather some information, but haven't really gotten serious yet
- Shopped for an electric vehicle, including a visit to at least one dealership to test drive
- I already have, or have had, a vehicle powered by electricity.

There is no substantive difference in the distribution of responses between 2014 and 2017 (Fig. 2). Any apparent change is within the sampling errors of the surveys. So long as the actual incidence of PEV ownership remains in the low single digit percentages, surveys such as these will be imprecise instruments for measuring change. It would have taken truly dramatic market growth from 2014 to 2017 to be ascertained by these surveys, so the apparent lack of change among those who already own or have owned a PEV is not what is most disturbing about these data. What is disturbing for the prospects of future PEV market growth is that no more households in 2017 had shifted toward any higher level of consideration; no more had become active shoppers, gathered any information, or even had the idea of a PEV occur to them.

Similar lack of difference in measures of knowledge, awareness, and evaluation of PEVs further support the argument of a general failure to grow consumer engagement across a broader population of households between 2014 and 2017. Figure 3 shows no difference in the percentage of car-owning households in California who can name a BEV for sale: barely a third can do so and this did not increase from 2014 to 2017.

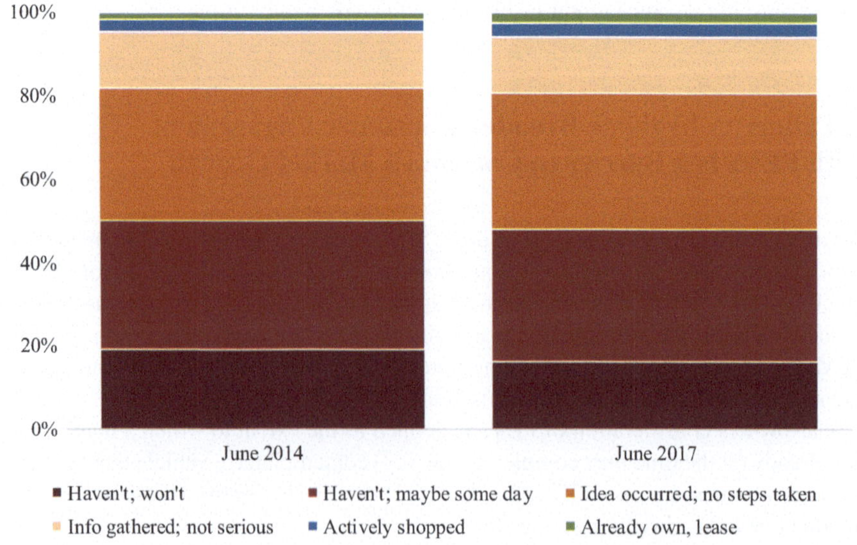

Fig. 2 PEV consideration among all car-owning households in California, percent. *Source* Kurani and Hardman (2018)

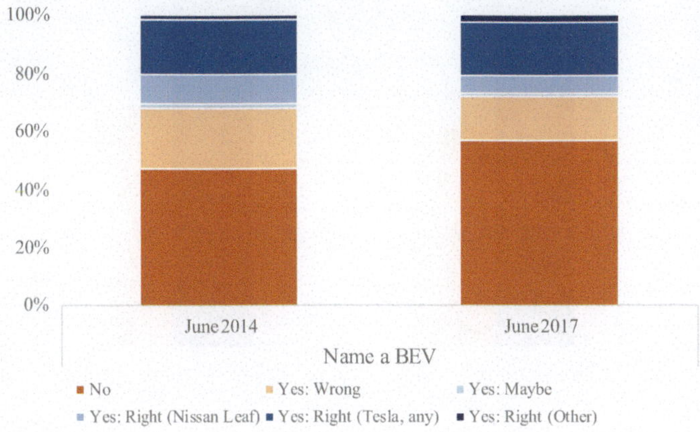

Fig. 3 Ability to name a BEV for sale in California, percent. *Notes* Ability to name a BEV is ascertained in two steps. Respondents are asked if they can name a battery electric vehicle presently for sale in the US; they answer yes or no. Only those who reply "yes" are then asked to name a manufacturer and model. These make-model answers are then classified as right or wrong. The data shown here apply the strictest of three sets of rules. If their make-model answer can in no way be construed as correct, they are classified as "Yes: Wrong." If they provide a vehicle name that comes as a BEV and ICEV or HEV but they do not specify the BEV variant, they are classified as "Yes: Maybe." The remaining three "Yes" answers classify the two most commonly offered correct names plus all other clearly correct responses grouped as "Yes: Right (Other)"

Despite new market entrants, BEV name recall did not extend much beyond the two best sellers. About half of respondents in both years state they are unable to name a BEV. Of those who say they can name one, many give incorrect names. Similar data for PHEVs are not shown, but the conclusions are the same; the ability of a sample of all car-owners to name PEVs for sale did not increase from 2014 to 2017—an interval over which the number of makes and models PEVs for sale (or that had been for sale) in California nearly doubled from 23 to 40.

As of the end of June 2014, California had spent approximately $140 million dollars on Clean Vehicle Rebates (CVRs) to PEV buyers. By the end of June 2017, this had increased to $440 million (CSE 2019). Similarly, the accrued value of the US federal tax credit to PEV buyers increased by hundreds of millions of dollars over the same interval. Despite these, awareness of such incentives did not increase from 2014 to 2017. To examine whether there has been an increase in awareness of PEV purchase incentives, we look at data from a different sample of only new-car buying households from December 2014 as well as the sub-set of new car buyers from the June 2017 sample of all car-owning households (Fig. 4). Even among those who actively shopped for and purchased any new car in California there was no increase in the percentage reporting awareness of incentives from either the US federal government or the California state government from 2014 to 2017.[1]

[1]Certainly, more people received CVRs and federal tax credits over this time. At the end of June 2014, approximately 67,000 CVRs had been paid to PEV buyers; by the end of June 2017 this had

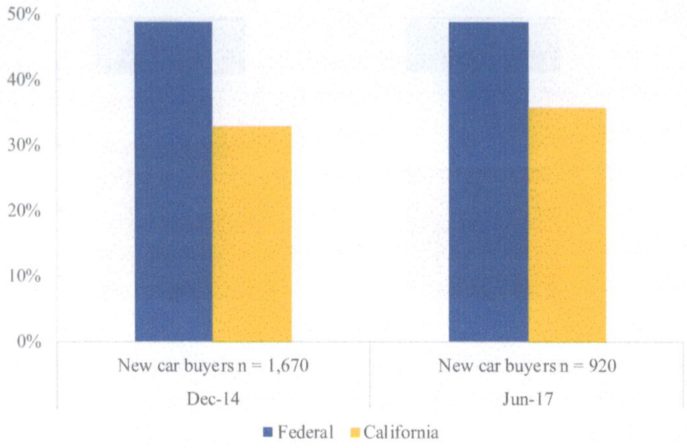

Fig. 4 Awareness of federal and California PEV purchase incentives, new car buyers and all-car-buying households, percent

Returning to the June 2014 and June 2017 surveys of all car-owning households, respondents were asked to offer their evaluations of six aspects of PEVs and PEV charging infrastructure by rating their agreement or disagreement with these statements:

- My household would be able to plug in a vehicle to charge at home.
- There are enough places other than my residence to plug in a car that runs on electricity.
- Vehicles that run on electricity take too long to charge their batteries.
- Vehicles that run on electricity do not travel far enough before needing their batteries to be charged.
- Vehicles that run on electricity cost more to buy than gasoline powered vehicles.
- Gasoline powered vehicles are safer than vehicles that run on electricity.

Answers were on a scale from −3 (strongly disagree) to +3 (strongly agree). Mean values and interquartile ranges (25th–75th percentile) for each item for both years are plotted in Fig. 5. (Items which include the interjection [not] had their signs switched so positive values in the figure always favor PEVs.) Differences between means for 2014 and 2017 which are statistically significant at alpha ≤ 0.05 are shown with an up or down arrow to indicate whether the change was less favorable (⇓) or more favorable (⇑) to PEVs.

In general, mean evaluations of PEVs and charging infrastructure either did not change or became slightly worse from 2014 to 2017. The means of three items are not significantly different: on average, the samples in both years slightly disagree with the statement there are enough places to charge PEVs and slightly agree PEVs take too long to charge and don't drive far enough before needing to charge. (The

increased to over 203,000. The point here is that this increase did not translate into any greater awareness across the broader population of new-car buyers.

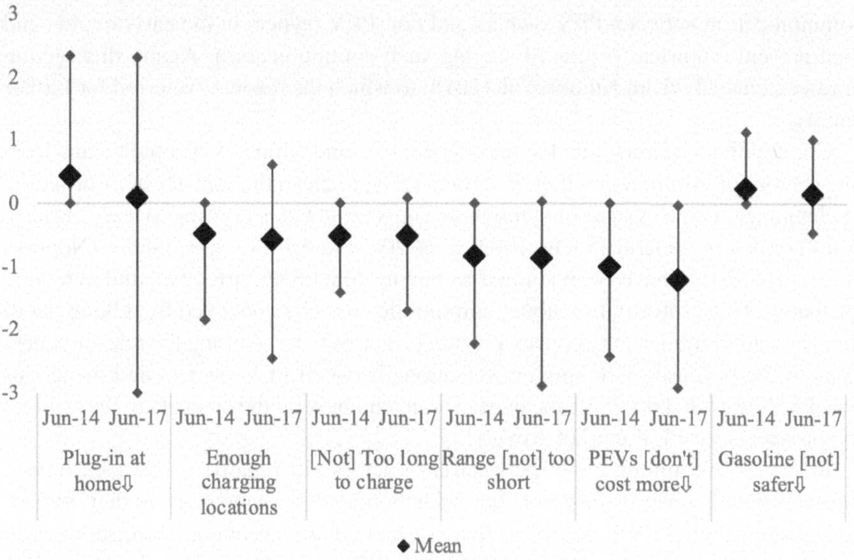

Fig. 5 Evaluations of PEVs, means and interquartile range; scale −3 = strongly disagree to +3 = strongly agree[a]. [a]All items recoded so that positive values favor PEVs. ⇩ Statistically significant decrease (less favorable) in mean value from 2014 to 2017

second two items are reverse coded.) For three items the means are less favorable to PEVs: on average, respondents in 2017 have a weaker agreement they can plug-in a PEV at home, stronger agreement PEVs cost more, and weaker disagreement that gasoline vehicles are safer than vehicles powered by electricity.

The interquartile ranges are larger for every item in 2017 than in 2014, and in particular, in the direction of negative evaluations for PEVs. For example, even though the mean level of (dis)agreement with the statement, "There are enough places other than their residences for people to plug in a car that runs on electricity" is unchanged from 2014 to 2017, the percentage of respondents who registered the greatest disagreement doubled from one-in-eight respondents in 2014 to one-in-four respondents in 2017. Even if on average the changes are slight between the two years, more respondents in 2017 express stronger disagreement with statements phrased in favor of PEVs.

Can PEV Owners Help Accelerate PEV Market Growth?

The survey data presented in the previous section show no growth in the percentage of car-owning households who have already considered PEVs and—at best—no more pervasive knowledge of PEVs and incentives or improved assessment of PEVs and charging availability. In this section we briefly describe possible theoretical roles of communication between PEV owners and non-PEV owners in the early market and then present empirical results of staging such communication. Again, this section draws extensively from Kurani et al. (2018) to which the reader is referred for further details.

Communication must involve the transfer of some "thing" via some means from one person (or group) to another; it is often taken to mean the transfer of information via language, i.e., a system of symbols or signs arranged according to a set of rules. Automobiles in general (Sachs 1983; Steg 2005) and PEVs specifically (Noppers et al. 2014, 2015) have been studied as having functional, affective, and symbolic attributes. Thus, information about automobiles may be conveyed by talking about them and automobiles may convey meanings themselves.[2] Among the relevant questions to the purpose of staging conversations between PEV owners and owners of non-PEVs are what do PEVs mean, to whom, and how do the answers to these affect the prospects for PEV market growth?

Research on communication between early actors with regard to a new idea, object, or practice and as-yet-to-be-actors has been theorized in many ways. In their review, Axsen and Kurani (2012) described four conceptual models of communication, each possibly operating within a specific context: diffusion, conformity, dissemination, and translation, operating within a reflexive context. They provided the following definitions:

- Diffusion is one-way, person-to-person sharing of information.
- Conformity is behavior influenced by perceptions of what others are doing.
- Dissemination describes efforts of organized social groups motivated to promote societal goods, i.e., collective action.
- Translation is the social negotiation of the meaning and benefits of a new technology or behavior.
- Reflexivity is a dynamic, continuous, self-awareness in which one defines and expresses oneself typically to create a sense of order, direction, and development.

Axsen and Kurani (2012) concluded that information flows between drivers of PEVs and owners of non-PEVs likely entail translation within a reflexive context. Based on this brief introduction, the following general propositions guided the design and conduct of workshops in which PEV owners and owners of non-PEVs were brought together:

1. If a larger number of people must buy PEVs to achieve the intended social benefits, then knowing what people who have already acquired PEVs and people who have not acquired PEVs talk about when they talk about PEVs may provide insights into whether and how more people can be encouraged or facilitated to acquire PEVs.
2. All motor vehicles embody a complex suite of symbols, encoding and conveying meanings including owners' identities. As PEVs are new, their meanings are still being negotiated and may be simply unknown to many people.

[2]Information can also be learned via experience; the feeling (affect) of acceleration may be more effectively conveyed by driving the car than by talking about it.

General Approach

Workshops were convened in three regions selected to cover a wide range of PEV market development in California circa 2014: Fresno, Sacramento, and Santa Clara County. Santa Clara County is the region with highest per capita PEV sales in California (5.5 PEVs per 1000 population), followed in decreasing order by Sacramento County (1.8 PEVs per 1000) and Fresno County (0.5 PEVs per 1000). PEV charging infrastructure development follows this pattern, Santa Clara County had by far the most extensive network, followed by Sacramento County and then Fresno County. Regional differences in the value of incentives was mixed. California offered drivers of PEVs solo-driver access to high occupancy vehicle (HOV) lanes. The relative prevalence of HOV lanes and traffic congestion confers higher possible value to people who live in Santa Clara County; HOV lane access had no value in Fresno County as there were no HOV lanes at that time. Conversely, while the Clean Vehicle Rebate (CVR) was available at that time to anyone in the state buying a PEV, PEV buyers in Fresno County were eligible for an additional financial purchase incentive that was larger than the CVR.

In all three regions, samples of PEV drivers were recruited first, then samples of non-PEV owners were recruited to broadly match the PEV owners on a set of socio-economic and demographic measures. Purposive sampling was used to increase the range of data observable and increase plausibility of identifying emerging themes (Erlandson et al. 1993). Sampling was done in three-stages: criterion, maximum variation, and convenience sampling. First, criterion sampling selects cases that meet particular criteria, in this case people who owned PEVs (and thus must have recently purchased a new car) and people who did not own PEVs (but who had recently purchased a new conventional or hybrid car). Second, maximum variation sampling covered the range of households identified in the first stage on socio-economic and demographic measures: income, age, gender, whether they had children living at home, and whether they were employed. Once this range of PEV owners was established, non-PEV owners were sampled to cover a similar range on the same measures. We did reverse this process for gender, that is we took the expected 50/50 gender split among all new car buying households and imposed it on the PEV-owners sample. In the case of gender, the population of buyers of all vehicles is split approximately evenly among women and men, but the population of early buyers of PEVs is strongly skewed towards men. Third, convenience sampling is based on access to a sample. We used convenience sampling (specifically, asking PEV drivers to recruit other PEV drivers) selectively to bolster recruiting in Sacramento and Fresno; this was not necessary in Santa Clara.

Descriptions of the three workshop samples are in Table 1. As participants were not randomly selected to represent a population, results reflect the experiences of specific owners of PEV and non-PEVs at specific places and times.

Battery electric vehicles (BEVs) represented in the study included the Fiat 500E, Ford Focus EV, Mitsubishi i-MiEV, Nissan Leaf, Toyota RAV4 EV, and Tesla Model S. These vehicles spanned what was then the spectrum of price, performance, luxury,

Table 1 Workshop samples

County	Number of participants	PEV/non-PEV split	Female/male split	Age range	Children in home, yes/no
Sacramento	20	10/10	10/10	23–73	10/10
Santa Clara	22	11/11	10/12	29–67	8/14
Fresno	17	8/9	9/8	27–66	8/9

driving range, and charging power. Plug-in hybrid electric vehicles (HEVs) included the Chevrolet Volt, Honda Accord Plug-In, and Toyota Prius Plug-In. These vehicles differed in their electric range from approximately 10 to 35 miles, but were more nearly similar in size, amenity, and cost than the BEVs.

Drawing on focus group design, the workshops capture data in a constructed social setting. Focus groups aid understanding of how individuals contextualize and categorize phenomena as part of a collective (Johnson 1996). Focus groups guide participants' interactions, promoting participants to explain or defend their views to each other (Owen 1997). Compared to focus groups, workshops may be similar in approach but typically include more participants, engaged in a sequence of events over longer periods of time, allowing for more interaction between participants, and involve creative activities.

The workshops were conducted in the first half of 2014. All were attended by the same four researchers. All were conducted in these six steps: (1) Introduction, (2) Agenda Creation, (3) Discussion (small, mixed PEV- and non-PEV owners) group discussion, (4) Discussion Summary, (5) Discussion (a second small, mixed group discussion), and (6) Concluding Discussions (split groups PEV-owner only and non-PEV owner only). Within this outline, the workshops were moderated in a semi-structured manner permitting long periods of uninterrupted conversation among participants.

Recognizing the limits on observable content imposed by the agenda created in Step 2, themes from the discussions in Steps 3, 4, 5, and 6 were identified in a three-step coding process: (a) open coding to locate themes and assign initial codes, (b) axial coding to examine, and if necessary, revise initial codes, and (c) selective coding, i.e., a search for examples to illustrate themes. Two researchers reviewed the thematically organized quotes repeating the three-step coding process to identify themes for Steps 3 and 4. For Steps 5 and 6, each small group was summarized by its moderator or notetaker and those summaries were compared and contrasted by all the researchers.

While identification of themes was necessary to categorize content, the purpose of the workshops required attention to what two or more people said to each other about those themes, especially exchanges between PEV and non-PEV owners. Therefore, text that had been coded for themes was then re-read in the context of what was said, by whom, prior to and following the coded text.

Workshop Results

Throughout, personal names are aliases appended with identifiers (PEV) or (non-PEV) to aid the reader. During Agenda Creation for each workshop, PEV costs and charging infrastructure were selected as the main topics of discussion. Beyond PEV costs, charging, and driving range, the non-PEV owner generally didn't have much else to ask until PEV owners introduced new subjects. The discussion of PEV costs and charging infrastructure are summarized here, followed by a summary of the closing discussions of the non-PEV owners' response to their workshop.

PEV Costs

Non-PEV owners were preoccupied with PEV purchase costs; they had no questions about purchase incentives because they did not know such existed. In response, PEV owners conveyed their accounts of costs, incentives, and benefits—private and public—of owning and driving PEVs. This from Landen (PEV) represents how PEV owners incorporated incentives into their accounts:

> The incentives [are] another big one because there are both state and federal rebates for EVs which will also offset the difference in price.

He goes on to discuss free PEV charging at his work place:

> They have on-site chargers which are free, so I can actually charge my car at work. It doesn't cost me a penny.

Because Landen can charge his PEV at work, he only occasionally charges it at home. Still, he describes cost savings at home as well:

> The cost to fill up an electric car is much less than the cost to fill up a gas car so the costs of running the car are much lower [than a gas car].

PEV owners also told of cost savings from lower maintenance. Calvin (PEV) said:

> My maintenance for [the three years of his PEV lease] is zero. There're no oil changes. You have to keep the fluids topped off and that's it.

The PEV owners in Fresno took pains to portray themselves as financially sensible car buyers, not politically liberal environmentalists. As with their peers in Santa Clara and Sacramento, their accounts included purchase rebates, tax credits, lower home electricity prices, savings from forgone spending on gasoline, and real or expected maintenance cost savings. Trent (PEV) explained:

> It was solely a financial decision…we did the math and I can own and operate that [PEV] for about 90 [dollars] a month…we're saving hundreds of dollars a month.

If PEV owners in Fresno spoke about local environmental benefits, they did so to reject such benefits as a reason they bought their PEV. Jessica (PEV) said,

> My impression was it's more the 'go green' people that drive these cars, but I'm not one of them...I'm not this huge tree hugger either. I'm very far from being liberal.

The non-PEV owners in all three workshops were uniformly surprised by these accounts of PEVs costing less than non-PEVs. This exchange from Fresno was typical:

Zoe (PEV):	*Through the California Air Resources Board I got $2500 and through the local valley air district I got $3000, so that's $5500.*
Shirley (non-PEV):	*Wait a minute. You got checks back?*
Trent (PEV):	*It's cash. 55 hundred bucks.*
Elizabeth (PEV):	*And that doesn't include the $7500 from the federal government.*

If the non-PEV owners in all three workshops were impressed with the cost savings afforded by incentives, they also worried incentives would no longer be offered when the time came they might consider buying a PEV.

PEV Charging Infrastructure

Non-PEV owners prompted the PEV owners to explain how they charge their vehicles, both at home and away. Away-from-home PEV charging both supports individual PEV drivers and signals (to PEV drivers) the larger transition to electric-drive vehicles. Reggie (PEV) described the opportunities for charging away from home to the Sacramento non-PEV owners,

> There's over 250 charging stations in Sacramento. Most people don't realize it. They put them in at restaurants, shopping centers, grocery stores, movie theatres, so you can utilize it.

Reggie refers to these stations in a way that signifies both the density of a network of charging opportunities and support (from in this case unnamed "they") for a transition to PEVs.

The PEV owners in Santa Clara noted a shift from free public charging toward billing for public charging; Derek said,

> The parking garages in downtown San Jose [Santa Clara County] ...They have charging stations in there, which used to allow you to charge for free...now you pay for them.

Still there were convenience benefits, often associated with the parking space itself as much or more than the electricity available at that parking space. Mark (PEV) described the situation at his workplace:

> Watching everybody else circling around and cursing...looking for a parking spot...I'm going to pull in right by the front door.

Fresno had by far the least developed public PEV charging of the three locations. PEV owners there lamented the lack of public charging, told of frustrating encounters with local businesses who declined to install PEV charging, and extolled the San Francisco Bay Area (including Santa Clara County) for its greater PEV charging infrastructure development.

Despite the visibility of PEV charging infrastructure to PEV owners, non-PEV owners appear simply to not have noticed it even in San Clara and Sacramento Counties. Regarding home charging, some non-PEV owners in all three workshops had questions about what kind of home charger is required, what happens if the battery is depleted while away from home, the benefits of using solar power to charge PEVs, and locations of chargers outside of their respective regions.

By the end of the discussions, the concept of a public charging station had been demystified for non-PEV owners. As Shirley (non-PEV) explained at the Fresno workshop,

> To me charging stations were out there—like the Jetsons whipping around with their cars in the air.[3] But it makes sense. It's just a refueling station for your car. So, I can think about it in that way and that makes more sense.

Overall Responses of Non-PEV Owners: Learning the Signs of PEVs

In their concluding conversations the overarching result discussed by non-PEV owners was their general lack of awareness of PEVs prior to their workshop. In Sacramento and Fresno, most of the non-PEV owners were simply unaware PEVs were for sale (three years after retail sales of PEVs started in California). By the end of all three workshops, many of the non-PEV owners discussed lack of awareness as a main hurdle to PEV sales. Given what they learned at the workshop, some non-PEV owners believed the lack of larger body styles was a related barrier. Brenda (non-PEV) put the two together:

> Maybe if there was more awareness, then more people would be buying. Then the demand will be greater, and there will be more variety…because if there was enough awareness there would be his truck [refers to another non-PEV owner who had expressed an interest in an electric pickup truck] and a minivan for me that I could drive, and I'd be saving a lot of money.

One outcome of the workshops was the awareness by non-PEV owners that they had not seen the signs of PEVs around them. Their lack of basic awareness of PEVs surprised the non-PEV owners and frustrated some of them. In Sacramento, three non-PEV owners had purchased a new car in the few months prior to their workshop

[3]"Out there" in American slang describes something far from normal, day-to-day experience. The Jetsons© was a cartoon television show produced in the US in the 1960s depicting a future in which everyone has flying cars—in addition to other imagined advances.

from a manufacturer that offered a PEV. At the end of their workshop, all stated they would have liked to have known PEVs were available. Elijah (non-PEV) said,

> We bought a four-cylinder [mid-size, gasoline-powered sedan] and we get great gas mileage. But if I would have realized all of the incentives, I may have thought more of going to the electric.

Olivia (non-PEV) in Santa Clara who had previously thought about a PEV stated a similar thought,

> Well, I seriously asked two dealers about [PEVs] and nobody told me anything…if I got the incentives and it was the same price as my car and I had researched EVs more…I would have considered it.

Sam (non-PEV), commenting on public PEV charging in Fresno (a region almost devoid of public charging at the time of the workshop),

> I haven't seen one [PEV charger]. And believe me I drive around. Believe me, about 200 miles a day in this town. I haven't seen one station. Yet, I haven't been looking. But I haven't seen one sign that says 'EV station here.' Or anything like this. Where are they?

The workshop shifted his perspective from *not seeing* PEV charging infrastructure to *seeing* there is no PEV charging.

PEV owners' accounts of cost savings made PEVs appealing to some non-PEV owners because in addition to the possibility of cost savings in a practical sense, the accounts made PEVs represent an idea many non-PEV owners valued: saving money. PEV-owners' accounts changed the meaning of PEVs from cars that are expensive to buy to cars that save money as cost savings accrue over time and maybe even at the point of purchase depending on incentives.

Further, for many of the non-PEV owners this switch of meaning regarding the cost of PEVs shifted PEVs away from a meaning they did not want: environmentalist. Shirley (non-PEV) explained,

> What this has shown me tonight is that these cars are not just for people who want to save the planet, lessen their carbon footprint. Like you said, there's math behind it. There's genuine savings. There's a reason that I can relate to.

There was concern the incentives that seemed essential to accounts of cost savings might not last. Darrel (non-PEV) said,

> They want the early adopters to do these things so they're offering incentives but when do they say, in a year or two years, 'No more free parking; no more HOV [high occupancy vehicle lane access for single occupant PEVs]'…because eventually [PEV owners are] going to be the majority and then there's no point to the HOV.

Non-PEV owners tended to question whether PEVs actually provide social, and in particular environmental, benefits—often in the form of the question, "Are PEVs really cleaner?" This question was asked about both air pollution and battery disposal. Still, at the end of their workshops some reflected on the desirability of achieving some social benefit or political goal—even if they remained unsure about PEVs ability to deliver these goals. This exchange among non-PEV owners in Fresno during their concluding discussion illustrates this shift in perspective:

Alfred (non-PEV): *Not being captive to the oil companies I think is a real benefit. Absolutely.*
David (non-PEV): *And I'm not an environmentalist wacko, again to use that term, but I like clean air. You know I like things being clean.*
Scott (non-PEV): *You said that too. I'm not crazy but I do like to be conscious.*
Mike (non-PEV): *Keep a low carbon footprint, like, why not?*

Discussion

Two meanings of PEVs that were clearly heard in all three workshops were (1) managing personal finances and (2) preserving the environment (using labels for meanings previously identified by Heffner et al. 2007). Coming into their workshops, the denotation "manage personal finances" meant "cost less" for PEV owners but "costs more" for non-PHEV owners. Non-PEV owners' preoccupation with questions of higher costs reveal an underlying assumption PEVs are expensive. PEV owners accounts—partial and informal balances of costs and benefits conveyed in stories about becoming and being PEV owners—were efforts by PEV owners to both portray themselves as financially prudent consumers and to position PEVs as less expensive than non-PEVs. By the end of the workshops non-PEV owners generally accepted these accounts. However, a sign that their skepticism and uncertainty about the cost of PEVs had not been entirely overcome was non-PEV owners worry that the incentives which were central to PEV owners accounts would end before non-PEV owners consider buying a PEV. The durability of the meaning of PEVs as smart way to "manage finances" is important to prospects for PEV markets to grow.

The "preserve the environment" meaning, especially as discussed in the Fresno workshop, illustrates the possibility of rejection of PEVs because of a symbolic meaning. Fresno is in a part of the state that is less likely to support government intervention on behalf of environmental goals. PEV owners in Fresno repeatedly positioned their PEV purchase as financially smart and rejected the idea that they would have to have been environmentalists to have bought a PEV. This dynamic was absent from the other two workshops. That some PEV owners would disavow environmental motives is instructive for proponents of the electric-drive transition. Broadcasting all denotations risks alienating potential buyers who would have responded favorably to some meanings but not others. It seems plausible the "relevant social groups" across which innovations spread are defined at least in part by the different meanings an innovation conveys to different people. Further, even some who might find the pro-environmental meanings of PEVs attractive may doubt PEVs really are better for the environment. Thus, even among those interested or at least curious about PEVs, there is a role for greater discussion of the transition itself, literally, why is a transition to electric drive vehicles a good idea and how do we know?

As practical travel tools, non-PEV owners started by doubting a PEV would allow them to accomplish their desired travel; both driving range and vehicle charging created doubt. PEV charging infrastructure is both a part of a system of electric mobility that helps to make PEVs more useful to their owners as well as an important signifier—again, to PEV owners. A growing PEV charging infrastructure provides three related signs of a growing PEV market. First, a single charger signals PEV charging at its locale. Second, any one charger may be perceived as adding to an increasing number of chargers signifying increasing density and extent of charging. Third, a single charger and a perception of an increasingly dense and extensive system of PEV charging may signify commitment by many actors, e.g., electricity suppliers, infrastructure network providers, other businesses, and governments, to continued support for the transition to electric drive. Those PEV owners who live in regions of high charging availability such as Santa Clara County and to a lesser extent Sacramento County used both the high aggregate numbers and specific locations of charging locations to assure non-PEV owners that PEV charging was widely available and that a transition to electric-drive is broadly underway. Those PEV owners who live where charging infrastructure deployment lags decried the absence of infrastructure necessary to support both their driving and an increasing number of PEVs. While rarely entirely settling non-PEV owners' disquiet about charging PEVs as practical, the PEV owners' accounts demystify the idea of charging.

For all this discussion of what PEV and non-PEV owners discussed in the workshops, the most important conclusion is that non-PEV owners did not see the signs of PEVs until they participated in these workshops with PEV owners. Despite small, non-random samples, at a minimum these workshops establish that whether information about PEVs was flowing to any substantively important degree from early actors to potential later ones necessary for a sustained transition remains a valid question. Outside constructed conversations such as the workshops reported here, processes of translating information about PEVs within reflexive contexts were likely not pervasive in California three years after the start of retail sales of PEVs. If we accept that owners of PEVs are necessary sources of information for others who are not-yet PEV buyers, at the time of the workshops there was still a clear necessity to instigate and facilitate communication between these two groups. The stark conclusion across the spectrum of PEV market development contexts represented by the three regions in this study is non-PEV owners were largely unaware, and thus not engaged in, processes of PEV market growth.

Conclusions

Results of large sample surveys and small sample workshops mutually reinforce the argument that continued growth of PEV markets faces a barrier in the form of the inattention to PEVs of the vast majority of car-owning and new-car buying households—even in a place widely regarded as a leader. Most car-owning households are not paying attention to PEVs or the idea of a transition to electric-drive. They are unable to recall names of BEVs or PHEVs even as the number of make-models offered for sale increases. Increasing public spending on vehicle purchase incentives doesn't appear to be producing widespread knowledge of incentives. At best, on average assessments of PEVs are not improving even as purchase prices decrease and driving range and charging opportunities increase. At worst, non-trivial percentages of the population of all-car owning households are becoming more negative in their evaluations of PEVs, especially as regards to purchase price and charging infrastructure.

This absence of difference between 2014 and 2017 supports the continued relevance of the results from listening to conversations between PEV owners and owners of non-PEVs staged in 2014. If PEV owners are to be relied upon to convey information to non-PEV owners, there is no data in the surveys that suggests that started to happen between 2014 and 2017. Conversations between PEV owners and non-PEV owners in the 2014 workshops revealed that non-PEV owners can be amenable to consideration of PEVs if basic pre-conditions are met. A reason to be interested and engaged in a transition to electric drive is only the first such condition. Awareness of the real possibility to buy a PEV may be the second. Untested assumptions by non-PEV owners about PEV price and performance are subsequent conditions. These conditions form a sequence of imagined problems that requires energy, motivation, and resources to overcome. As the assessment of many non-PEV owners is their existing cars and trucks are fine, they lack motivation to invest time, energy, attention, money or any other resources to solve what would be their imagined sequence of "the problems with PEVs" if they had imagined that list in the first place. Thus, the first step to overcoming the attention barrier would seem to be many more people need to find their personal motivation to engage with PEVs.

The connection between individual actions, such as those relayed in the accounts of PEV-owners during the workshops, and societal goals was explored by Kurani et al. (2007). They discussed the intersection between personal narratives and social discourses in terms of Giddens' (1991) "life politics". Updating their conclusions to the present, PEV owners give accounts of their lives "in which they have connected their choices about automobiles to local, regional, and global concerns". Telling such stories "more broadly, to more people" as was done on a limited scale in the workshops strengthens those personal narratives. A complementary strategy to bottom-up story telling by individual consumers is for society to engage compelling societal discourses about why a transition to electric-drive is important. These discourses aren't only for political leaders' speeches but would be supported and repeated by institutional actors across public and private sectors. The purpose of both approaches is not to find a single unifying narrative, but to help diverse people find their reason why they would want a PEV.

References

Alliance of Automobile Manufacturers. (2018). *Advanced technology vehicle sales dashboard.* Data compiled by the Alliance of Automobile Manufacturers using information provided by IHS

Markit. Data last updated August 23, 2018. Retrieved June 7, 2018 from https://autoalliance.org/energy-environment/advanced-technology-vehicle-sales-dashboard/.

Axsen, J., & Kurani, K. S. (2012). Social influence, consumer behavior, and low-carbon energy transitions. *Annual Reviews of Environment and Resources, 37*, 311–340. https://www.annualreviews.org/doi/full/10.1146/annurev-environ-062111-145049n.

Center for Sustainable Energy. (2019). *California Air Resources Board clean vehicle rebate project, rebate statistics*. Data last updated April 16, 2019. Retrieved June 7, 2019 from https://cleanvehiclerebate.org/rebate-statistics.

Erlandson, D. A., Harris, E. L., Skipper, B. L., & Allen, S. D. (1993). *Doing naturalistic inquiry: A guide to methods*. Newbury Park, CA: Sage.

Giddens, A. (1991). *Modernity and self-identity*. Stanford, CA: Stanford University Press.

Heffner, R. R., Kurani, K. S., & Turrentine, T. S. (2007). Symbolism in California's early market for hybrid electric vehicles. *Transportation Research Part D, 12*, 396–413. https://doi.org/10.1016/j.trd.2007.04.003.

International Energy Agency. (2017, June 8). *EV30@30, press release*. Beijing, China. https://www.iea.org/media/topics/transport/3030CampaignDocumentFinal.pdf.

International Energy Agency. (2018). *Electric vehicles: Tracking clean energy progress*. Accessed September 8, 2018. http://www.iea.org/tcep/transport/evs/.

Johnson, A. (1996). It's good to talk: The focus group and the sociological imagination. *The Sociological Review, 44*, 517–538. https://journals.sagepub.com/doi/10.1111/j.1467-954X.1996.tb00435.x.

Kurani, K., & Hardman, S. (2018). Automakers and policymakers may be on a path to electric vehicles; consumers aren't. *Greenlight Blog*. Institute of Transportation Studies, University of California, Davis. https://its.ucdavis.edu/blog-post/automakers-policymakers-on-path-to-electric-vehicles-consumers-are-not/.

Kurani, K. S., Turrentine, T. S., & Heffner, R. R. (2007). Narrative self-identity and societal goals: Automotive fuel economy and global warming policy. In D. Sperling & J. Cannon (Eds.), *Driving climate change: Cutting carbon from transportation*. Academic Press. https://www.sciencedirect.com/science/article/pii/B9780123694959500142.

Kurani, K. S., Caperello, N., TyreeHageman, J., & Davies, J. (2018) Symbolism, signs, and accounts of electric vehicles in California. *Energy Research and Social Science, 46*, 345–355. https://www.sciencedirect.com/science/article/pii/S221462961830118X.

Noppers, E. H., Keizer, K., Bolderdijk, J. W., & Steg, L. (2014). The adoption of sustainable innovations: Driven by symbolic and environmental motives. *Global Environmental Change, 25*, 52–62. https://doi.org/10.1016/j.gloenvcha.2014.01.012.

Noppers, E. H., Keizer, K., Bockarjova, M., & Steg, L. (2015). The adoption of sustainable innovations: The role of instrumental, environmental, and symbolic attributes for earlier and later adopters. *Journal of Environmental Psychology, 44*, 74–84. https://doi.org/10.1016/j.jenvp.2015.09.002.

Owen, H. (1997). *Open space technology: A user's guide* (2nd ed.). San Francisco: Berrett-Kooehler Publishers.

Sachs, W. (1983). Are energy-intensive life-images fading? The cultural meaning of the automobile in transitions. *Journal of Economic Psychology, 3*, 347–365.

Steg, L. (2005). Car use: Lust and must. Instrumental, symbolic and affective motives for car use. *Transportation Research Part A, 39*, 147–162.

Gasoline Savings from Electric Vehicles in the US

Tamara L. Sheldon and Rubal Dua

Abstract Without the option to purchase plug-in electric and/or hybrid vehicles, conventional counterfactuals used in literature may underestimate the fuel savings from clean vehicle adoption, thus overestimating the costs of securing associated environmental benefits. Using a nationally representative sample of new car purchases in the U.S., a vehicle choice model-based counterfactual approach is proposed in this chapter that allows for the prediction of what consumers would purchase if these clean vehicles were unavailable. The cost of demand-side policies in the form of financial incentives to encourage plug-in electric vehicle adoption is estimated.

Policy relevant insights:

- In the US, gasoline consumption under a no clean vehicle scenario increases by 1.7%, compared with a 1.1% increase based on a conventional counterfactual.
- Many pivotal buyers would instead purchase premium brands and larger vehicles, leading to an increase in the share of light trucks, which are subject to less stringent, but more difficult to meet standards.
- Assuming a vehicle lifetime of 16 years, the conventional counterfactual overestimates the cost of gasoline savings from clean vehicle adoption incentives by $1.16 (27%) per gallon compared with the choice model-based counterfactual.

Keywords Fuel efficiency · Clean vehicles · Air pollution · Transportation policy

Introduction

The goal of many transportation policies is to increase clean vehicle market share. However, evidence suggests that consumers most likely to purchase advanced clean

---

T. L. Sheldon
Department of Economics, University of South Carolina, 1014 Greene St., Columbia, SC 29208, USA

R. Dua (✉)
King Abdullah Petroleum Studies and Research Center (KAPSARC), P.O. Box 88550, Riyadh 11672, Saudi Arabia
e-mail: rubal.dua@kapsarc.org

M. Contestabile et al. (eds.), *Who's Driving Electric Cars*, Lecture Notes in Mobility, https://doi.org/10.1007/978-3-030-38382-4_4

vehicle, such as plug-in electric vehicles (PEVs), are early adopters and environmentally conscious (e.g., Sheldon et al. 2017; Dua et al. 2019; Sheldon and Dua 2018). These consumers may therefore be more likely to 'trade up' from a hybrid to a PEV, resulting in considerably lower gains than replacing a standard, internal combustion engine (ICE) vehicle with a PEV. On the other hand, luxury PEV consumers may otherwise purchase high emitting, high performance vehicles. Achieving greenhouse gas (GHG) and pollution reduction goals depends not only on clean vehicle market share but also on which vehicles are taken off the road as a result of clean vehicle sales. Failure to account for such counterfactuals may result in significant over- or underestimates of environmental benefits of clean vehicles. Here we estimate a detailed vehicle choice model to provide a better counterfactual than the existing literature for the US market.

Complicating the analysis of a counterfactual fleet is the impact of U.S. National Highway Traffic Safety Administration's Corporate Average Fuel Economy (CAFE) and Environmental Protection Agency's recent GHG standards, which regulate carbon dioxide emissions from automobiles under the Clean Air Act. The CAFE and GHG standards set specific sales-weighted fuel economy and GHG emissions targets for automakers with the goal of improving energy use and reducing emissions. PEV sales help auto manufacturers meet CAFE standards. In a counterfactual world without PEVs, auto manufacturers would still have to meet CAFE and GHG targets, meaning that in theory, fleet fuel economy could not decrease below the targets, at least in the medium and long run. Nevertheless, since CAFE standards vary by vehicle size, i.e., separate standards for cars and light trucks, a change in the vehicle class mix could still impact fleet fuel economy. Additionally, gasoline savings from PEV adoption depends not only on which ICE vehicles are taken off the road, but also on how much those vehicles are driven.

In this chapter, we fill the gap in the literature on environmental benefits of clean vehicles by predicting a counterfactual fleet of light-duty vehicles in the US and estimating fuel savings from clean vehicle adoption. Our nationally representative sample of nearly 275,000 new vehicle purchases in the U.S. in 2015 includes household level demographic and attitudinal variables. Linking these data to a database of vehicle characteristics, we estimate an innovative vehicle choice model that incorporates consumer heterogeneity, resulting in more precise market share estimates. Using this model, we predict national vehicle sales at the make-model level assuming unavailability of (1) PEVs and (2) PEVs and hybrid vehicles (HEVs). We use these predictions to construct counterfactual fleet fuel economy and gasoline consumption for both scenarios.

Results suggest that if PEVs, which account for 0.81% of the 2015 market share, were not available, fuel economy of cars would decrease nearly 1% and that of light trucks would drop by 0.23% for a total decline in fleet fuel economy of 0.60%. If PEVs and HEVs, which jointly account for 3.38% of the 2015 market share, were not available, fuel economy of cars would decrease by 2.49% and that of light trucks would drop by 0.37% for a total decline in fleet fuel economy of 1.68%. Since many would-be PEV and HEV buyers would otherwise purchase a larger vehicle, an absence of PEVs and HEVs would lead to a shift in the vehicle class mix towards light trucks, which are subject to less stringent fuel economy standards. Lastly, since PEV and HEV consumers also tend to drive more miles, in the absence of clean vehicles gasoline consumption rises by 1.71%, greater than the decrease in fleet fuel economy.

Together, these results imply that clean vehicle technology has still led to a significant reduction in gasoline consumption. Finally, we estimate greater increases in fuel economy and decreases in gasoline consumption relative to counterfactuals relied upon in existing literature ('conventional counterfactuals'). A simple calculation estimates the cost of the gasoline savings resulting from PEV adoption incentives in the US to be $4.31 per gallon, assuming a vehicle life of approximately 16 years (Davis et al. 2013). While relatively expensive, it is significantly less than the $5.47 per gallon we estimate using conventional counterfactuals.

Background

Federal, state, and local governments in the U.S. offer a variety of incentives to promote clean vehicle adoption. There is a federal tax credit for PEVs and a dozen states offer additional financial incentives to PEV buyers in the form of rebates (Sheldon and Dua 2019a), tax credits, and sales tax exemptions. Local incentives include subsidized battery recharging stations, preferred parking, and access to high occupancy vehicle lanes. These policies are generally intended to decrease carbon dioxide emissions as well as to reduce local air pollution such as sulfur dioxide and nitrogen oxides.

An emerging literature attempts to quantify environmental benefits of PEVs. Archsmith et al. (2015) estimate adoption of a PEV results in total present day environmental benefits of $425 (in western states) in terms of a reduction in GHG emissions. Holland et al. (2016) estimate environmental benefits of PEVs, including both reductions in local air pollution as well as reductions in GHG emissions. They find benefits as large as $2800 per PEV, depending on the carbon intensity of electricity generation, which varies by region. To estimate these pollution reductions, both studies must rely on a counterfactual to PEVs. For example, Archsmith et al. (2015) use average values for midsize vehicles to calculate their counterfactual gasoline mileage. Holland et al. (2016) use the ICE vehicle that is most similar to each PEV model in the analysis. The counterfactual for the Ford Focus PEV is the conventional Ford Focus and the counterfactual for the Tesla Model S is the BMW 740.

Zivin et al. (2014) estimate marginal emissions of electricity demand to assess impacts of PEVs on carbon emissions. They also find that PEVs are associated with lower carbon emissions in western states than ICEs, but higher emissions in other regions, since PEVs tend to charge during off-peak hours when marginal emissions are larger. The authors compare PEVs to two alternatives: a 'comparable' economy car and a hybrid. The comparable ICE has the average fuel economy, 31 mpg, of the Toyota Corolla, Honda Civic, Chevrolet Cruze, and Ford Fiesta. The HEV counterfactual is the Toyota Prius.

Lacking knowledge of alternative purchase decisions, these are logical counterfactuals. However, estimated environmental benefits from a PEV purchase may be overstated if the consumer would otherwise purchase an HEV or highly fuel efficient vehicle. Likewise, estimated environmental benefits from a PEV purchase may be understated if the consumer would otherwise purchase a higher-emitting larger or premium vehicle.

To encourage the development of clean vehicle technologies, the CAFE/GHG standards are less stringent for manufacturers who produce alternative fuel vehicles such as PEVs. Jenn et al. (2016) find that each alternative fuel vehicle sold in place of a conventional vehicle therefore weakens the fleet efficiency standards, resulting in an increase of up to 60 tons of carbon dioxide and 7000 gallons of gasoline consumption. Existing work to estimate environmental benefits of PEVs tends to overlook the point made by Jenn et al. (2016) that if there were no PEVs, the counterfactual fleet would still have to comply with CAFE/GHG standards. For example, if PEVs were no longer available, the average fuel economy of midsize sedans might increase or else the auto manufacturers might use price induced sales shift strategies in order to be in compliance.

Data

The primary dataset, which was leased from Strategic Vision Incorporated, is from a survey of households that purchased a new vehicle in model year 2015 (October 2014–September 2015). It is a representative sample of the U.S. national new vehicle market and includes approximately 275,000 observations along with weights that correspond to the ratio of the number of buyers for each make and model in the national market to the number of respondents for the same make and model in the survey. The survey collects information on household characteristics and attitudes. It also includes information on the new vehicle purchased, including the vehicle identification number (VIN) obtained by Strategic Vision Incorporated from either the automakers or IHS Polk. Household characteristics include respondent's age, education level, household income and number of miles the respondent intends to drive his or her new vehicle each month. Although the dataset is not state-wise representative, we do know each respondent's state of residence as well as whether he or she lives in a metropolitan, suburban, rural, or farming area.

Although some households self-report new vehicle characteristics, to ensure accuracy of vehicle attributes, we match new vehicles to their characteristics on 11-digit VIN using Edmunds database of vehicle characteristics. Vehicle characteristics include manufacturer's suggested retail price (MSRP), fuel economy (EPA unadjusted/laboratory values),[1] horsepower, fuel capacity, curb weight, wheel base, track width, and body type. For PEVs, fuel economy is reported in miles per gallon (mpg)

[1]In practice, fuel economy tends to be lower (and fuel consumption greater) than in laboratory tests. Thus, the overall real-world savings would be lower in absolute terms (and associated costs

equivalent, which is the average distance traveled per unit of energy consumed normalized to the energy content of a gallon of gasoline. EPA calculations assume that in terms of energy use, 1 gallon of gasoline is equivalent to 33.7 kW-h of electricity. For PEVs, the data also include electric range. With regard to fuel economy, for the vehicle choice model, this is calculated as the harmonic mean of city fuel economy and highway fuel economy, which are weighted by 55% and 45%, respectively. This is consistent with the EPA's unadjusted/laboratory combined fuel economy calculations, which are used for compliance standards and for the fuel economy labels that consumers see. However, in our counterfactual analysis of fleet fuel economy, we use city and highway weights of 43% and 57%, respectively, which is consistent with EPA methodology for adjusted fleet fuel economy calculations, and is the best estimate of real world performance. For more details see EPA (2015).

In our analysis, we use several constructed variables. For ICEs, range is calculated as the product of fuel capacity and fuel economy. Range is equal to the EPA stated electric range for BEVs. For PHEVs, range is calculated as the product of fuel capacity and fuel economy, plus the electric range. Performance is defined as horsepower over curb weight.

Table 1 shows summary statistics of the new vehicles purchased in the nationally representative sample, as well as mean monthly vehicle miles traveled (VMT). For VMT calculations, we replace missing observations with the sample-weighted mean monthly VMT of 1116 miles. In this chapter, when we refer to PEVs, we refer to both battery electric vehicles (BEVs), which are all-electric, as well as plug-in hybrid vehicles (PHEVs), which have both an electric battery and an internal combustion engine.

Vehicle Choice Model

The purpose of our vehicle choice model is to predict vehicle choices when certain alternatives are unavailable. A standard conditional logit model exhibits the independence of irrelevant alternatives (IIA) property, meaning that the odds of choosing one vehicle over another are independent of the choice set for all other pairs. For example, if the PEV alternative is removed from the choice set, the probability that had previously been allocated to choosing the PEV is split equally across remaining alternatives. This could lead to unrealistic substitution patterns, especially if PEV consumers are more likely to purchase an HEV or an SUV, for example.

The foundation of our model is the following mixed logit model. Consumer n chooses the vehicle that maximizes her utility of choosing vehicle i, U_{ni}. Utility is a linear function of a vector of vehicle characteristics, X_i, as follows:

$$U_{ni} = \mathbf{X}_{\mathbf{i}}'\boldsymbol{\beta}_n + \varepsilon_{ni}. \tag{1}$$

higher) than our estimates. However, relative differences between our estimates and conventional counterfactuals would remain.

Table 1 Summary statistics

	Mean	St. dev.
Price ($)	30,581	12,335
Fuel economy (mpg)	25.06	8.96
Horsepower	232	85
Fuel capacity (gal)	18.4	5.3
Curb weight (lbs.)	3831	896
Wheel base (in.)	113	13
Track width (in.)	63.7	3.2
Range (mi)	426	77
HEV (%)	2.57	
PHEV (%)	0.29	
BEV (%)	0.52	
Convertible (%)	0.74	
Coupe (%)	3.33	
Full-size (%)	0.18	
Hatchback (%)	5.01	
Minivan (%)	2.98	
Pickup (%)	12.55	
SUV (%)	32.79	
Sedan (%)	34.06	
Wagon (%)	8.35	
Monthly VMT	1116	843

Assuming ε_{ni}'s are independently distributed Type-I extreme value errors, the probability of individual n selecting vehicle i can be written as:

$$\pi_{ni} = \int \frac{\exp(\mu_n \mathbf{X}_{\mathrm{i}}' \boldsymbol{\beta})}{\sum_{j=1}^{J} \exp(\mu_n \mathbf{X}_{\mathrm{j}}' \boldsymbol{\beta})} f(\boldsymbol{\beta}|\boldsymbol{\theta}) \partial \boldsymbol{\beta}, \tag{2}$$

where μ_n is a scale parameter commonly assumed to equal 1 and where $f(\beta|\theta)$ is the density function of β.

In contrast to the conditional logit model that assumes fixed parameters, the mixed logit allows for random parameters. When we estimate the mixed logit model, we estimate the distribution of the random parameters. However, the mixed logit does not inform how consumer characteristics correlate with the parameter distributions. While in theory we could interact the mean and/or variance of mixed logit parameters with consumer characteristics, such models tend not to converge to a well-defined maximum value due to numerical instability. We encountered similar issues attempting to estimate latent class models on this large dataset.

In order to more accurately identify heterogeneity in preferences across consumers, we utilize a 'mesh' strategy where we divide the sample into subsamples (or 'meshes') and for each subsample, estimate a mixed logit on aggregate market shares for that subsample. Estimating separate choice models for different consumer segments identified through clustering is common in literature (Axsen et al. 2015). However, since categorical variables such as location and socio-economic group are used in this study, a simple categorical segmentation was used instead of clustering. The choice sets are at the make-model-engine type level, where engine type is either ICE, HEV, BEV, or PHEV. For example, the conventional Ford Focus and the Ford Focus Electric appear as separate alternatives. This results in approximately 400 alternatives.

Table 2 shows the three variables that we use to divide the sample into meshes: (1) socio-economic group, which is a function of age, income, and education, (2) pro-environmental attitude, and (3) residential location. We generate a separate 'mesh' for each socio-economic by pro-environment by residential location group of consumers, for a total of 90 meshes. Table 3 shows summary statistics of the number of observations and weighted observations in each mesh. For computational ease, we aggregate household level data to market share level within each mesh, summing frequency weights for each make-model-engine observation such that the sample remains representative. Since each household within a mesh that chooses a particular make-model-engine is identical with respect to variables used in the estimation, estimating the mixed logit model on aggregate market shares using frequency weights is equivalent to estimating the model on a representative agent.

For most mesh estimates, vehicle characteristics in the utility function include price, fuel economy, range, performance, engine type indicators (HEV, PHEV, or BEV, with ICE as the omitted category), make indicators, and body type indicators (convertible, coupe, full-size car, hatchback, minivan, pickup, SUV, and wagon, with sedan as the omitted category). We exclude ultra-luxury brands to avoid skewing our estimates, since though representing a tiny market share, their prices may be outliers.

Table 2 Mesh categories

Socio-economic group	Pro-enviro[a]	Residential location
Age < 35/Income > \$70K/College	No or Missing	Metropolitan City
Age < 35/Income > \$70K/No College	Yes	Suburban Community of a Large City
Age < 35/Income ≤ \$70K/College		Small Town or Rural City
Age < 35/Income ≤ \$70K/No College		Farming Area
Age 35+/Income > \$70K/College		Missing
Age 35+/Income > \$70K/No College		
Age 35+/Income ≤ \$70K/College		
Age 35+/Income ≤ \$70K/No College		
Missing		

[a]Individual indicated he or she agrees somewhat or strongly with 'I would pay significantly more for an environmentally friendly vehicle'

Table 3 Mesh sizes

	Observations	Frequency (weighted)
Minimum[a]	5	111
Second min[a]	12	731
25th percentile	378	15,308
Median	1174	57,390
75th percentile	2967	138,907
Maximum	38,981	1,733,866

[a]The smallest mesh includes observations with age < 35/income > \$70K, no college, pro-environmental, and missing residence. The second smallest mesh is similar except college educated. Given the limited sample sizes for these two meshes, for each we utilize the estimates from the most similar mesh. That is, the mesh that has identical characteristics except rather than missing residence, includes observations from suburban residences

These brands are also not represented by most meshes. These ultra-luxury brands include Alfa Romeo, Aston Martin, Bentley, Ferrari, Jaguar, Lamborghini, Lotus, Maserati, McLaren, and Rolls-Royce. Due to collinearity amongst the variables, we construct a variable that is the product of fuel economy, range, and performance, scaled by price. This variable is specified to have a random parameter. To better characterize body type substitution patterns, the SUV indicator (which is the second most popular body type after sedan) is also specified to have a random parameter. A minority of mesh estimates do not converge when attempting to estimate this model, particularly for meshes with fewer observations. In these cases, we simplify the model first by considering SUV a fixed rather than a random parameter, and, if convergence is still not achieved, then by excluding the make indicators. Table 4 summarizes the three utility functions utilized in our analysis. Lastly, due to collinearity between the BEV indicator and other attributes, in a first stage we fit the utility function excluding the BEV and Tesla make indicators. Then, in a second stage we re-estimate the utility function constraining the fuel economy-range-performance-price coefficient to equal that from the first stage and including the BEV and Tesla indicators.

For PEV prices, we adjust MSRP by state and national incentives. Specifically, for PHEV and BEV purchase observations from states with incentives, we reduce MSRP by the maximum incentive amounts shown in Fig. 1 obtained from DeShazo (2016). These state incentives range from \$500 to \$7500, with most falling in the \$2000–\$3000 range. The federal tax credit varies by vehicle, with a maximum amount of \$7500 (DOE-EPA). A household cannot claim more than it pays in federal income tax. For example, if a household purchases a PEV eligible for a \$7500 tax credit but only pays \$5000 in federal taxes, it will get to offset the full \$5000 of federal taxes but will not receive an additional rebate or benefit on top of that. The PEV tax credit does not roll over and thus cannot be used to offset future taxes. Due to the complexity of estimating household federal taxes owed and the lack of necessary data to do so (e.g., precise pre-tax income, mortgage, student loan information, number

Table 4 Utility function

Utility function	Fixed parameters	Random parameters	Number of meshes
1	Makes, Bodies, HEV, PHEV, BEV	SUV, (Fuel Economy * Range * Performance)/Price	58
2	Makes, Bodies, HEV, BEV, PHEV	(Fuel Economy * Range * Performance)/Price	22
3	Bodies, HEV, BEV, PHEV	(Fuel Economy * Range * Performance)/Price	10

Note Body indicators (Bodies) include convertible, coupe, full-size car, hatchback, minivan, pickup, SUV, and wagon, with sedan as the omitted category. Make indicators (Makes) are included for all makes except ultra-luxury brands, which are omitted from the sample as outliers. These ultra-luxury brands include Alfa Romeo, Aston Martin, Bentley, Ferrari, Jaguar, Lamborghini, Lotus, Maserati, McLaren, and Rolls-Royce. Due to collinearity between the BEV indicator and other attributes, in a first stage we fit the utility function excluding the BEV and Tesla make indicators. Then, in a second stage we re-estimate the utility function constraining the fuel economy-range-performance-price coefficient to equal that from the first stage and including the BEV and Tesla indicators

of dependents, etc.), we make the simplifying assumption that all PEV purchases in the data are eligible for the maximum federal tax credit. Given the observation that most PEV buyers are high income, we feel this is a reasonable assumption. Therefore, in addition to any state level subsidies, we adjust PEV prices downwards by the maximum amount of the federal tax credit available for each model.[2]

In an effort to move inventory, many vehicle manufacturers have been offering cash rebates and special lease offers for their PEVs on top of government incentives. While these vary across vehicle, manufacturer, and location, the cash rebates average around 9% of MSRP (Campbell et al. 2017). Therefore, for all PEVs (other than Tesla), we reduce MSRP by 9% before deducting federal tax credit and any state incentives. When we collapse individual data to aggregate market share, we take the weighted average price across individuals.

After estimating mixed logit models for each mesh, we use estimation results to predict consumer vehicle choices separately for each mesh. Using consumer weights, we aggregate the separate mesh predictions to market shares of the representative national sample. Table 5 shows the model predictions for BEV, PHEV, and HEV market shares are identical to the actual market shares in the sample.

Table 5 Actual versus predicted market shares

	BEV (%)	PHEV (%)	HEV (%)
Actual	0.52	0.29	2.57
Predicted	0.52	0.29	2.57

[2]See https://www.fueleconomy.gov/feg/taxevb.shtml.

Counterfactual Analysis

We predict consumer vehicle choice under three different scenarios: (0) the actual set of available vehicles is available, (1) PEVs are not available, and (2) neither PEVs nor HEVs are available. There are two major assumptions underlying our counterfactual analysis. Our vehicle choice model models consumer choice conditional upon having decided to purchase a new vehicle. Currently, we are unable to include an outside option to not purchase a new vehicle.[3] Thus, our counterfactual analysis assumes that the total number of new vehicle purchases is the same- i.e., when PEVs and/or HEVs are not available, the same population still decides to purchase a new vehicle. This is a reasonable assumption because although a household might delay purchasing a new vehicle as they anticipate the arrival of a particular model (e.g., the Tesla Model 3), in a counterfactual world devoid of PEVs or HEVs, a household would not have these models to anticipate.

The second assumption is that in the counterfactual scenarios, prices of available vehicles are unchanged- i.e., we only model the demand side, not the supply side. Langford et al. (2015) model both the supply and demand sides of the new vehicle market for their counterfactual analysis. They find that if hybrids were not available from 2001 to 2008, equilibrium prices of close substitutes would be lower. However, Table 3 of their paper shows a maximum price decrease of $18. Such small price changes are unlikely to significantly change our counterfactual analysis results.

Table 6 shows market share estimates for various vehicle types under each scenario. In Scenario 1 when PEVs are not available, less than 5% of the would-be PEV buyers (who make up 0.81% of the total market) purchase HEVs instead. More would-be PEV buyers purchase SUVs (11%) and pickups (11%) instead. Overall, more than a quarter of the would-be PEV buyers switch to a light truck, leading to an increase in overall light truck market share (from 48.24 to 48.45%) and a decrease in passenger car share (from 51.76 to 51.55%). These trends are exacerbated in Scenario 2 when HEVs are also unavailable, with light truck share increasing to 49.36% and passenger car share decreasing to 50.64%.

Table 7 shows market shares of different makes in each of the three scenarios. From Scenario 0 to Scenario 1, where PEVs are no longer available, market shares of makes with popular PEV models such as Chevrolet (maker of the Volt), Ford (maker of the C-MAX Energi, Fusion Energi, and Focus Electric), and Toyota (maker of the Prius Plug-in) decline. Nissan's predicted market share actually increases from Scenario 0 to Scenario 1, despite the popularity of the Nissan Leaf, which is no longer available in Scenario 1. This suggests that some would-be PEV buyers switch to other non-PEV Nissan models. Similarly, Cadillac and Mercedes-Benz market shares increase slightly and BMW's remains the same despite several PEVs of these makes no longer being available in Scenario 1. This is consistent with larger increases in market share of several premium brands, including Audi and Lexus. Jeep and Subaru, whose popular models are SUVs, also have larger increases in market share.

[3]This would require knowing how many people did not purchase a new vehicle in each mesh category.

Table 6 Market shares, fuel economy, and gasoline consumption of counterfactual scenarios

	Scenario 0	Scenario 1	Scenario 2
	Predicted	No PEVs	No PEVs or HEVs
BEV (%)	0.52	2.61	
PHEV (%)	0.29		
HEV (%)	2.57		
SUV (%)	32.7	32.8	33.2
Convertible (%)	0.8	0.8	0.8
Coupe (%)	3.4	3.4	3.5
Full size car (%)	0.2	0.2	0.2
Hatchback (%)	5.3	5.1	5.0
Minivan (%)	3.0	3.0	3.1
Pickup (%)	12.6	12.7	13.0
Sedan (%)	33.6	33.6	32.6
Wagon (%)	8.5	8.5	8.6
Passenger car share (%)	51.76	51.55	50.64
Light truck share (%)	48.24	48.45	49.36
MPG passenger cars	28.81	28.55	28.09
MPG light trucks	19.58	19.53	19.51
MPG fleet	23.47	23.33	23.08
Annual billion gal of gasoline	7.14	7.18	7.26

Together, the top panel of Tables 6 and 7 suggest that while some would-be PEV buyers switch to HEVs and other small cars, many instead switch to larger vehicles including SUVs and light trucks, as well as premium makes, which tend to have lower fuel economy.

Similar patterns emerge for Scenario 2 in Table 7. When HEVs are also unavailable, Toyota and Honda, makers of popular HEVs, have significant market share declines. Market shares of makers of larger vehicles increase (e.g., GMC, Jeep, and Subaru). However, unlike Scenario 1, market shares of non-premium brands such as Chevrolet and Nissan increase more than those of premium brands. This suggests that while many would-be PEV and HEV buyers substitute toward larger vehicles, substitution toward premium brands is driven by would-be PEV buyers.

For each scenario we use the predicted fleet to calculate the vehicle-weighted fleet fuel economy, as shown in the bottom panel of Table 6. In Scenario 0, predicted fleet fuel economy is 23.47 mpg. In Scenario 1, when PEVs are removed from the choice set, predicted fleet fuel economy declines 0.60% to 23.33 mpg. In Scenario 2, when

Table 7 Brand market shares (percent) of counterfactual scenarios

	Scenario 0	Scenario 1	Scenario 2
	Predicted	No PEVs	No PEVs or HEVs
Acura	0.83	0.84	0.85
Audi	1.30	1.32	1.35
BMW	2.11	2.11	2.16
Buick	1.36	1.37	1.40
Cadillac	0.99	1.00	1.02
Chevrolet	9.00	8.98	9.20
Chrysler	1.56	1.57	1.62
Dodge	2.73	2.74	2.80
FIAT	0.27	0.26	0.27
Ford	10.96	10.91	10.96
GMC	3.07	3.10	3.17
Honda	10.66	10.60	10.06
Hyundai	4.46	4.49	4.51
INFINITI	0.85	0.86	0.85
Jeep	5.31	5.34	5.44
Kia	3.92	3.93	3.98
Land Rover	0.04	0.04	0.04
Lexus	2.38	2.41	2.34
Lincoln	0.65	0.65	0.66
MINI	0.41	0.41	0.43
Mazda	2.20	2.23	2.29
Mercedes-Benz	2.20	2.21	2.26
Mitsubishi	0.50	0.50	0.51
Nissan	7.93	7.96	8.15
Porsche	0.74	0.73	0.75
Ram	2.64	2.65	2.72
Scion	0.39	0.39	0.40
Smart	0.04	0.03	0.04
Subaru	4.19	4.23	4.34
Tesla	0.16	0.00	0.00
Toyota	13.67	13.62	12.86
Volkswagen	2.17	2.17	2.20
Volvo	0.34	0.35	0.36

both PEVs and HEVs are removed from the choice set, predicted fleet fuel economy declines 1.68% from Scenario 0 to 23.08 mpg. To put these changes in perspective, recall that in 2015, PEVs accounted for 0.81% of the market and HEVs for 2.57%. It is also worth noting that given the relatively small market share of PEVs and HEVs, our counterfactuals rely on changes in the tails of the distribution of vehicle purchases. While our large sample size provides substantial power, uncertainty and errors are to be expected.

For Scenarios 1 and 2 we also calculate a 'conventional' counterfactual generated similar to counterfactuals currently used in the literature and policy analysis to show how our results differ when we use a more sophisticated choice model to generate the counterfactuals. The conventional fuel economy counterfactuals assume that if BEVs, PHEVs, or HEVs are unavailable, they are replaced by a vehicle with the average fuel economy of that BEV/PHEV/HEV's size (e.g., compact, mid-size, or full-size) and class (e.g., car, pickup, or SUV). In Scenario 1, the conventional counterfactual predicts a 0.48% decrease in fleet fuel economy relative to the 0.60% estimate. In Scenario 2, the conventional counterfactual predicts a 1.12% decrease in fleet fuel economy relative to the 1.68% estimate.

Our model predicts larger decrease in fleet fuel economy mainly because the conventional counterfactuals do not allow would-be PEV drivers to switch vehicle class. Most PEVs are compact or midsize cars. The conventional counterfactuals assume that if the PEVs or HEVs are not available, these consumers would choose compact or midsize ICEs. However, the vehicle choice model suggests that in the absence of their preferred PEV or HEV, many consumers would instead purchase a larger vehicle (or a premium vehicle), which typically have lower gas mileage than compact or midsize ICEs. This difference suggests that conventional counterfactuals such as those used in existing literature underestimate the fuel economy improvements from PEVs and HEVs.

Using the predicted fleet and reported monthly miles driven we also calculate total annual gasoline consumption of the new vehicle fleet for each scenario, as shown in the bottom panel of Table 6. Annual gasoline consumption in Scenario 1 is 0.61% greater than Scenario 0 (7.18 vs. 7.14 billion gallons). Annual gasoline consumption in Scenario 2 is 1.71% greater than Scenario 0 (7.26 vs. 7.14 billion gallons). In other words, we predict that if PEVs were unavailable, the fleet would consume 0.61% more gasoline, and if HEVs also were unavailable, the fleet would consume 1.71% more gasoline.

Annual gasoline consumption values for Scenarios 1 and 2 are calculated using the conventional fuel economy counterfactuals and the sample-weighted mean VMT of 1116 miles per month (as existing studies, lacking individual data, typically rely on population averages). In Scenario 1, the conventional counterfactual predicts a 0.49% increase in annual gasoline consumption relative to our 0.61% estimate. In Scenario 2, the conventional counterfactual predicts a 1.14% increase in annual gasoline consumption relative to our 1.71% estimate.

Thus, when accounting for heterogeneous vehicle preferences and substitution patterns, predicted gasoline savings from PEV and HEV adoption are 25% and 50% greater, respectively, than the conventional counterfactuals suggest. This difference

Table 8 Miles driven per month by fuel type

	Mean	St. dev.
All	1116	843
ICE	1114	832
HEV	1160	984
BEV	1181	1437
PHEV	1167	1340

is due to the larger decrease in fleet fuel economy in the predicted scenarios relative to the conventional counterfactuals. If more PEV drivers switch to light trucks instead of midsize cars, for example, then gasoline consumption would be greater.

We predict that gasoline consumption increases by slightly more than fleet fuel economy decreases (0.61% relative to 0.60% and 1.71% relative to 1.68%). This is because PEV and HEV drivers tend to drive more miles. Table 8 shows the number of miles that households intend to drive their newly purchased vehicle, broken down by vehicle type. The average BEV and PHEV drivers' average monthly VMT are 5–6% greater than the average ICE driver, and the average HEV driver's VMT is more than 4% greater than the average ICE driver. Thus, when a would-be PEV or HEV driver instead purchases an ICE, this vehicle not only gets lower gasoline mileage, but is also driven more than the average ICE.

Fuel Savings from PEV Adoption Incentives

Many states offer financial incentives for PEV adoption in addition to the federal tax credit of up to $7500. Using the estimates from our vehicle choice model, we adjust prices by removing the state and federal incentives previously added and predict PEV sales. Absent these incentives, PEV market share decreases from 0.81 to 0.59%. This suggests that (0.81–0.59)/0.81 = 27% of PEV sales, and therefore 27% of the gasoline savings from Scenario 1, are a result of PEV incentives. We define the total policy cost as the sum of all state and federal incentives, assuming all PEVs received incentives for which they were eligible. Dividing the total policy cost by gasoline savings resulting from the policy (27% of gasoline savings from Scenario 1) equals $69 per gallon of annual gasoline savings. Assuming a vehicle lifetime of approximately 16 years (Davis et al. 2013), this cost decreases to $4.31 per gallon. This is nearly $1.16 per gallon less expensive than the $5.47 per gallon we calculate using the conventional counterfactual. Nevertheless, despite the significant environmental benefits from PEV adoption, demand-side policy costs appear to be relatively expensive (Sheldon and Dua 2019b).

Caveat: CAFE/GHG Standards

As shown in Table 6, we predict that if PEVs and HEVs were not available in 2015, average fuel economy for the fleet of new vehicles would be lower and gasoline consumption higher. One caveat is that these may only represent short run effects because, without PEVs to sell, some vehicle manufacturers may adjust supply (e.g., putting fuel savings technology on more models or lowering the price of more fuel efficient models) in order to comply with CAFE/GHG standards. Such supply responses would increase the fleet fuel economy in the long run.

We calculate 2015 CAFE standards and compliance of our nationally representative sample based on official methodology (EPA-NHTSA 2012). The fuel economy values in our data are adjusted, unlike the laboratory performance test results used to evaluate regulatory compliance. Gillis et al. (2016) estimate that a CAFE test score of 54.5 mpg would correspond to an EPA consumer label of 40 mpg. Accordingly, we use a rough conversion factor of $54.5/40 = 1.3625$ to convert our fuel economy values for use in a simple back-of-the-envelope CAFE compliance calculation. We find that while several automakers are non-compliant in 2015, none of the compliant automakers would change to non-compliant in either of our counterfactual scenarios. Therefore we can interpret the fuel savings from PEVs and HEVs as additional to those from CAFE standards. However, it is important to note that the 1.3625 is a rough average conversion factor and does not take into account the nuances of CAFE compliance calculation. A more granular look at the separate fleets while incorporating adjustment factors for alternative fuel vehicles, such as flex-fuel vehicles, would be needed for a full-scale evaluation of the CAFE compliance, which is beyond the scope of this study.

Even if some manufacturers became non-compliant in the counterfactual scenarios, fleet fuel economy would still be lower because many would-be PEV buyers would purchase light trucks instead, as shown in Table 6. CAFE/GHG standards are size-based standards and are less stringent for larger vehicles. A shift in the vehicle mix toward light trucks would result in a larger share of vehicles being subject to less stringent, but difficult to meet, standards. Furthermore, CAFE/GHG standards only regulate fleet fuel economy, not gasoline consumption. If would-be PEV drivers with high VMT purchase less fuel efficient vehicles, their increase in gasoline consumption will not be offset by lower VMT drivers buying slightly more fuel efficient vehicles as a result of CAFE/GHG standards.

Caveat: The Rebound Effect

Our estimates of changes in gasoline consumption implicitly assume that drivers of PEVs drive the same number of miles, regardless if they drive the PEV or alternative ICE vehicle. In other words, they assume zero rebound effect. If there is a rebound effect, these drivers may increase their vehicle miles travelled when adopting the

PEV, suggesting that miles travelled on the ICE substitute could be lower, which would reduce the gasoline savings from PEVs. However, if there is a rebound effect, it is likely to be small.

Gillingham (2014) estimates a rebound effect of 9% for Californian households in the period 2000–2006. Small and Van Dender (2007) find that the rebound effect for motor vehicle travel in the U.S. has been declining over time and is decreasing with income. With income at 1997-2001 levels they estimate a short- and long-run rebound effect of 3.1% and 15.3%, respectively. Literature on consumer residential energy demand also suggests the rebound effect is declining with income (Henly et al. 1988; Reiss and White 2005). Since PEV drivers tend to be high income earners, any rebound effect is likely to be on the low end.

In an analysis of the 2009 federal Cash for Clunkers program, West et al. (2017) find that while households did purchase more fuel efficient vehicles as a result of the scrappage program, despite a lower cost per mile due to an increase in fuel economy, households did not respond by increasing miles driven. Gillingham et al. (2015), who estimate a gasoline price elasticity of −0.1, find that this elasticity is driven by low fuel economy vehicle drivers. They find that higher fuel economy vehicles' miles traveled are highly inelastic to changes in the gasoline price and that newer vehicles are also less responsive. Since PEV consumers tend to have high incomes and since we analyze new vehicle sales, the consumers who are 'switching' in the counterfactual case are likely to have inelastic price elasticities of demand for VMT. As such, any potential rebound effect is likely to be very small.

Conclusion

We find that if PEVs and HEVs were not available, fleet fuel economy would decrease. Many clean vehicle buyers would instead purchase larger vehicles, leading to a shift in the vehicle class mix. Would-be PEV buyers would also substitute toward premium brands, which tend to have lower fuel economy. Also, since PEV and HEV consumers tend to drive more miles, in the absence of clean vehicles, gasoline consumption rises by slightly more than the decline in fleet fuel economy. Together, these results imply that clean vehicle technology has led to a significant reduction in gasoline consumption and will continue to do so in the near future despite potential leakage from CAFE and GHG regulations. Our estimates of fuel economy improvements and gasoline savings from PEVs are larger than conventional back of the envelope calculations, suggesting that recent estimates of environmental benefits of PEVs may be underestimated. Finally, our results imply that PEV adoption incentives cost $4.31 per gallon of gasoline saved, assuming a vehicle lifetime of 16 years, suggesting need for improvement in the cost-effectiveness of current demand-side measures to support PEV adoption.

References

Archsmith, J., Kendall, A., & Rapson, D. (2015). From cradle to junkyard: Assessing the life cycle greenhouse gas benefits of electric vehicles. *Research in Transportation Economics, 52,* 72–90.

Axsen, J., Bailey, J., & Castro, M. A. (2015). Preference and lifestyle heterogeneity among potential plug-in electric vehicle buyers. *Energy Economics, 50,* 190–201.

Campbell, R., Zhou, Y., Lin, Z., & Ward, J. (2017). Analysis of manufacturer plug-in electric vehicle incentives. In *Transportation research board annual meeting compendium of papers*.

Davis, S., Diegel, S., & Boundy, R. (2013). *Transportation energy data book: Edition 34*. US. DOE 2013. Available from https://cta.ornl.gov/data/editions/Edition34_Full_Doc.pdf.

DeShazo, J. R. (2016). Improving incentives for clean vehicle purchases in the United States: Challenges and opportunities. *Review of Environmental Economics and Policy, 10*(1), 149–165.

DOE-EPA. *Federal tax credits for all-electric and plug-in hybrid vehicles*. Available from https://www.fueleconomy.gov/feg/taxevb.shtml.

Dua, R., White, K., & Lindland, R. (2019). Understanding potential for battery electric vehicle adoption using large-scale consumer profile data. *Energy Reports, 5,* 515–524. https://doi.org/10.1016/j.egyr.2019.04.013.

EPA. (2015). *Light-duty automotive technology, carbon dioxide emissions, and fuel economy trends: 1975 through 2015*. Available from https://www.epa.gov/fuel-economy-trends/trends-report.

EPA-NHTSA. (2012). *2017 and later model year light-duty vehicle greenhouse gas emissions and corporate average fuel economy standards*. Available from https://www.federalregister.gov/documents/2011/12/01/2011-30358/2017-and-later-model-year-light-duty-vehicle-greenhouse-gas-emissions-and-corporate-average-fuel.

Gillingham, K. (2014). Identifying the elasticity of driving: Evidence from a gasoline price shock in California. *Regional Science and Urban Economics, 47,* 13–24.

Gillingham, K., Jenn, A., & Azevedo, I. M. L. (2015). Heterogeneity in the response to gasoline prices: Evidence from Pennsylvania and implications for the rebound effect. *Energy Economics, 52,* S41–S52.

Gillis, J., Brobeck, S., & Cooper, M. (2016). *Automakers are on the road to meeting fuel efficiency standards*. Available from https://consumerfed.org/wp-content/uploads/2016/04/2016-Fuel-Economy-Report-April-25–2016.pdf.

Henly, J., Ruderman, H., & Levine, M. D. (1988). Energy saving resulting from the adoption of more efficient appliances: A follow-up. *The Energy Journal, 9*(2), 163–170.

Holland, S. P., Mansur, E. T., Muller, N. Z., & Yates, A. J. (2016). Environmental benefits from driving electric vehicles? *American Economic Review, 106*(12), 3700–3729.

Jenn, A., Azevedo, I. M. L., & Michalek, J. J. (2016). Alternative fuel vehicle adoption increases fleet gasoline consumption and greenhouse gas emissions under United States corporate average fuel economy policy and greenhouse gas emissions standards. *Environmental Science and Technology, 50*(5), 2165–2174.

Langford, R. P., & Gillingham, K. (2015). Market benefits from hybrids. *Working paper*.

Reiss, P. C., & White, M. W. (2005). Household electricity demand, revisited. *The Review of Economic Studies, 72*(3), 853–883.

Sheldon, T. L., DeShazo, J. R., & Carson, Richard T. (2017). Demand for battery-electric and plug-in hybrid vehicles: Policy lessons for an emerging market. *Economic Inquiry, 55*(2), 695–713.

Sheldon, T. L., & Dua, R. (2018). Gasoline savings from clean vehicle adoption. *Energy Policy, 120,* 418–424. https://doi.org/10.1016/j.enpol.2018.05.057.

Sheldon, T. L., & Dua, R. (2019a). Assessing the effectiveness of California's "replace your ride". *Energy Policy, 132,* 318–323. https://doi.org/10.1016/j.enpol.2019.05.023.

Sheldon, T. L., & Dua, R. (2019b). Measuring the cost-effectiveness of electric vehicle subsidies. *Energy Economics, 84,* 104545. https://doi.org/10.1016/j.eneco.2019.104545.

Small, K. A., & Van Dender, K. (2007). Fuel efficiency and motor vehicle travel: The declining rebound effect. *The Energy Journal, 2007,* 25–51.

West, J., Hoekstra, M., Meer, J., & Puller, S. L. (2017). Vehicle miles (not) traveled: fuel economy requirements, vehicle characteristics, and household driving. *Journal of Public Economics, 145,* 65–81.

Zivin, G., Joshua, S., Kotchen, M. J., & Mansur, E. T. (2014). Spatial and temporal heterogeneity of marginal emissions: Implications for electric cars and other electricity-shifting policies. *Journal of Economic Behavior & Organization, 107,* 248–268.

The Present and Future Market for PEVs in Canada: Evidence from a Mixed-Method Research Program

Jonn Axsen and Zoe Long

Abstract Understanding the current and future market for plug-in electric vehicles (PEVs) requires an understanding of consumers, technology suppliers, policy, and the interplay between. We illustrate a mixed-method research program that explores this interplay, using the case of Canada. Our approach is "reflexive" in two ways. As researchers, we connect insights from several research methods, including qualitative interviews, quantitative surveys, choice models, technology adoption models, and policy evaluation. For consumers, we prompt reflexivity by helping them to learn about PEVs, and how the technology may (or may not) connect with their lifestyle. Consumer data were collected from samples of "Pioneers" (the earliest buyers of PEVs), and "Mainstream" new-vehicle buyers. Results show that Mainstream consumers have low awareness of PEVs, though one-quarter demonstrate interest in purchasing one after learning about it. Interview and survey results indicate diverse consumer perceptions and motivations, including orientations towards new technology, the environment, and practicality. We combine these survey data with technological constraint data (supply and charging access) to construct a technology adoption model, which can help explain limited market share (1–2% PEV sales) and simulate how policy might increase future sales. Finally, we demonstrate how insights from this research can be used to evaluate PEV-supportive policies.

Keywords Consumer research · Plug-in electric vehicles · Consumer behaviour · Technology forecasting · Policy evaluation

Introduction

Despite much effort and policy support, alternative fueled vehicles have repeatedly failed to displace gasoline and diesel powered passenger vehicles in the vast majority of countries (Melton et al. 2016). In recent years, plug-in electric vehicles (PEVs) have shown promise in several markets, including "pure" battery electric vehicles (BEVs) and plug-in hybrid electric vehicles (PHEVs) that can be powered by gasoline

---

J. Axsen (✉) · Z. Long
Simon Fraser University, Burnaby, BC, Canada
e-mail: jonn_axsen@sfu.ca

M. Contestabile et al. (eds.), *Who's Driving Electric Cars*, Lecture Notes in Mobility,
https://doi.org/10.1007/978-3-030-38382-4_5

and plugged into recharge. In 2017, PEV sales comprised 0.9% of new light-duty vehicle sales in Canada and have risen to 2.1% in the first half of 2018 (Klippenstein 2018)—a new market share that is similar to many other developed countries (though, a few regions have achieved much greater sales rates, as covered in other chapters of this book).

Understanding the current and potential future market for PEVs requires an understanding of policy, consumers, technology suppliers, and the interplay between. Because increased PEV uptake may be signaling the beginning of a transition to widespread electric mobility, many stakeholders want to better understand PEV buyers, including their preferences, motivations, and potential PEV usage patterns. Specifically, national and sub-national policymakers have sought to understand how best to incentivize and support further PEV market growth; automakers have sought to understand what PEV designs and attributes will be valued by consumers; and electric utilities have sought to understand how widespread PEV adoption might impact electrical load. This chapter summarizes a research program that explores several aspects of the potential PEV market, including comparison between current PEV owners (who we refer to as "Pioneers") and "Mainstream" conventional new vehicle buyers. In efforts to anticipate future sales, it is arguably important to identify and to study the "Potential Early Mainstream" consumers—the likely next adopters of PEVs. Following modeling studies which suggest that plug-in electric or other alternative fuel vehicles must make up a majority of market share by 2040 to meet 2050 greenhouse gas reduction targets (International Energy Agency 2015; Williams et al. 2012), PEV acceptance among Potential Early Mainstream buyers may be necessary to displace fossil fuels and decarbonize the passenger vehicle sector.

We use the case of Canada, where individual provinces demonstrate a variety of policy and consumer contexts. The movement of people and goods presently accounts for about one-quarter of Canada's greenhouse gas (GHG) emissions, with passenger vehicles alone generating 16% (Environment and Climate Change Canada 2017). Studies in Canada and the US suggest that meeting national and provincial GHG targets would require about 80–90% of passenger vehicle sales be those which do not produce GHG emissions by 2050 (Bahn et al. 2013; Greene et al. 2014; Sykes and Axsen 2017). With Canada's current electric grid, a PEV could reduce emissions by 45–98% compared to a conventional gasoline vehicle (Axsen et al. 2015b). These reductions will become even more substantial as Canada's electricity grids continue to move towards low-carbon generation.

In this chapter we use Canada to demonstrate an approach to PEV consumer research that is "reflexive" in at least two ways. First, as researchers, we seek to connect insights from a variety of research methods, including qualitative interviews, quantitative surveys, latent class choice models, technology adoption models, and policy evaluation. This can be called a "mixed-method" approach in that we integrate insights from both quantitative and qualitative research (Creswell 2011; Teddlie and Tashakkori 2011). Secondly, we support reflexivity in consumers by helping participants to learn about and reflect upon PEVs, and how the technology may (or may not) connect with their lifestyle. We have developed these methods to aid with the inherent difficulty in trying to anticipate consumer response to new technologies, especially when consumers have little to no experience with that technology. We presently integrate these findings around the following research objectives:

- Qualitatively explore consumer perceptions, preferences, and motivations regarding adoption and usage of PEVs (using semi-structured interviews);
- Quantitatively analyze consumer demand, perceptions, preferences, and motivations concerning PEVs (using representative survey samples);
- Simulate the role of technology constraints and policy in current and future PEV sales (using a technology-adoption model); and
- Develop a simple framework to evaluate PEV-supportive policy (using insights from all the above).

In this chapter we provide a description of this reflexive, mixed-method research approach, and illustrate it by reporting results from several related research projects. We first provide a rationale and overview of the program, as well as contextual details regarding the case of Canada, before working through sections on each of the four objectives.

Rationale for a Reflexive, Mixed-Method Research Approach

Anticipating consumer demand for emerging or not yet existing technologies is one of the great challenges of market research (Hauser et al. 2006). The consumer may have little or no awareness of the technology, misinformation about the technology, or no opportunity to experience and reflect upon the technology. For such products, consumers have significant uncertainty in estimating product usefulness, they may face difficulty in trying to translate the new attributes (or attribute levels) into personal benefits, and realization of benefits may require changes in consumption or use patterns (Hoeffler 2003). To be successful in the market, adoption and use of such products are likely to require consumer learning, and could in turn induce other changes in behavior (Urban et al. 1996). Forecasts of consumer response and subsequent market share will depend on a variety of complex factors, including changes in technology, diffusion of information, discovery of new uses, reduction in price, change in competition, and growth of related infrastructure (Al-Alawi and Bradley 2013; Brand et al. 2017). For these reasons, we have developed a reflexive approach to studying consumers and to integrating insights across a mixed-method research program more generally.

While **reflexivity** is defined differently in a variety of contexts, generally it describes "situation[s] in which individuals are responding to information about themselves" (Turrentine and Kurani 1998, p. 212). Reflexivity has also been described as a sort of "feedback" on success and failures that helps the individual to assess progress in an activity or learning experience (Greenblat 1981). To conceptualize the consumer, we draw from Giddens' (1991) notion of reflexivity within the context of modernity: under conditions of modernity, individuals and institutions lack the certainty and direction provided in traditional societies, and therefore individuals must interact with other people, institutions, and even objects to meet basic needs and to trial, establish, and reinforce self-identity. This process is "reflexive" in the sense that the individual is continually interacting with their environment, including new technologies. Individuals are more likely to become aware of, actively learn about, and seek experience with products they see as aligning with aspects of their current or their desired self-concept and lifestyle (Axsen and Kurani 2012). Continuing with the notion of reflexivity within modernity, the consumer is not just an "actor" or "respondent" that plays a role and enacts their preferences—but rather, they are a "participant" in discovering and learning about new products and behaviours, and how these may relate to their lifestyle and self-concept. Hence, we utilize the term **Reflexive Participant** to describe an approach or perspective to studying consumers.

Taking this notion further, the research process itself can be described as reflexive, where the researcher is also participating in the learning process with the participant (Bourdieu 1977; Bourdieu and Wacquant 1992). In this chapter we describe how several projects in our research program link together in this regard, including different data collection methods (interviews and large-scale surveys), different methods of engagement with consumers (open-ended questions, stated-choice experiments, design space exercises), and different models (discrete choice models, technology-adoption models, energy-economy models). Because these studies are meant to yield policy relevant insights, the researchers are also interacting with different policymakers and stakeholders over time, which helps to identify and refine research objectives for subsequent projects.

Overview of the Research Program

Figure 1 provides a simple conceptualization of how the various methods link and interact together. The figure is not meant to be exhaustive in its coverage of research methods, nor is it meant to imply a necessary order (e.g., top to bottom). In fact, the arrows are meant to convey the reflexive iterations between different methods

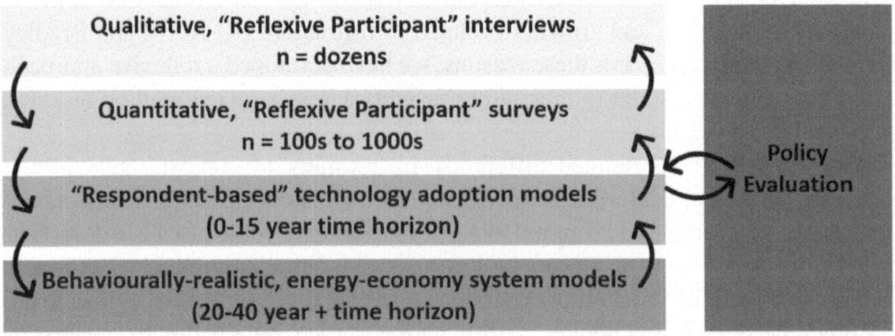

Fig. 1 Overview of our multi-method "reflexive" approach to research

and projects, which can interact in a number of ways. As examples: results from one project can help to generate research questions or hypotheses to explore in another; incomplete explanations from one study or method can help to identify gaps that could be better filled by another method; and consistency in findings across methods can help to improve confidence in those findings. We illustrate this process throughout the chapter using the Canadian Plug-In Electric Vehicle Study (Axsen et al. 2015b), which involved consumer data collection from 2013 to 2015 in Canada.

While the broader research program has been iterating between different methods over the past decade, here we begin by explaining how we collect data directly from consumers through qualitative interviews and quantitative surveys. In-person, semi-structured interviews (usually with about 20–30 households) can further the understanding of respondent preferences and provide important insights into individual (and household) motivations and decision-making processes. Semi-structured interviews are useful for identifying participants' perceptions because the method allows and encourages complex and unanticipated answers while capturing some of the wider context and meaning behind responses that can be lost in large-scale quantitative surveys (Guba and Lincoln 1994; McCracken 1988). Further, we include interactive elements (part of the "structure" of semi-structured interviews), which give participants the opportunity to think through and develop their responses—a process that itself can also be observed and documented as part of the analysis. Examples include stated response exercises, where participants construct and select a particular PEV design, and explain why they selected that particular design. Of course, one must be careful in generalizing results from a small-sample qualitative exploration to the broader population. For this reason, qualitative research is considered hypothesis generating (rather than hypothesis testing) and exploratory in nature (McCracken 1988).

As a complement to such interviews, we implement quantitative surveys to collect data from much larger samples of new vehicle-buying households (usually with representative samples of 100–1000s of respondents). The Reflexive Participant Approach to survey research draws from the earlier work of Turrentine and Kurani (1998) and Lee-Gosselin (1996), which has since been adapted to several surveys on PEVs (Axsen and Kurani 2009, 2013a, b). This approach is characterized by several components, including:

- in-depth collection of respondent background (e.g., context, travel patterns, technological awareness, and vehicle history);
- a technology primer to help the respondent to learn about the technology;
- a task or exercise to trigger reflection upon the technology, in relation to one's lifestyle; and
- a preference eliciting response exercise, for example a stated choice experiment and/or design space exercise, completed after the reflexive experience (i.e., within one to two weeks after), and personalized to match the respondent's context.

As with the interview method, the intention of the Reflexive Participant Approach to surveys is to help respondents learn about and reflect on how the case technology is compatible (or not) with their lives. The interactive exercises in these surveys also help respondents construct and express their preferences for PEVs, which we assume were likely non-existent prior to beginning the survey. The insights gained from consumers via these in-depth, reflexive interviews and surveys can be used to inform technology-adoption models (as well as energy-economy models, not addressed here—see Mau et al. (2008) and Axsen et al. (2009) for more details). Incorporating rich, empirical consumer data into these models is useful for accurately representing consumer behaviour, which can lead to more realistic estimates and forecasts of future demand given specific technology, policy, and market assumptions. Such models can be well-suited for examining how prospective policies or combination of policies can impact adoption of new technologies, such as PEVs.

The Case of Canada: PEV Sales and Policy

Before summarizing our results, we provide a brief context to our case study of Canada. At the time of writing this Chapter (in 2018), PEVs sales in Canada have continued to grow, though still remain low relative to the 2040 or 2050 sales or GHG goals noted above. Between 2011 and 2017, the number of PEVs on the road increased from 1500 to 39,000 (Klippenstein 2018). Similarly, the number of publicly accessible PEV chargers has increased from 0 in 2011 to 5841 in 2017 (International Energy Agency 2018). Uptake has significantly varied by province; in 2015, PEVs represented between 0 and 2% of new vehicles sales across Canadian provinces (Table 1). Most of the sales (95%) were concentrated in Canada's most populous provinces, namely Quebec, Ontario, and British Columbia (with a total of over 40,000

Table 1 Canadian and provincial 2015 PEV new market share and cumulative sales (2011–2017)

Provinces	Electric vehicle new market share (2015) (%)	Electric vehicle sold (2011–2017)
Canada	0.9	47,788
British Columbia	2.0	8662
Alberta	0.3	1189
Saskatchewan	0.1	113
Manitoba	0.2	234
Ontario	0.7	16,644
Quebec	1.4	20,639
New Brunswick	0.1	123
Nova Scotia	0.1	133
Prince Edward Island	0.0	16
Newfoundland and Labrador	0.0	31

Adapted from Axsen et al. (2016b)

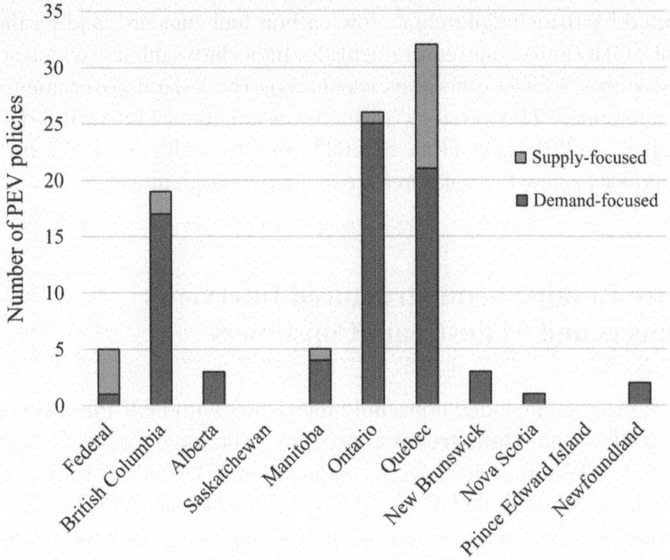

Fig. 2 Number of PEV-supportive policies, by provincial and federal government in Canada (Melton et al. 2017)

BEVs and PHEVs), which account for 75% of the total population. As of 2017, thirty-four PEV models were available for sale Canada (in at least some part), which is a substantial improvement from the initial three models offered in 2011. Sales of PHEVs and BEVs are approximately equal in Canada (as of 2017). The Chevrolet Volt dominates PHEV stock, while BEV stock is split among several models (in order of total sales): the Tesla Model S, Nissan Leaf, Tesla Model X, and Chevrolet Bolt.

Canada has a patchwork of PEV-supportive policies (Fig. 2), which are mostly **demand-focused** and in place at the provincial level (see Melton et al. (2017) for more details). Demand-focused policies focus on encouraging consumers to purchase ZEVs, for example by offering financial or non-financial incentives, making PEVs more attractive relative to other technologies. For example, three provinces (British Columbia, Ontario and Quebec) currently have financial incentives for PEVs (though as of writing this chapter, Ontario is in the process of canceling their incentives). These same three provinces also allow PEV drivers to access high occupancy vehicle lanes, and the whole country is implementing carbon pricing.

Canada has relatively fewer **supply-focused policies** (Fig. 2), defined as those that generally encourage or require auto manufacturers to sell PEVs, for example by specifying a minimum share of vehicles sold to be zero-emissions vehicles or requiring that vehicles sold in a region meet a fleet average emissions intensity. Quebec is the only province that has implemented the most direct supply-focused policy, a Zero-Emissions Vehicle (ZEV) mandate. It requires that 22% of total light-duty vehicle sales be ZEVs by 2025 (Government of Quebec 2017). PEV uptake is also supported by British Columbia's low-carbon fuel standard, and by the federal government's GHG emissions requirements for light-duty vehicles (which are largely harmonized with US GHG emissions standards). The federal government has also developed a national "ZEV strategy", which was announced in April 2019 and sets national targets for PEV sales (10% by 2025, 30% by 2030, and 100% by 2040) as well as introduces a new federal purchase incentive programme.

Qualitative Results: Semi-structured Interviews with Pioneers and Mainstream Consumers

Our research program included household interviews with PEV **Pioneers**, who currently own a PEV, and **Mainstream** consumers, who have recently bought a new vehicle but not a PEV. Interviews were conducted in 2013 with 22 Mainstream consumers (Axsen et al. 2017), and from 2014–15 with 17 Pioneers (Axsen et al. 2018). Each interview was between one and two hours long and mostly took place in participants' homes. In these studies, we explored consumer perceptions of the PEVs at the time of the interview. We (the interviewers) did not comment on the technology itself or the "correctness" of participants' perceptions or understanding of these technologies.

The objective of these interview studies was to examine consumers' PEV awareness, perceptions, and motivations regarding PEVs. Analysis followed a framework that categorizes perceptions of PEV attributes (or benefits) as functional, symbolic, or societal (Table 2) (Axsen and Kurani 2012). Functional attributes, defined as perceptions of what the technology does for the individual (e.g., performance, cost), can play a strong role in consumer decision making. However PEV interest is typically driven by other factors as well. Symbolic attributes, defined as perceptions of what message or messages the technology conveys (e.g., a vehicle that represents being ethical, or being intelligent), also can influence consumers choices. PEVs can also be perceived as having societal attributes or benefits, defined as perceptions about what the technology does for society (e.g., reducing air pollution or GHG emissions).

Table 2 Categorization and examples of PEV attributes (adapted from Axsen and Kurani 2012)

Attribute type	Description	Examples
Functional	Technology's use and operation	High performance (e.g. fast acceleration) Affordable (e.g. low operating costs)
Symbolic	Expression of meaning or self-identity	Represents being ethical "Looks" intelligent
Societal	Impact on society more broadly	Reduce air pollution and greenhouse gas emissions Increase energy security

Insights from PEV Pioneers

Interviews with PEV Pioneers revealed a diverse set of perceptions and 18 distinct motivations pertaining to purchase and use of the PEV. As expected, Pioneers generally displayed a high level of awareness about PEVs, including sophisticated and well-formed perceptions which were overall positive. Although motivations that pertained to functional attributes were most common (10 of the 18), we identified four symbolic and four societal motivations. The most frequently mentioned motivations pertained to cost minimization and practicality (functional attributes), environmental protection (societal attributes), as well as embracing new technology and being a PEV ambassador to others (symbolic attributes). Less frequently mentioned motivations included reduced oil dependence and support for innovative companies. Most participants stated that they have adjusted their vehicle usage since purchasing a PEV.

All Pioneer participants reported frequent usage of public charging infrastructure, with the most frequently mentioned motivations being convenience, followed by the symbolism of being an ambassador—the latter of which has not been identified in previous research. Interestingly, despite their high reported usage of public charging, most participants (13 of 17) mentioned that they did not feel that infrastructure was necessary for their typical driving needs. This particular finding may be unique to North America and other regions with a high level of home charging access (e.g., where most new car buyers own their home, which is a single-family dwelling with a garage).

Using these interview data, we constructed four segments of Pioneer participants which differ by their engagement in pro-environmental and technology-oriented lifestyles (Fig. 3):

- "Tech Enthusiasts" demonstrated high levels of engagement in a technology-oriented lifestyle, such as purchasing the newest technologies, and expressed little to no engagement in pro-environmental practices.
- "High-tech Greens" were engaged in both technology-oriented and pro-environmental lifestyles and were more likely to engage in activities related to pro-environmental technology.
- "Low-tech Greens" had high levels of engagement in a pro-environmental lifestyle, such as eating a vegetarian diet, but little to no engagement in technology-oriented practices. These participants emphasized the motivations of environmental protection and reduced oil dependence for purchasing a PEV.
- Those in the "Practical" segment did not engage in either pro-environmental or technology-oriented lifestyles, instead expressing engagement in a family-, recreation-, or social-oriented lifestyle. These participants were most likely to emphasize practical motives for buying their PEV, such as cost savings or convenience.

A major insight from this work is the importance of heterogeneity in terms of lifestyle, perceptions, and motivations among PEV Pioneers. This contrasts with

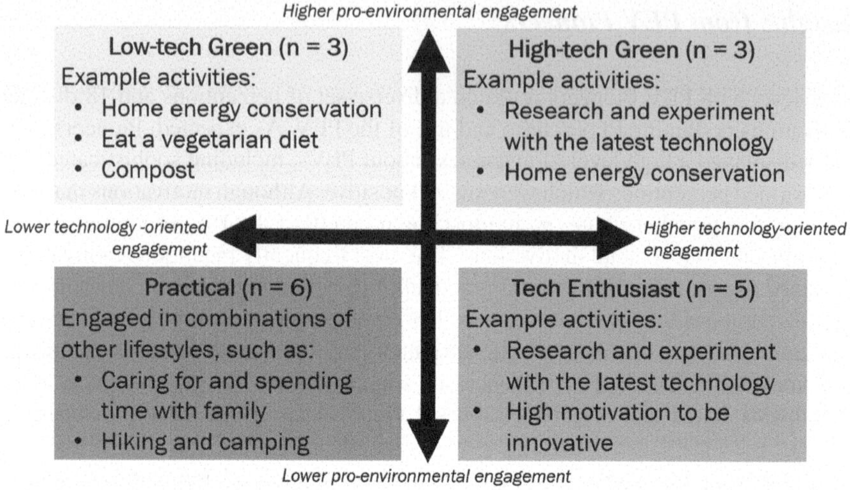

Fig. 3 Four lifestyle segments of PEV pioneers. Adapted from Axsen et al. (2015b)

some common perspectives, such as research following the Diffusion of Innovations approach (Rogers 2003) that tends to assume the earliest buyers of new technologies are homogenous, and mostly (or entirely) motivated by their trait of being "innovative". In this case of PEV Pioneers in Canada, we find evidence that suggests a significantly heterogeneous set of consumers' motivations, and of consumers.

Insights from Mainstream Consumers

In contrast, interviews with Mainstream consumers revealed overall misunderstanding and low awareness of PEVs. Specifically, 19 of 22 households interviewed were unaware of PHEVs as a distinct type of vehicle, confusing the description with that of conventional hybrids (in terms of whether the vehicle could be fueled with gasoline, electricity, or both). We categorized households by their awareness, familiarity, and understanding of PEVs (Fig. 4). Fourteen households were categorized as having "Basic" knowledge: they were aware that hybrids and BEVs existed but not that PHEVs exist and did not understand how hybrids are refueled (believing they can be recharged using external electricity). Five households were characterized as having "Moderate" knowledge, understanding how hybrids are refueled, but still were unaware of PHEVs. Only three households were identified as having "High" knowledge, indicated by their awareness of PHEVs, understanding of how they are refueled, and how they differ from hybrids and BEVs.

The interview protocol then helped to explain these basic concepts to participants to permit further discussion of consumer perceptions and interests relating to

<table>
<tr><th rowspan="2">Category of electric vehicle knowledge</th><th rowspan="2">Households</th><th colspan="3">Knowledge of vehicle type</th></tr>
<tr><th>Hybrids</th><th>Plug-in hybrid vehicles (PHEV)</th><th>Battery electric vehicles (BEV)</th></tr>
<tr><td>Basic knowledge</td><td>Fay
The Mathews
Andreas
The Chens
Lei
Violet
Veronica
The Parks
The Morrettis
Christine
Al
The Dimirovics
Margaret
Clair</td><td>Experience, but low familiarity</td><td rowspan="2">Not aware that this type of vehicle exists</td><td rowspan="3">No experience, but aware that this type of vehicle exists</td></tr>
<tr><td>Moderate knowledge</td><td>The Youngs
Sandra
Omar
Kevin
The Nicolovs</td><td rowspan="2">Experience and familiarity</td></tr>
<tr><td>High knowledge</td><td>Daryl
Liz
The Fengs</td><td>Experience and familiarity</td></tr>
</table>

Fig. 4 Interviewed households categorized by their PEV knowledge (Axsen et al. 2017)

PEVs. Unlike the Pioneer households, many perceptions of PEVs held by Mainstream respondents were negative or uncertain. Five distinct themes of perceptions of PEV attributes also emerged:

- Range (Functional): Twelve households expressed that the range of a BEV would be insufficient for their needs—for example, that a destination they wished to visit would be beyond the presumed maximum range of the available BEVs.
- Charging (Functional): Many participants (15) perceived that the challenge of finding convenient places to charge away from home would be a drawback of PEV ownership—however three participants perceived that it would be a benefit to recharge at home.
- Cost (Functional): Most participants (19) perceived that PEV adoption would result in saving money on fuel costs, though eleven participants were unsure about the operating costs of PEVs.
- Novelty (Functional and Symbolic): Some households (10) believed PEV technology is unproven due to its newness (functional), whereas some (9) households expressed a more symbolic perception of PEVs being different or “strange”, compared to conventional vehicles.

- Environment (Societal): Most participants (19) stated that PEV adoption would help to reduce pollution (though no participants explicitly mentioned climate change), and four households expressed concern that PEV batteries might be an environmental drawback.

Quantitative Results: Large-Scale Surveys with Mainstream Consumers

As noted earlier, our surveys follow the Reflexive Participant Approach, in part to bridge insights from the qualitative studies to a quantitative context. In 2013, we implemented a three-part web-based survey that followed the Reflexive Participant Approach, as part of the Canadian Plug-in Electric Vehicle Study. The final sample consisted of 1754 respondents (including Mainstream consumers and PEV Pioneers) from English-speaking Canada. Following the Reflexive Participant Approach, the survey included three parts:

- Part 1 of the survey was a web-based questionnaire that collected background details on each respondent's vehicle fleet, energy usage, and PEV-related knowledge and perceptions.
- Part 2 was a mail-out and web-based survey package that included several components, including:
  - A technology primer with basic information on the vehicle technologies and charging scenarios to be discussed in the next part of the survey, as well as a travel diary and home charging assessment. As noted, qualitative research repeatedly demonstrates the low familiarity and high confusion among Mainstream respondents. The primer explained the differences between conventional, hybrid, and different types of plug-in electric vehicles, and provided explanations for the vehicles' various features (e.g., approximate driving range, recharging time). It also explained the format of Part 3 of the survey.
  - A 3-day driving diary detailing timing and trip distances, destinations, type of parking locations and proximity of those locations to an electrical outlet, and potential charging opportunities.
  - A more detailed home charging readiness questionnaire, where respondents located and assessed the potential to install a charger at their home parking spot.
- Part 3 was another web-based survey, completed two weeks after Part 1, that elicited respondents' interests and preferences for PEV technologies via two stated response methods, both of which were customized to each respondent based on their intended next vehicle purchase to increase the realism of the response exercise:

- The **design space exercises** provided respondents with a "space" or design envelope—a series of design options that respondents could select from to create their preferred vehicle design given a particular context (following Axsen and Kurani 2009, 2013b; Kurani et al. 1994, 1996). Each respondent first selected a currently available "base" vehicle that represented the body type, price, and fuel economy of the vehicle they would buy next if they were limited to a conventional gasoline vehicle. Design options included conventional, hybrid, PHEV, and BEV versions of the vehicle. Attributes such as range and home charging could be improved at extra cost.
- The **stated choice experiment** was similar, but instead of showing a design space, it presented respondents with "choice sets" including the four drivetrains (conventional vehicle, hybrid, PHEV, and BEV), each differentiated by a set of pre-defined "packages" of vehicle attributes.

In general, the survey revealed that Mainstream respondents have low PEV familiarity and awareness (Axsen et al. 2016a). Only a minority of respondents reported that they were either "familiar" or "very familiar" with the Nissan Leaf (17%), Chevrolet Volt (20%) and Toyota Prius (34%) (Fig. 5). The majority of respondents also demonstrated confusion with the operation of each of the three vehicle models. In Part 1 of the survey, respondents were asked about how the Toyota Prius (hybrid), Nissan Leaf, and Chevrolet Volt were fueled (either gasoline only, gasoline and electricity, or electricity only). Only 20% of respondents successfully indicated that the Toyota Prius is fueled only by gasoline, and one-third successfully identified the correct fueling method for the Chevrolet Volt (gasoline and electricity) and the Nissan Leaf (electricity only). More recent survey work (conducted in 2017) finds

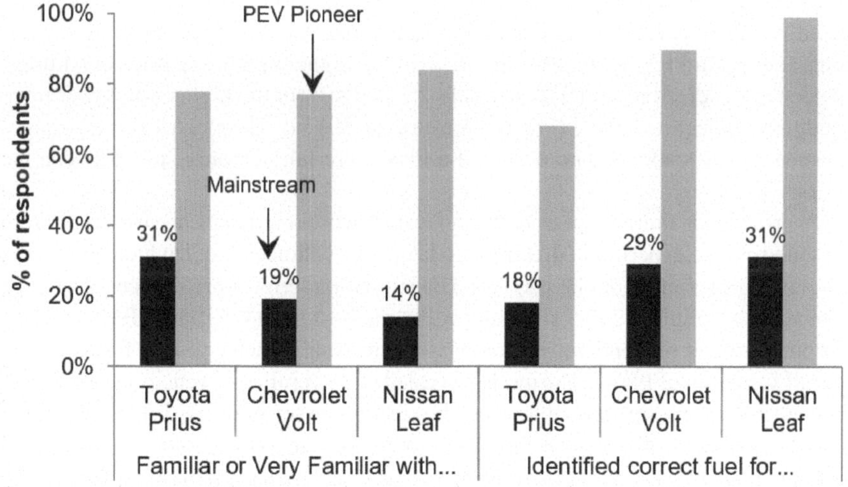

Fig. 5 PEV familiarity and refueling understanding of Mainstream respondents and PEV pioneers. Adapted from Axsen et al. (2016a)

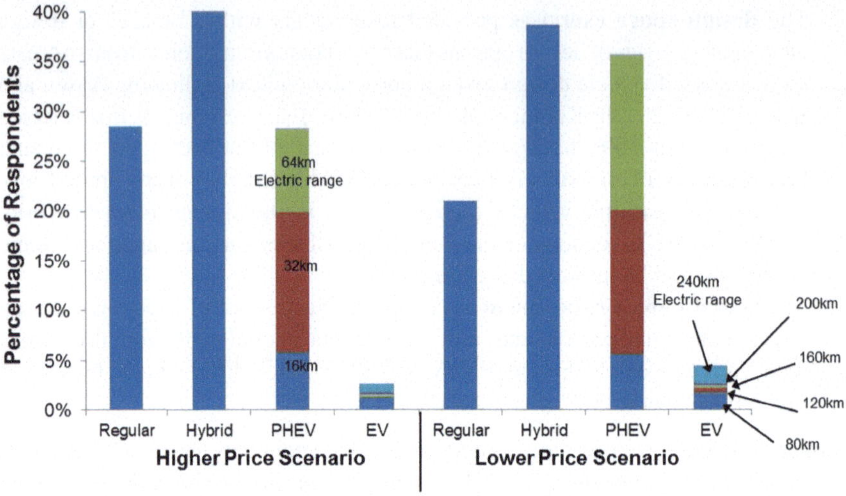

Fig. 6 Proportion of respondents selecting each vehicle type in the higher and lower price scenarios (Axsen et al. 2015b)

similar levels of PEV awareness, familiarity, and fuel type understanding among Mainstream consumers in Canada (Long et al. 2019). Thus, we see no evidence that consumers' awareness of PEVs has increased in the four years since the collection of this 2013 data. In contrast, Pioneer survey respondents have extremely high PEV awareness (Fig. 5).

As noted, the survey helped educate respondents about PEVs and their unique attributes (i.e., during Parts 1 and 2 of the survey), so that they could complete the stated response exercises. In the design space exercise, over one-third of respondents selected a PEV, with 28% and 36% (in the higher and lower price conditions, respectively) selecting a PHEV and only 2% and 4% (in the higher and lower price conditions) selecting a BEV (Fig. 6) (Axsen et al. 2015a). The electric driving ranges selected by respondents who designed a PEV were fairly evenly split between the options.

To provide an alternate perspective on respondent preferences, we used data from the stated choice experiment to estimate a latent class choice model, which divides the sample into segments (or classes) based on similar patterns of preferences. This type of model is also suitable for estimating willingness-to-pay (WTP) for PEVs and their attributes among respondents in each class, a measure which is useful for assessing overall valuation of PEVs. The model revealed that overall preferences for PEVs and their attributes vary substantially across the sample (Axsen et al. 2015a). We identified five classes which differ according to their interest in vehicle types, valuation of fuel savings, lifestyle, and environmental concern (Fig. 7). The largest differences between the classes are overall vehicle preferences, though other differences exist as well. The full details are presented by Axsen et al. (2015a), though we provide

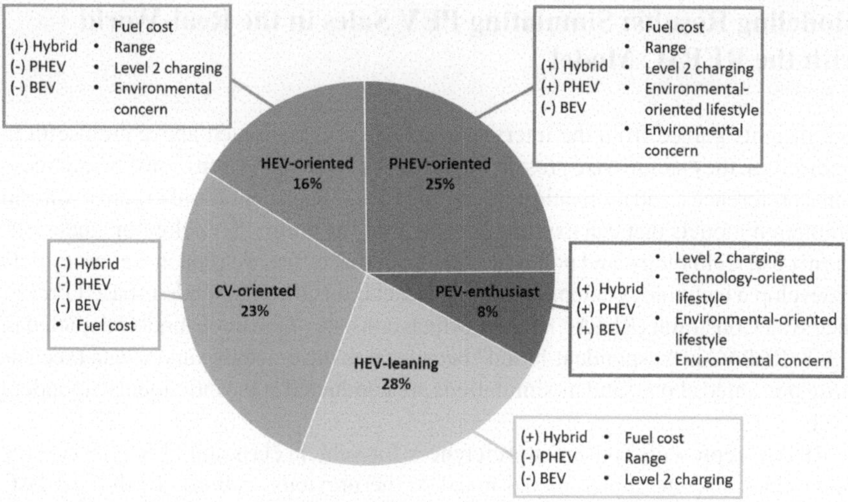

Fig. 7 Description of five latent classes of respondents (each box summarizes attributes that were significantly valued by respondents in that class)

summaries here of the classes that are most relevant to our objectives (focusing on the two classes that best represent the "Potential Early Mainstream" respondents):

- The "PEV-enthusiast" class (8% of the sample) has extremely high WTP values for HEVs, PHEVs, and BEVs, indicating that respondents in this class have very strong positive preferences for PEVs (hence the "PEV-enthusiast" label).
- The "PHEV-oriented" class (25% of sample) generally prefers PHEVs and is on average willing to pay an extra $15,000 for such a design.
- The remaining classes are described as being "HEV-oriented" (16%), "HEV-leaning" (28%), and "CV-oriented" (23%). All three classes have negative or non-significant WTP values for PHEVs and BEVs.

Figure 7 summarizes several other patterns in respondents' attribute valuation and personal characteristics. Respondents in most classes value having Level 2 charging at home. The classes differed by sensitivity to fuel costs—for example, the PHEV-oriented class places the highest value on fuel cost savings. Interestingly, fuel cost is not valued by the PEV-enthusiast class, suggesting that those respondents' PEV interest is not financially motivated. PEV-enthusiast and PHEV-oriented respondents are the most likely to engage in environmental-oriented lifestyles and to have high levels of environmental concern. PEV-enthusiast respondents are most likely to also engage in technology-oriented lifestyles. This PEV-enthusiast class may be indicative of the extreme nature of "very early" PEV buyers in North America who generally have very different characteristics, preferences, and motivations relative to later buyers. In contrast, the PHEV-oriented class positively values PHEVs and HEVs (but not BEVs), and values fuel cost savings while also exhibiting environmental concern and lifestyle engagement.

Modeling Results: Simulating PEV Sales in the Real-World with the REPAC Model

The insights gained from the interviews and surveys presented above are useful in themselves: they summarize present (at the time of data collection) consumer perceptions, preferences, and motivations regarding PEVs. Such data can also help to inform simulation models that can explore the potential for future PEV adoption under different cost, technology, and policy scenarios. In that effort, we used these survey data to develop a technology-adoption model. We call this the REspondent-based Preference And Constraint (REPAC) model, which consists of the three models depicted in Fig. 8. REPAC is "respondent-based" because it is informed by survey data (such as those presented above) and its simulations are conducted at the individual respondent level.

REPAC represents consumer preferences for vehicle class and PEV type (via the latent class discrete choice model noted in the previous section). To this, REPAC applies real-world constraints, specifically: lack of PEV familiarity, lack of home charging, and limited supply of PEVs (including PEV model variety and dealership availability). The latent class model simulates the probability that each respondent will choose a conventional vehicle, hybrid, PHEV, or BEV in the absence of constraints on the choice. The output of this sub-model can be thought of as unconstrained demand (or latent demand)—preferences that are not actually realized as PEV demand due to constraints. The vehicle model describes the cost and performance of the archetypal vehicles (one for each drivetrain) that we use to populate the choice model simulation. Finally, the constraints model represents the factors that prevent the unconstrained demand of each respondent from being realized as sales.

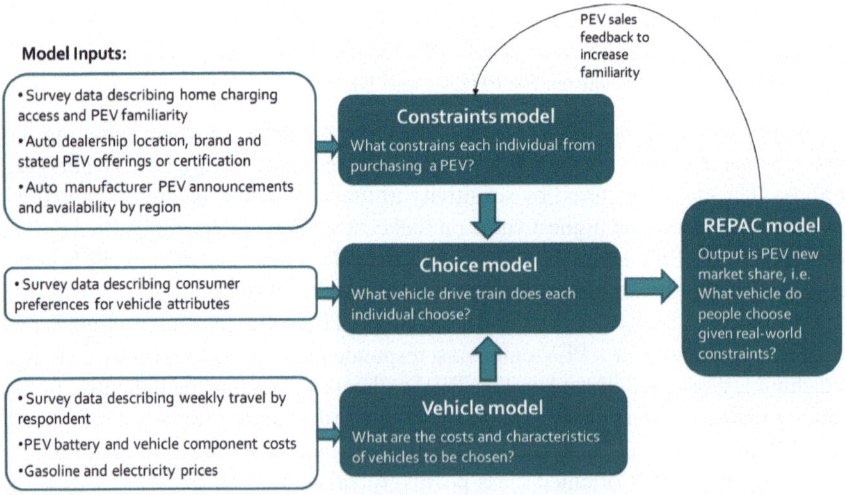

Fig. 8 Structure of the REPAC market share forecast model (Wolinetz and Axsen 2017)

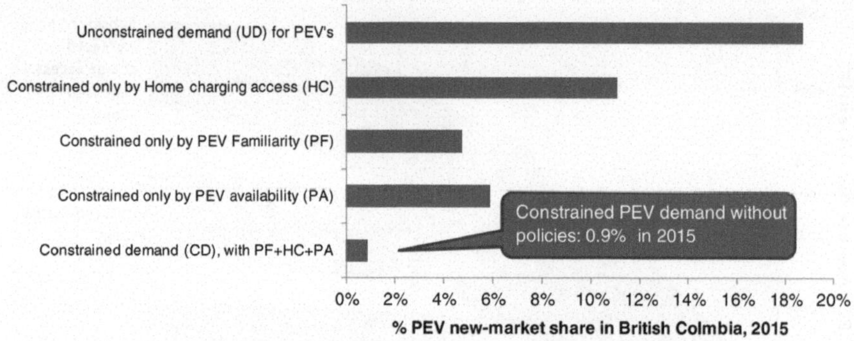

Fig. 9 Impact of REPAC constraints on PEV new vehicle market share, 2015 (Wolinetz and Axsen 2017)

Figure 9 illustrates the individual and combined effects of the three constraint categories applied in REPAC for the model year 2015, using data from the survey and latent class model presented above (Wolinetz and Axsen 2017). Unconstrained demand (or latent demand) for PEVs was found to be 19% of new market share. Applying only the home charging constraint (where 54% of respondents have reliable home charge access) reduced PEV demand to 11%. Applying only the PEV familiarity constraint (where 22% of respondents were familiar with PEVs prior to taking the survey) reduced the PEV new market share to 5%. Applying only the PEV supply constraint reduced market share to 6%. Applying all three constraints yielded a constrained demand of 0.9% in 2015, which was very similar to the actual market share in British Columbia in that year—roughly 1% prior to the introduction of a PEV subsidy (Stevens 2016).

We also applied this model to help policymakers anticipate the effects of different policies on PEV adoption, for instance in five to 15 years into the future. Figure 10 depicts the modeled PEV new market share trajectory in three policy scenarios from 2015 to 2030. In the "Weaker" demand-focused policy scenario, a short-term PEV subsidy (2015–2020) results in a PEV new market share between 2 and 12% by 2020. After the subsidy is removed in 2021, new market share falls slightly to between 1 and 10% by 2030. The "Stronger" demand-focused policy scenario (maintaining the subsidy to 2030 and increasing home charging access) increases PEV new market share to 4–17% in 2020 and 17–28% by 2030. Finally, the "Strong supply- and demand-focused policy" scenario adds the supply-focused policy that generally approximates the effects of policies like the ZEV mandate. This final scenario demonstrates the impact of an aggressive supply-focused policy, inducing PEV availability to increase on a trajectory that eliminates supply constraints by 2030 (increasing the variety of PEV models to 20 in 2020 and 60 in 2030, spread equally across all vehicle classes, available from all dealerships). In this scenario, the PEV new market share forecast reaches 38–49% by 2030.

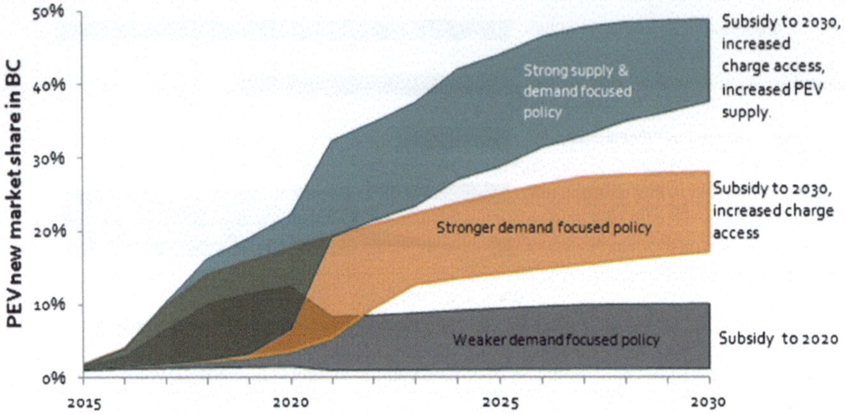

Fig. 10 PEV new market share under policy scenarios, with shaded areas representing uncertainty (Wolinetz and Axsen 2017)

Informing Policy Evaluation: The PEV Policy Report Card

We believe there are a number of useful policy implications arising from the insights gained from these interview, survey, and modeling results (here and elsewhere in our research program and in the literature more broadly). However, policymakers often note that such academic results, while rigorous, are difficult for governments to access and translate into policy. In order to help bridge this gap, we developed a PEV policy evaluation framework (Melton et al. 2017). The framework boils these insights down to one goal: PEVs capturing at least 40% of new passenger vehicle market share by 2040 (in line with deep GHG abatement targets). The framework uses available research and literature (such as that summarized above) to translate different policy types into 2040 market share "points" based on policy stringency and duration. To aid in the communication of this analysis, we assigned an overall letter grade (A, B, C, D or F) to a given region based on the suite of policies in place, which is why we call it the "PEV Policy Report Card". We applied this framework to Canada's 10 provinces based on policies in place in November 2016, effectively assigning report cards to each province and to Canada as a country.

Our framework considered eight categories of PEV policies. The five demand-focused policies included: financial incentives, high occupancy vehicle lane access, public (non-home) charging infrastructure deployment, PEV-ready building codes, and carbon pricing. The three supply-focused policies considered were zero-emission vehicle (ZEV) mandates, fuel economy standards, and low-carbon fuel standards. We did not evaluate other policies because their local market share impact is likely small (e.g., voluntary programs) or particularly uncertain (e.g., research and development support). For each policy type, we identified a "benchmark" stringency and duration based on a judgmental estimate of the maximum politically acceptable level in North America. In some cases, the benchmark was based on already implemented

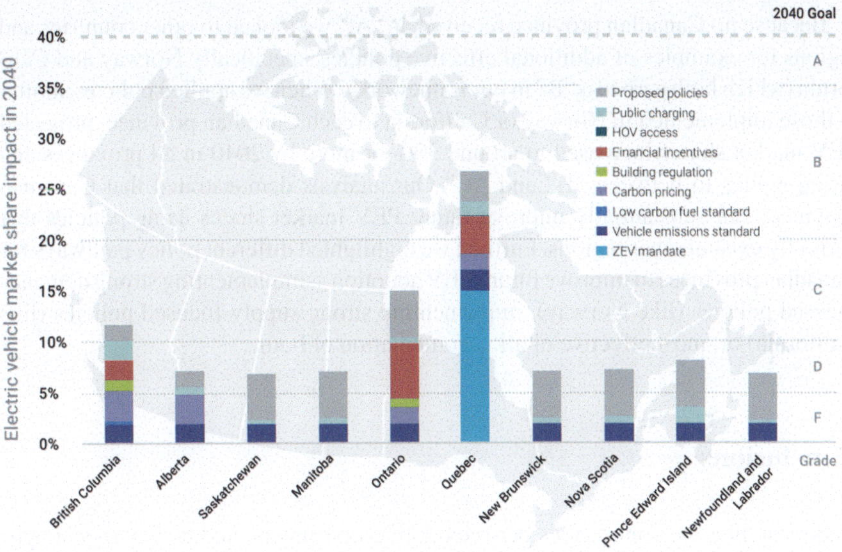

Fig. 11 Evaluation of PEV policies, by province (Axsen et al. 2016b)

policies, while in other cases the benchmark was taken from the literature. We then drew from available literature to estimate the effect of that benchmark stringency on the "40% by 2040" PEV sales target we assumed, in terms of new market share "points". To estimate the total effect of a given region's portfolio of PEV-supportive policies on 2040 market share, we summed the estimated market share points for each individual policy. While technology-adoption models (such as REPAC) have been developed that can explore such policy interactions, the "PEV Policy Report Card" was intentionally designed to be a more accessible tool for policymakers.

Our evaluation found that all provinces fall short of the 2040 goal of 40% PEV market share (Fig. 11). Six provinces received an "F" because current policies are unlikely to encourage 2040 PEV new market share to rise above 5%: Saskatchewan, Manitoba, New Brunswick, Nova Scotia, Prince Edward Island, and Newfoundland and Labrador (together representing 13% of the Canadian population). Quebec ranked highest among Canadian provinces, with a 2040 market share of 23% or a "B-". This score was in large part due to its ZEV mandate, which accounts for 15 market share points—the highest impact of any single policy in Canada. Policies in British Columbia, Ontario and Alberta were expected to have lesser impacts, with PEV new market share points of 10% or less by 2040. The most significant policies in these provinces included financial incentives (British Columbia and Ontario) and carbon pricing (British Columbia and Alberta). Consequently, British Columbia and Ontario received a "C-", while Alberta received a "D". Finally, by prorating each province's score according to the size of its vehicle fleet, we determined a net grade of "C-" for Canada as a whole (not shown).

Because no Canadian province received an "A," we looked to other countries and regions for examples of additional effective policies, specifically Norway and California (which both earned a "B" in our framework). When we applied policies similar to those implemented in Norway or California to each Canadian province, projected PEV market shares increased to around 30% or more by 2040 in all provinces and raised grades to between "B" and "A." Our analysis demonstrated that Canadian provinces can substantially improve future PEV market shares using policies that currently exist in other regions. Further, we highlighted different policy pathways for Canadian provinces to improve future PEV adoption—implementing strong demand-focused policies (like Norway), implementing strong supply-focused policies (like California) or, most effective of all, a combination of both.

Conclusions

Understanding the connections between policy, consumers, and technology suppliers is critical to understanding the current and potential future market for PEVs. Our research program begins with understanding the consumer, using a mixed-method, reflexive approach, which aims to help both the researcher connect insights from a variety of research methods and the consumer to learn about the technology. Results from interviews and surveys of PEV Pioneers and Mainstream new vehicle buying households have revealed limited awareness, a wide variety of perceptions and motivations regarding PEVs, and a high degree of heterogeneity across consumers. Interestingly, we observe a strong potential for PEV demand – once Mainstream new car buyers learn about PEV technology (mainly the attributes of PHEVs), then at least 15-20% (or more) express an interest in purchasing one, given realistic price premiums.

These behavioural insights (preferences and constraints) were integrated into a technology-adoption model, REPAC. Modeling simulations help to explain the currently limited PEV market in Canada and demonstrate how policy can build the market in the next 10 to 15 years. Finally, our "PEV Policy Report Card" illustrated how current policies are insufficient to grow PEV market share to levels compatible with 2050 climate goals.

We believe that taking these methods and results together can make the case for the value of mixed-method, reflexive research. While academic models tend to reward the publication of individual studies (typically using only one method), the integration of multiple methods can be quite complementary, with the skilled use of one method making up for deficiencies in another, while also building upon one another's results to identify consistent findings and persistent gaps.

References

Al-Alawi, B., & Bradley, T. (2013). Review of hybrid, plug-in hybrid, and electric vehicle market modeling studies. *Renewable and Sustainable Energy Reviews, 21,* 190–203.

Axsen, J., Bailey, J., & Castro, M. A. (2015a). Preference and lifestyle heterogeneity among potential plug-in electric vehicle buyers. *Energy Economics, 50,* 190–201. https://doi.org/10.1016/j.eneco.2015.05.003.

Axsen, J., Cairns, J., Dusyk, N., & Goldberg, S. (2018). What drives the Pioneers? Applying lifestyle theory to early electric vehicle buyers in Canada. *Energy Research and Social Science, 44,* 17–30. https://doi.org/10.1016/j.erss.2018.04.015.

Axsen, J., Goldberg, S., & Bailey, J. (2016a). How might potential future plug-in electric vehicle buyers differ from current "Pioneer" owners? *Transportation Research Part D: Transport and Environment, 47,* 357–370. https://doi.org/10.1016/j.trd.2016.05.015.

Axsen, J., Goldberg, S., Bailey, J., Kamiya, G., Langman, B., Cairns, J., et al. (2015b). *Electrifying vehicles: Insights from the Canadian plug-in electric vehicle study*. Burnaby, BC: Simon Fraser University.

Axsen, J., Goldberg, S., & Melton, N. (2016b). *Canada's electric vehicle policy report card*. Burnaby, BC: Simon Fraser University.

Axsen, J., & Kurani, K. (2009). Early U.S. market for plug-in hybrid electric vehicles. *Transportation Research Record: Journal of the Transportation Research Board, 2139,* 64–72. https://doi.org/10.3141/2139-08.

Axsen, J., & Kurani, K. S. (2012). Interpersonal influence within car buyers' social networks: Applying five perspectives to plug-in hybrid vehicle drivers. *Environment and Planning A, 44*(5), 1057–1065.

Axsen, J., & Kurani, K. S. (2013a). Connecting plug-in vehicles with green electricity through consumer demand. *Environmental Research Letters, 8*(1). https://doi.org/10.1088/1748-9326/8/1/014045.

Axsen, J., & Kurani, K. S. (2013b). Hybrid, plug-in hybrid, or electric—What do car buyers want? *Energy Policy, 61,* 532–543.

Axsen, J., Langman, B., & Goldberg, S. (2017). Confusion of innovations: Mainstream consumer perceptions and misperceptions of electric-drive vehicles and charging programs in Canada. *Energy Research and Social Science, 27,* 163–173. https://doi.org/10.1016/j.erss.2017.03.008.

Axsen, J., Mountain, D. C., & Jaccard, M. (2009). Combining stated and revealed choice research to simulate the neighbor effect: The case of hybrid-electric vehicles. *Resource and Energy Economics, 31*(3), 221–238.

Bahn, O., Marcy, M., Vaillancourt, K., & Waaub, J. P. (2013). Electrification of the Canadian road transportation sector: A 2050 outlook with TIMES-Canada. *Energy Policy, 62,* 593–606. https://doi.org/10.1016/j.enpol.2013.07.023.

Bourdieu, P. (1977). *Outline of a theory of practice*. London, UK: Cambridge University Press.

Bourdieu, P., & Wacquant, L. (1992). *An invitation to reflexive sociology*. Chicago, IL: University of Chicago Press.

Brand, C., Cluzel, C., & Anable, J. (2017). Modeling the uptake of plug-in vehicles in a heterogeneous car market using a consumer segmentation approach. *Transportation Research Part A: Policy and Practice, 97*(Supplement C), 121–136. https://doi.org/10.1016/j.tra.2017.01.017.

Creswell, J. W. (2011). Controversies in mixed method research. In N. K. Denzin & Y. S. Lincoln (Eds.), *The SAGE handbook of qualitative research* (4th ed.). London, UK: SAGE Publications.

Environment and Climate Change Canada. (2017). *Canadian environmental sustainability indicators: Greenhouse gas emissions Gatineau*. QC: Government of Canada.

Giddens, A. (1991). *Modernity and self-identity: Self and society in the late modern age*. Stanford, CA: Stanford University Press.

Government of Quebec. (2017). Analyse d'impact réglementaire du projet de règlement d' application de la Loi visant l'augmentation du nombre de véhicules automobiles zéro émission au Quebec afin de réduire les émissions de gaz à effet de serre et autres polluants.

Greenblat, C. S. (1981). Seeing forests and trees: Gaming-simulation and contemporary problems of learning and communication. In C. S. Greenblat & R. D. Duke (Eds.), *Principles and practices of gaming-simulation*. Beverly Hills, CA: Sage.

Greene, D., Park, S., & Liu, C. (2014). Analyzing the transition to electric drive vehicles in the U.S. *Futures, 58,* 34–52. https://doi.org/10.1016/j.futures.2013.07.003.

Guba, E. G., & Lincoln, Y. S. (1994). Competing paradigms in qualitative research. In N. K. Denzin & Y. S. Lincoln (Eds.), *Handbook of qualitative research*. SAGE: Newbury Park, CA.

Hauser, J., Tellis, G. J., & Griffin, A. (2006). Research on innovation: A review and agenda for "marketing science". *Marketing Science, 25*(6), 687–717. https://doi.org/10.2307/40057216.

Hoeffler, S. (2003). Measuring preferences for really new products. [Article]. *Journal of Marketing Research (JMR), 40*(4), 406–420.

International Energy Agency. (2015). *Energy and climate change: World energy outlook special report*. Paris, France: OECD/IEA.

International Energy Agency. (2018). *Global EV outlook 2018: Towards cross-modal electrification*. Paris, France: OECD/IEA.

Klippenstein, M. (2018). Canadian EV sales. www.tinyurl.com/CanadaEVsales. Accessed July 6, 2018.

Kurani, K. S., Turrentine, T., & Sperling, D. (1994). Demand for electric vehicles in hybrid households: An exploratory analysis. *Transport Policy, 1,* 244–256. https://doi.org/10.1016/0967-070X(94)90005-1.

Kurani, K. S., Turrentine, T., & Sperling, D. (1996). Testing electric vehicle demand in 'hybrid households' using a reflexive survey. *Transportation Research Part D: Transport and Environment, 1,* 131–150. https://doi.org/10.1016/S1361-9209(96)00007-7.

Lee-Gosselin, M. (1996). Scope and Potential of interactive stated response data collection methods. In *Conference on Household Travel Surveys: New Concepts and Research Needs, Transportation Research Board Conference Proceedings* (pp. 115–133). Irvine, CA.

Long, Z., Axsen, J., & Kormos, C. (2019). Consumers continue to be confused about electric vehicles: Comparing awareness among Canadian new car buyers in 2013 and 2017. *Environmental Research Letters, 14*(11), 114036.

Mau, P., Eyzaguirre, J., Jaccard, M., Collins-Dodd, C., & Tiedemann, K. (2008). The 'neighbor effect': Simulating dynamics in consumer preferences for new vehicle technologies. *Ecological Economics, 68,* 504–516. https://doi.org/10.1016/j.ecolecon.2008.05.007.

McCracken, G. (1988). *The long interview*. Newbury Park, CA: SAGE.

Melton, N., Axsen, J., & Goldberg, S. (2017). Evaluating plug-in electric vehicle policies in the context of long-term greenhouse gas reduction goals: Comparing 10 Canadian provinces using the "PEV policy report card". *Energy Policy, 107,* 381–393. https://doi.org/10.1016/j.enpol.2017.04.052.

Melton, N., Axsen, J., & Sperling, D. (2016a). Moving beyond alternative fuel hype to decarbonize transportation. *Nature Energy, 1*(3). https://doi.org/10.1038/nenergy.2016.13.

Rogers, E. (2003). *Diffusion of innovations* (5th ed.). New York, NY: Free Press.

Stevens, M. (2016). Electric vehicles sales in Canada: 2015 final numbers.

Sykes, M., & Axsen, J. (2017). No free ride to zero-emissions: Simulating a region's need to implement its own zero-emissions vehicle (ZEV) mandate to achieve 2050 GHG targets. *Energy Policy, 110,* 447–460. https://doi.org/10.1016/J.ENPOL.2017.08.031.

Teddlie, C., & Tashakkori, A. (2011). Mixed methods research: Contemporary issues in an emerging field. In N. K. Denzin & Y. S. Lincoln (Eds.), *The SAGE handbook of qualitative research* (4th ed.). London, UK: SAGE Publications.

Turrentine, T. S., & Kurani, K. S. (1998). Adapting interactive stated response techniques to a self-completion survey. *Transportation, 25,* 207–222.

Urban, L., Weinberg, D., & Hauser, R. (1996). Pre-market forecasting of really-new products. *Journal of Marketing, 60*(1), 47.

Williams, J. H., Debenedictis, A., Ghanadan, R., Mahone, A., Moore, J., Iii, W. R. M., et al. (2012). 2050: The pivotal role of electricity. *Science (New York, N.Y.), 335*, 53–60.
Wolinetz, M., & Axsen, J. (2017). How policy can build the plug-in electric vehicle market: Insights from the respondent-based preference and contraints (REPAC) model. *Technological Forecasting and Social Change, 117,* 238–250. https://doi.org/10.1017/CBO9781107415324.004.

Europe

Norway—The World Leader in BEV Adoption

Erik Figenbaum

Abstract Norway is the leading per capita BEV market in the world. The share of BEVs in the fleet exceeded 5% in December 2017 with another 2.5% being PHEVs. The market shares of BEVs and PHEVs in 2017 were 20% and 18% respectively. The total market share of BEVs and PHEVs in 2018 was well above 40% and in 2019 exceeded 50%. These vehicles have easily been adopted by Norwegian households. 4 out of 5 BEV owners, however, own another vehicle, giving them the option to swap vehicles when the range is too short. A strong incentive package has been introduced since 1990. The political targets which these incentives are intended to support have developed over the years. While these incentives were initially introduced to allow for experimentation with electric vehicles, and subsequent industrialization efforts, they are now geared towards meeting climate policy targets of reducing transportation sector greenhouse gas emissions. These incentives induced 20 years of BEV niche market experimentation and led to a fleet of 3000 BEVs by 2009. The long-term stability of BEV policies, the public awareness of BEVs created by their owners, the competences developed in the niche markets, and the strong incentive package, became a window of opportunity for the traditional passenger vehicle importers. When BEVs became available from their brands from 2010, they could import and sell them with immediate success using their vast national dealer networks.

Keywords User experiences · Multi Level Perspective (MLP) · Policy and incentives · Market development

Introduction

Norway is the leading per capita BEV market in the world. The share of BEVs in the fleet exceeded 5% in December 2017 with another 2.5% being PHEVs (Figenbaum 2018). The market shares of BEVs and PHEVs in 2017 were 20% and 18% respectively. The market in 2018 has been even stronger, as shown in Fig. 1, and by the end of May 2018 the BEV share climbed up to 6% with a total of 160,000 BEVs on

E. Figenbaum (✉)
Institute of Transport Economics, Oslo, Norway
e-mail: efi@toi.no; erik.figenbaum@toi.no

M. Contestabile et al. (eds.), *Who's Driving Electric Cars*, Lecture Notes in Mobility,
https://doi.org/10.1007/978-3-030-38382-4_6

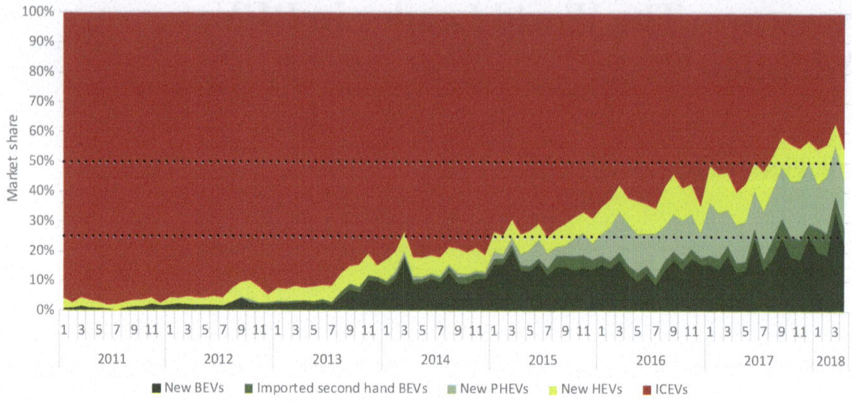

Fig. 1 Passenger vehicle technology monthly market shares 2011–2018 (Figenbaum 2018)

Norway's roads. ICEV ownership in Norwegian households peaked in 2015 and has decreased ever since (Figenbaum 2018). Since the end of 2018 1 out of 10 vehicles has a plug and can run fully or partially on electricity.

The average BEV and PHEV is driven approximately 15,700 km/year, which is equal to ICEVs that are a 2010 model year or newer (Figenbaum and Kolbenstvedt 2016). PHEVs have been estimated to have a 50% all-electric driving share of the total km driven (Figenbaum and Weber 2016) in Norway. Therefore, all the BEVs and PHEVs combined were driven about 260 million electric km in May 2018. They are used extensively in daily traffic, especially in and around cities. The toll ring around Oslo city had for instance a 16% BEV share of light duty vehicles in April 2018, but in the morning rush hours the share reached 24% (Dagsavisen 2018). BEVs and PHEVs are thus starting to make a profound impact on the environment and the transportation system.

CO_2-emissions from passenger cars in Norway had been steadily increasing since 1990, despite the growing share of BEVs in the fleet, and efficiency improvements in diesel and gasoline vehicles. It is only from 2015 that a sharp downward trend in emissions can be observed, as seen in Fig. 2. One of the reasons for the lack of emission reduction is the overall growth in the total number of vehicles in Norway (Figenbaum 2018). The growth rate has not been affected by the introduction of BEVs and PHEVs.

Consumers own 85% of the BEVs in the fleet (NPRA 2018). This chapter therefore focuses on how consumers have interacted with BEVs in Norway. It starts off with a description of the data sources used in the analysis of consumer behaviour. It then provides an overview of specific conditions in Norway that have influenced the adoption rate of BEVs. This is followed by a historical overview of the adoption process of BEVs since 1990, in order to gain a better understanding of how consumers fit into the greater societal transition towards electromobility. We then explore the effects and effectiveness of the Norwegian BEV policies and incentives. Subsequently we present the practical user experiences with BEVs, leading into a

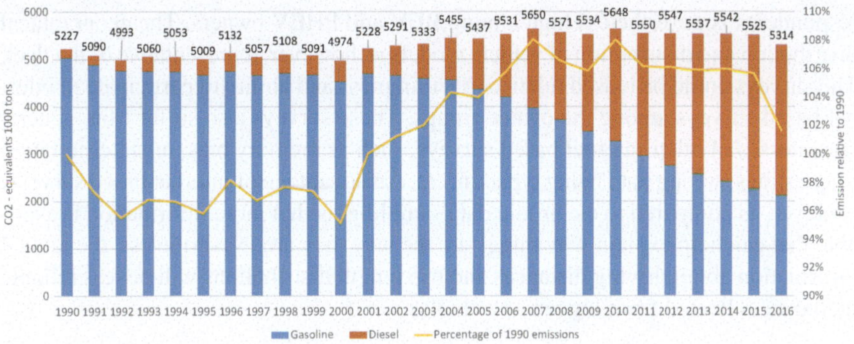

Fig. 2 Greenhouse gas emissions from passenger vehicles in Norway. 1000 ton CO_2-equivalents. *Source* SSB (2018c)

discussion of the potential for future BEV adoption, and we look at how BEVs influence other societal targets. Lastly, we discuss the parts of the Norwegian experience that are the most relevant to other countries.

Data Sources—Consumer Adoption of BEVs

Two national internet-based surveys have been carried out, one in 2014 and one in 2016, by the Institute of Transport Economics. These surveys were designed to gain a basic understanding of a wide range of issues related to consumer choice of and preference for BEVs compared to ICEVs and PHEVs, and how these vehicles are operated. The 2014 survey of 1721 BEV and 2241 ICEV owners was carried out as part of the international Era-Net Electromobility + project COMPETT (Figenbaum et al. 2014). A 2016 survey of 3111 BEV, 2065 PHEV and 2241 ICEV owners was carried out with the support from the Norwegian Public Roads administrations EMIROAD project (Figenbaum and Kolbenstvedt 2016). The BEV respondents in both surveys were recruited among members of the EV Association, a consumer membership organization for BEV owners. All BEV buyers receive a free year of membership in this association when buying their new BEV, courtesy of the dealers. The EV association members can, thus, be considered to be a representative sample of BEV owners in Norway. The ICEV owners were in 2014 recruited via e-mail from a regional sample of members of the National Motorist Organization NAF. The region covered Oslo and an area stretching 100 km west of Oslo. In 2016, a national representative sample of members from NAF was drawn. All private owners of PHEVs received in 2016 a postcard with information about where to find and fill out the 2016 survey.

These surveys provide a broad overview of how BEVs are bought, used and operated and what users think about their performance. The main weakness of these surveys is the fairly low response rate of 17–25%, although the total number of respondents is nevertheless high among BEV and PHEV owners. The geographical distribution and distribution of vehicle models is, however, representative of the fleet. The surveys could be biased relative to willingness and ability to participate in internet based surveys among the owner groups. These surveys also suffer from general weaknesses of other internet based surveys, such as memory bias, misinterpretation of questions, wrong data being typed in, etc. An advantage of internet based surveys, however, as compared to electronic data sampling, is that more information is available that can help explain why things are the way they are. Nevertheless, the factual information about driving distances and trip length distributions will be less reliable than data collected from loggers in vehicles.

Favorable Conditions, but Rough Winters

Norway has some particularly favourable conditions for an early market uptake of BEVs and PHEVs. The electricity is clean and non-controversial. 96% comes from hydroelectric power plants mainly built before 1980, another 2% is wind power. Most Norwegians heat their houses fully or partially with electricity during the winter. Most households therefore have a high power robust grid connection. Power cables to other European countries stabilize the main grid and secure access to electricity in dry years. Residential electricity grid connections can support 3.6 kW home charging of vehicles (NVE 2016), and in some areas up to 11 kW.

Electricity is cheap thanks to the large share of hydroelectric power. High taxes on fossil fuels make gasoline and diesel expensive. As a result, compared to running a car on diesel fuel, BEV owners in Norway save the most on energy cost in the whole of Europe (Figenbaum and Kolbenstvedt 2015).

The Norwegian power generation, industry and several other sectors are part of the EU greenhouse gas emission trading system (EU ETS). Dedicated national climate policies for these sectors are therefore not needed. The Norwegian electricity mix has anyway a close to zero carbon content. Transportation is therefore the main remaining sector to decarbonize besides agriculture.

Norway does not have a car industry. Up to the 1960s, vehicles were rationed to keep imports low and preserve the balance of trade. When the rationing was lifted, registration taxes on new vehicles were introduced to limit and control sales, and to generate income for the government. Today, the taxes on vehicles are still substantial, and are used to influence the types of vehicles people buy, and the type of energy these vehicles use. The tax has changed several times over the years, from a tax based on the value of the car, to the current tax which is proportional to the vehicle's curb weight, CO_2-emission and NO_X-emission. The 2018 tax rates are shown in Table 1.

Table 1 Registration taxes in Norway (National budget 2018)

CO_2 (g/km)	Euro/(g/km)	NO_X (mg/km)	Euro/(mg/km)	Curb weight (kg)	Euro/kg
<40	−110.27	>0	7.093	0–350	0
40–74	−93.72			351–1200	2.651
75	0.00			1201–1400	6.605
76–100	91.47			1401–1500	20.641
101–130	99.55			>1500	24.006
131–200	268.60				
>200	344.98				

There is a tax deduction for vehicles with CO_2 emissions below 75 g/km. The total registration tax is the sum of the three elements and cannot be negative.[1] In addition to these taxes, PHEVs and ICEVs are also levied the 25% Value Added Tax (VAT) on the sales value of the vehicle minus the registration tax. The high taxes made substantial purchase incentives for BEVs possible, by introducing exemptions from the registration tax and the VAT. Both these incentives have a low administrative cost for the government.

There are also high taxes on fuels for transportation. The fuel tax consists of a CO_2 tax and a "road use" tax. The latter should cover the societal cost of transportation, such as road construction and maintenance, queues, accidents, pollution and noise. For electricity there is an "electricity tax" (per kWh), and users pay for the transmission cost (per kWh), and the cost of a connection to the grid (an annual fee). The VAT on fuels and electricity is 25%. The energy prices are compared in Fig. 3.

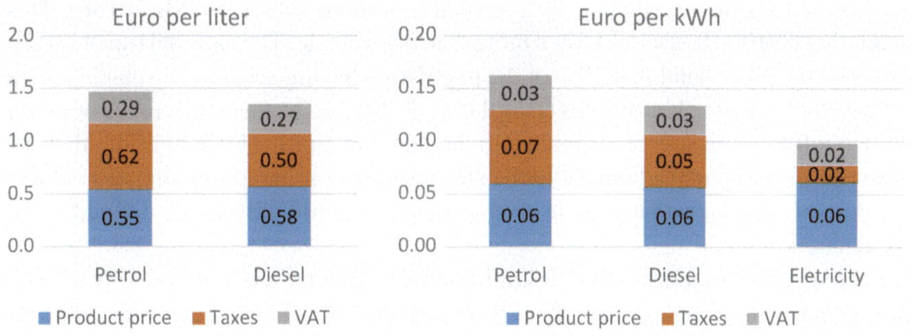

Fig. 3 Taxes and product prices of energy carriers in Norway (SSB 2018a; Drivkraft Norge 2018)

[1] The total tax for a vehicle weighing 1300 km, emitting 100 g CO_2/km and 100 mg NOx/km would be 91.477 * 25 + 100 * 7.093 + 850 * 2.651 + 100 * 6.605 = 5910 Euros. If that vehicle's CO_2-emission had been 50 g/km, the tax would be only 1374 Euros due to the negative CO_2 tax below 75 g/km.

Electricity has about the same product price per kWh as gasoline, but the taxes are much lower. In addition, electric vehicles use much less energy per km than ICEVs.

Since the 1990s road tolls have become a major source for financing new roads in most parts of the country (Figenbaum 2017). These tolls can accumulate up to 500–1000 euros for a typical commuter over a year, and even up to 3000–4000 for some (Figenbaum 2018). BEVs have since 1997 been exempted from all these toll roads. BEVs have also since 1999 been exempted from municipal parking charges, and have been charged less on some national main road ferry services since 2008. These incentives have been revised and BEVs can now be charged for up to 50% of the ICEV rates for parking (since 2017), road tolls (since 2019) and ferries (since 2018).

The Norwegian BEV incentives are summarized in Table 2. The only PHEV incentive is a low or even zero registration tax due to their low CO_2-emissions, and a deduction of 26% (reduced from 01.07.2018) of their weight prior to calculating the weight tax.

The mean household income in Norway is 2.3 times the EU28 average, whereas the median is 2.4 times the EU28 average (Eurostat 2018a). The cost of living is 1.4 times the EU28 average (Eurostat 2018b). The high registration taxes therefore do not prevent people from buying new vehicles, as seen by the vehicle ownership rate being the same as the EU average at 506 and 505 vehicles per 1000 inhabitants (Eurostat 2018c).

In households with more than one vehicle, the range and charge time drawbacks of electric vehicles will be less important. Figenbaum and Kolbenstvedt (2016) found that such households form an ideal BEV user group, and have been the main adopters of BEVs so far. Multi vehicle ownership is common among families in all areas of Norway apart from the densest parts of cities, where vehicle ownership in general is low. According to Statistics Norway, 53% of Norway's 2.25 million households in 2012 owned 1 vehicle, whereas 29% owned 2 or more vehicles (SSB 2018b). This means that 650 000 households own more than one vehicle. The national travel survey shows on the other hand that 49% of the population having access to a vehicle live in multi-vehicle owning households (Hjorthol et al. 2014). The main difference between the two information sources is related to the fact that multi vehicle households have more people living in the household than the average households owning one vehicle has. Another reason is that vehicle access could be slightly differently defined in the two datasets.

Low speeds on main roads (80 km/h) and frequent 60 km/h zones, as well as a limited network of motorways (90–110 km/h), lead to actual average travel speeds of only 70–80 km/h between most major Norwegian cities (Figenbaum 2018). The low average travel speed helps preserve range by keeping energy consumption low. In countries with higher actual speeds the "experienced utility" of BEVs could be lower, as the energy consumption increases rapidly and the range is quickly reduced at higher speeds (Galvin 2017; Fetene et al. 2017).

Winters are tough in Norway. Temperatures in some places reach below −20 °C. In the areas where most people live, the winter temperatures would however normally be in the range of 0 to −15 °C. The low temperatures lead to a high energy consumption

Table 2 Relative advantage of different BEV incentives in Norway (Figenbaum 2018)

Incentives	Intro year	BEV buyers—relative advantage	Future plans
Fiscal incentives: reduction of purchase price/yearly cost gives competitive prices			
Exemption from registration tax	1990/1996	The tax is based on ICEV emissions and weight. Example taxes: VW Up 3000 €. VW Golf: 6000–9000 €	To be continued until 2020
VAT exemption	2001	Vehicles competing with BEVs are levied a VAT of 25% on sales price minus registration tax	To be continued until 2020
Reduced annual vehicle license fee	1996/2004	BEVs and hydrogen vehicles 52 € (2014-figures). Diesel rate: 360–420 € with/without particulate filter	To be continued indefinitely
Reduced company car tax	2000	The company-car tax is lower but BEVs are seldom company cars	This incentive was up for revision in 2017/18, but remained unchanged
Exemption from reregistration tax	2018	Tax on change of ownership of ICEVs/PHEVs. 0–3 year-old vehicles + 1200 kg: 610 Euros, 4–11 years 370 Euros	
Direct subsidies to users: reduction of variable costs and help solving range challenges			
Free toll roads	1997	In Oslo users save 600–1000 €/year. Some places savings exceed 2500 €	Law revised so that rates for battery electric vehicles in toll roads and ferries will be decided by local governments, up to a maximum rate of 50% of the ICEV rate
Reduced fares on ferries	2009	Similar to toll roads saving money for those using car ferries	
Financial support for normal chargers	2009	Reduce investors risk, reduce users range anxiety, expand usage	A national plan for charging infrastructure shall be developed
Financial support for fast chargers	2011	More fast-charging stations influences BEV km driven and market shares	ENOVA[*] support to establish fast charging along major corridors. City fast charging left to commercial actors
User privileges: reduction of time costs and providing users with relative advantages			

(continued)

Table 2 (continued)

Incentives	Intro year	BEV buyers—relative advantage	Future plans
Access to bus lanes	2003/2005	BEV users save time driving to work in the bus lane during rush hours	Local authorities have given the authority to introduce restrictions if BEVs delay buses
Free parking	1999	Users get a parking space where these are scarce or expensive and save time looking for a space	Local authorities has seen 2017 been given the authority to introduce rates up to 50% of the ICEV rate
Free charging (some places)		Not regulated by national law, but often bundled with free municipal parking	Local authorities and parking operators decides whether this incentive will continue

*ENOVA, a Government agency, supports the introduction of energy saving and climate gas reduction technologies in Norway. In the transportation sector it supports charging infrastructure and refuelling infrastructure for alternative fuels

being required to heat the cabin (Haakana et al. 2013), and all vehicles use winter tyres leading to higher rolling resistance and thus higher energy consumption per km. The low temperatures also increase the air drag. Although the terrain is snow covered in the winter season, main roads are normally kept snow free. However, during heavy snow fall the roads will temporarily be snow covered, leading to an even higher energy consumption (Haakana et al. 2013). The length of the winter season varies between north and south and inland versus the coastal strip. The typical winter season is from mid-November to mid-April.

A Multi-level Perspective on BEV Adoption

Theoretical Approach and Main Results

Before looking at how BEVs are used and who the owners are, a closer look at how the BEV development in Norway came about will be useful. It will frame the role of consumers in the larger societal transition towards the adoption of BEVs. A suitable research framework for this purpose is the Multi-level Perspective (MLP) developed by Geels (2012) and others. The MLP aims to explain the diffusion of technologies as an interplay between actors and practices at three levels (Geels 2012):

1. Niches, often demonstration programs, involving users, buyers and niche producers, where new technologies can be tested out and improved.

2. Regimes with actors such as car producers, fuel stations, consumer organizations, NGOs and established practices of vehicle-based mobility.
3. The landscape with influential exogenous factors such as oil price, international politics etc.

These "levels" refer to heterogeneous configurations of increasing stability, constituting a nested hierarchy, in which regimes are embedded within landscapes, while niches exist inside or outside regimes. A new technology, developed initially in niche markets, can potentially, via a window of opportunity, break into an existing regime or replace an existing regime when developments in the global landscape put pressure on that regime.

The following sections utilize the MLP-framework to examine closely what happened in the three decades from 1990 to 2018, based on the analysis of Figenbaum (2017). In that analysis a layer of national governance was introduced between the landscape and the regime, as the Norwegian vehicle taxes and policies have a very large influence on the types of vehicles that are sold in Norway.

Local Testing and Industrialization Efforts—1990 to 1999

The BEV era in Norway started already in 1990, although no one understood it at the time. When the first BEV was imported in 1990, the importer, an NGO, managed with help from publicity generated by the pop group a-ha, to get a registration tax exemption for that vehicle. This event marks the start of the BEV era in Norway as seen in Fig. 4.

The Ministry of Finance stated in a letter to the NGO that the exemption was granted to allow for the experimentation with BEVs in the market place. This exemption has been effective for all BEVs since then, and has had a profound impact on the

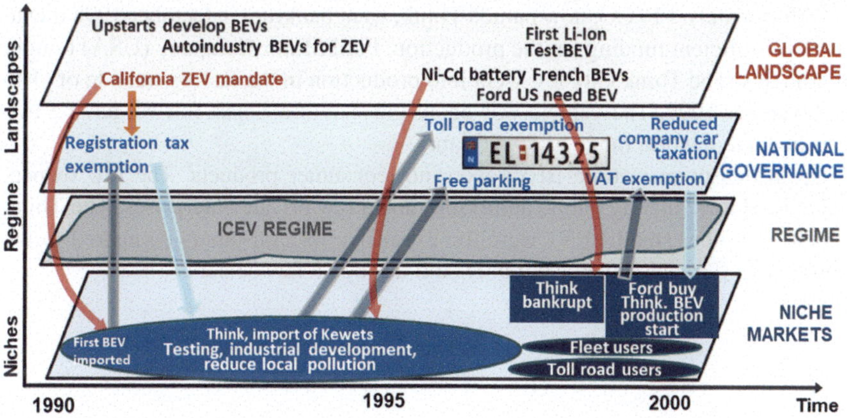

Fig. 4 MLP framework. BEV development 1990–2000. Adapted from Figenbaum (2017)

subsequent BEV market development. Without that exemption BEVs would have been unsellable in Norway at that point in time, as the registration tax would have doubled the sales price of the vehicle. This imported vehicle was a low volume conversion of an ICEV and was therefore expensive even without taxes. Soon after, the first BEV importer was established, importing Kewet Eljets from Denmark.

The Norwegian EV Association was established as an industry organization, with the utility Oslo Energi and the Municipality of Oslo as its main supporters. At this point in time there were no specific national policies for BEVs apart from the registration tax exemption, and the Association worked to improve conditions for BEVs.

California's ZEV mandate required manufacturers to sell BEVs, which led to the industrial development of BEVs and enabling technologies. A Norwegian BEV developer (PIVCO) emerged, inspired by these developments. PIVCO presented a prototype of a plastic body BEV at the Olympics held in Norway in 1994. The vehicle ran around at temperatures down to −20 °C, demonstrating that BEVs could work under Norwegian conditions. Soon after Calstart tested a small number of PIVCO BEVs in their station car program in California. A production intent version was developed from 1996 with the support of Lotus Engineering (UK), and major suppliers such as Siemens (drive-system) and Saft (NiCd batteries). In France, the industrialization of BEVs using robust NiCd batteries was undertaken by Renault, Peugeot and Citroën, with support from the French government. The automotive regime in Norway was not involved in significant BEV activities in this period, although a few BEVs were imported by Peugeot, Citroën and Renault importers.

Oslo Municipality wanted PIVCO to establish production in Oslo. Their target was therefore to create a local market for BEVs. They pushed in collaboration with the EV Association for local incentives such as exemptions from the toll ring around Oslo and free parking, which both required changes to national laws. Free road tolls were introduced in Oslo in 1997 and free parking in 1999, and then nationwide. BEV specific vehicle number plates were introduced from 1999 to make the control of such incentives easier.

Unfortunately, PIVCO, now named Think, went bankrupt in October 1998 due to a lack of sufficient funding for the production. Ford Motor Company (USA) bought the remains. The Think City EV went into production in the factory outside of Oslo in November 1999. The vehicle was about the size of a Smart fortwo, having two seats and a real world range of 50–70 km.

Up to this point in time BEVs were not consumer products. The few owners that existed were mainly public authorities and a few private enterprises. The Think vehicle and other small BEVs were too expensive for consumers compared to the utility they offered and they were only sold through a few dealers.

Optimism, Dismay and Renewed Hope—2000 to 2009

With Ford owning Think, the decade started with optimism for the future prospects of BEVs in Norway. Ford Motor Company owned Think between 1999 and 2002 and politicians saw a potential for Norway to become a vehicle producer. Politicians expanded the BEV friendly policies with the introduction of the exemption from the VAT from 2001 as seen in Fig. 5. The exemption came about because the Think City became so expensive by the time it reached the market that it did not sell well. Ford therefore also started to develop a lower cost higher volume vehicle intended for launch in 2003, designed for Californian and European markets.

Some Norwegian Ford dealers started selling the Think BEV. In total about 1000 vehicles were produced and sold while Ford owned Think. Ford decided to pull out of Think in 2003 when the Californian ZEV mandate changed, and they did not have to sell BEVs there anymore, which was the main reason why they had bought Think in the first place. The changes in California seem to have led to a broad collapse in BEV developments across the globe with governments losing interest in the technology. Think was sold to a UK company (Frazer-Nash) that proved unable to get the new model into production and Think went bankrupt a few years after Ford pulled out.

Think thus missed the opportunity that the opening of the bus lanes to BEVs from 2003 created. The road authorities decided to throw minibuses out of the bus lanes because commuters used them to save time on their commutes to work. The BEV community used this opportunity to lobby for BEVs to be allowed into the bus lane. They managed to get a test of BEV access to the bus lanes in the Oslo area. A couple of years of testing sufficed to establish that BEVs in bus lanes did not delay buses. The incentive became national, creating a small consumer BEV buying niche. There was also a rapid expansion of the number of toll road projects across the country in this period, leading to niches of consumers owning BEV around these projects. Due to the great amounts of time saved on their commutes when using bus lanes

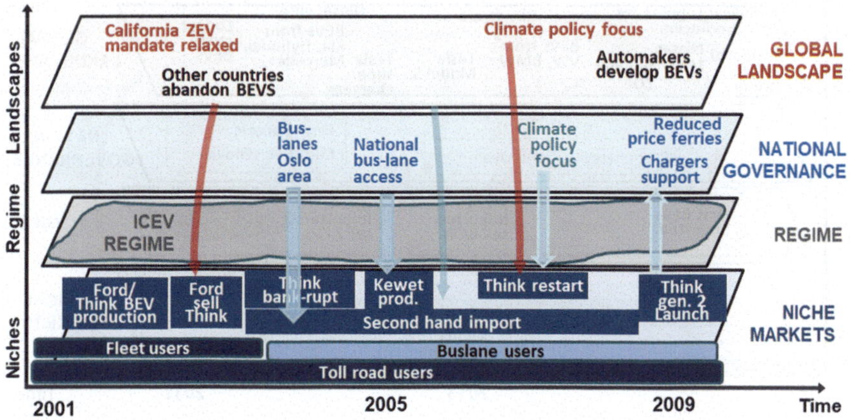

Fig. 5 MLP framework. BEV development 1990–2000. Adapted from Figenbaum (2017)

and the large toll road savings in some areas of Norway, consumers now became the main ownership group. The Norwegian Kewet importer established Norwegian production of the Buddy EV (based on the Kewet) from 2005. BEVs were from 2003 imported second hand from countries that had abandoned their BEV policies and test programs. These two main vehicle channels kept the BEV consumer market alive, gradually building up a consumer owned BEV fleet that reached 3000 in 2009.

Think re-emerged with a new production model in 2008, backed by Norwegian and international investors inspired by the global focus on climate policy and greenhouse gas emission reduction. BEVs were from 2008 firmly integrated as part of Norway's climate policy. The government supported the BEV market and the renewed industrial efforts with a new incentive: lower prices on national ferry services. The 2009 financial crisis delayed Think's production ramp up. Think needed new investors, and when Finnish Valmet bought into Think the production moved to Finland, leading to further production delays. A 5 million Euro national program to install public chargers was initiated to counter the 2009 financial crisis.

On the global landscape, not yet visible, the first incumbent vehicle manufacturers initiated BEV industrialization programs that would eventually lead to a flow of models with a greater consumer appeal coming on the market in the next decade.

The Modern BEV Jump Starts the Market—2010 to 2018

The international investments in BEV production on the global landscape resulted in the Mitsubishi I-Miev (and its siblings) and the Nissan Leaf entering the Norwegian market from 2011 as seen by the sequence of events in Fig. 6. The importers rolled out these vehicles across their entire dealer network with immediate success. Buyers, of which over 85% were consumers (Figenbaum and Kolbenstvedt 2016), could take

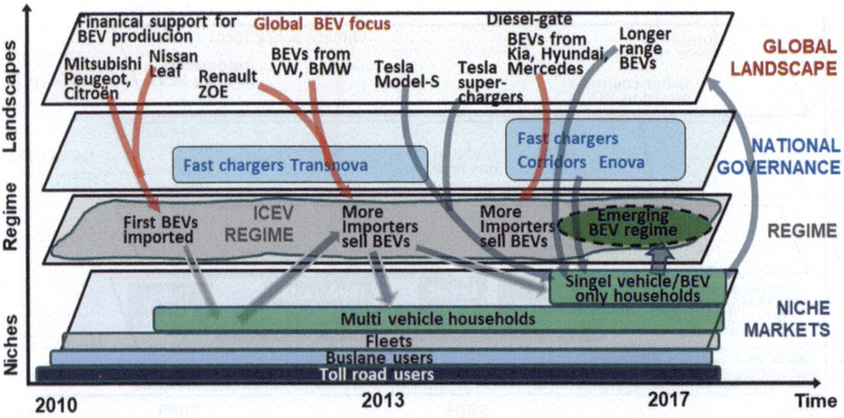

Fig. 6 MLP framework. BEV development 1990–2000. Adapted from Figenbaum (2017)

immediate advantage of all the incentives that were in place. These incentives also made the salesperson's job easier (Assum et al. 2014; Kolbenstvedt and Assum 2018).

The buyers were consumers wanting to save time driving in the bus lane and save money on the toll road exemption, with the assurance of a product delivered by an incumbent vehicle manufacturer through an existing vehicle dealer. Soon multi-vehicle households in general adopted BEVs, especially after 2013, when the price of BEVs came below the cost of buying an ICEV (Figenbaum 2018). A process of peer to peer communication between consumers further increased sales (Figenbaum and Kolbenstvedt 2016). Tesla Model S came on the market from 2013 demonstrating that long range driving and larger BEVs are viable. With the build out of the reliable Supercharger network in Norway, Tesla also demonstrated that long range driving was possible in practice, as seen by their driving pattern in the 2016 user survey (Figenbaum and Kolbenstvedt 2016).

After 2013, the EU target to reduce new vehicle CO_2-emission to below 95 g/km by 2021, and even further down in later years, combined with improvements in batteries and reduced battery cost, led to a massive development of BEVs in the auto-industry. The international driving forces were strengthened further after the diesel scandal hit the automotive sector in late 2015. Therefore, new vehicle importers soon followed the success of Mitsubishi and Nissan, launching a number of BEVs that expanded the market to increasingly larger groups of consumers.

BEVs now diffused from cities and into surrounding municipalities, and gradually to consumers living in rural areas (Figenbaum 2017, 2018), as seen in Fig. 7. The population in rural areas depend more on vehicle based transport, as public transport cannot be operated economically in low population density areas. BEVs are therefore

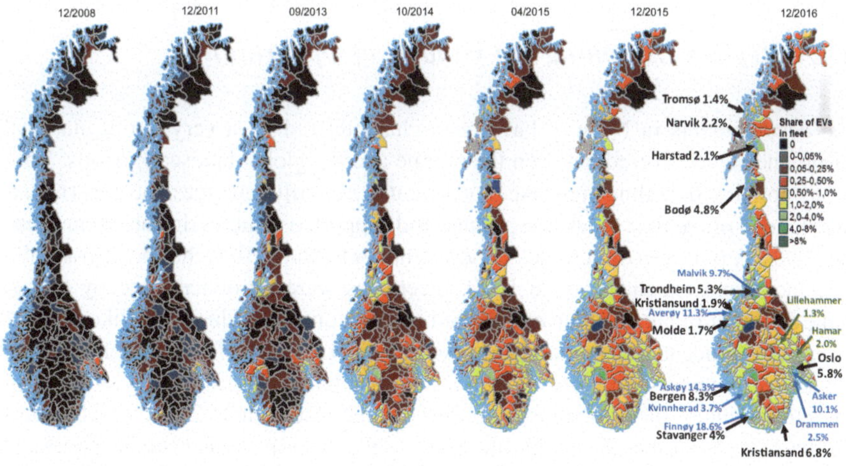

Fig. 7 BEV diffusion across Norway. Share of the total fleet by the 428 municipalities (428 in total, showed as individual areas coloured according to BEV share of the fleet). Adapted from Figenbaum (2017, 2018)

also an option in such areas, and by the end of 2016 BEVs were present almost everywhere in Norway, as seen in Fig. 7.

The large market expansion of BEVs in Norway in this period was almost entirely driven by consumer adoption. BEVs are now treated as any other vehicle option by dealers, subject to their traditional customer needs analysis in the sales process (Kolbenstvedt and Assum 2018). Private fleets have, however, been slower in adopting BEVs, in part because the main BEV incentive, the VAT exemption, has no value for enterprises.

The Norwegian EV association emerged as a pure consumer organization (Figenbaum 2017). A clever move was when all vehicle importers decided to give away a year of free membership in the EV association to those who bought a BEV (Figenbaum 2017). As BEV sales rapidly expanded, this built up a powerful lobby and user support organization. By 2018 the EV Association had over 20 full time employees working daily to improve conditions for consumers owning BEVs.

National programs to build out fast chargers made life with BEVs gradually easier for consumers after 2011. By the end of 2018 it was possible to drive and fast charge a BEV between major cities and along major roads in southern Norway, and partly in the northern parts (Figenbaum 2018). The first consumers could now get by with only owning a BEV, and the market expanded. By January 2018, 89% of the population knew someone owning a BEV, 66% had been a passenger in one, 34% had driven one, and only 22% had never been inside a BEV (Elbilbarometer 2018).

This Norwegian experience demonstrates that a move from niche activities to full market penetration can be fast when the substantial resources of the incumbent actors are mobilized, and consumers see real benefits of adoption.

ICEV Regime Exploiting the Window of Opportunity

The developments in Norway have been characterized by a very strong national governance level layered between the regime and the global landscape levels. With strong policies, including high taxes and incentives of different types, the government have fended off vehicle types that pollute, and supported vehicles that are clean since the 1990s. A policy to direct sales of new vehicles towards BEVs has been gradually introduced and expanded since 1990. Incentives were introduced one by one to make a market impact and have remained in place once introduced. Unlike in other countries the incentives are mainly tax exemptions, i.e. a loss of government income, not a government expenditure as with a purchase bonus program. Due to a lack of attractive vehicles these incentives produced small results up to 2009, which means that the loss of income was negligible. Some of the incentives had been introduced in national laws, such as the free toll roads and free parking. It is a more cumbersome process to change laws than budgets, so these incentives have been well protected.

The long lasting niche activities in Norway raised the consumer awareness of BEVs over a very long time period. There were already 3000 existing BEV owning consumers in place when the auto-industry launched their BEVs after 2010. A substantial window of opportunity thus lay open when the modern BEVs with 4–5 seats and full comfort and safety features were introduced, offering much higher value for money for consumers than earlier BEVs. Sales took off immediately among consumers when the importers used their resources and extensive dealer networks. Due to the incentives and falling prices, BEVs had by 2013 become the cheapest vehicles to buy in the segments where they were present (Figenbaum 2017), leading to even higher market shares. BEVs came within the reach of families owning more than one vehicle which could rather effortlessly put these vehicles into their daily transportation routines. Sales escalated further through positive consumer peer-to-peer communication (Figenbaum 2017; Figenbaum and Kolbenstvedt 2016).

The stable policies with large incentives and increasing awareness among consumers made the BEV uptake go much faster in Norway than in other countries. The stability of the BEV policy is rather remarkable when one also takes into account the fact that the rationale for the policy has shifted substantially over the years. First the rationale was to allow experimentation, then to foster industrialization, and finally climate policy became the dominant political motivation from 2007.

Without the exemptions from the registration tax from 1990 and the VAT from 2001, the story of BEVs in Norway would however have been very different.

Extreme Incentives Give Extreme Results

The total cost of ownership (TCO) for BEVs is in 2019 substantially lower than for ICEVs. The shift in TCO in favour of BEVs occurred already in 2013. The reason is that BEVs are exempted from all purchase taxes, whereas ICEVs are taxed with both the registration tax and the 25% VAT. A compact sized PHEV avoids the registration tax due to the deduction in weight tax and the low CO_2-emission.

An example of the purchase price and annual cost differences[2] of BEVs, PHEVs and ICEVs is shown in Fig. 8 for three powertrain versions of the VW Golf. As can be seen, BEV owners pay less for their vehicle, save energy cost, pay reduced annual tax, and avoid costs for oil change, compared to owners of ICEVs and PHEVs. These factors, amount to a consumer advantage of 2000 Euros per year over an ICEV.

Figenbaum and Kolbenstvedt (2015) found, as a matter of fact, that the annual energy cost savings of a BEV compared to an ICEV in Norway is the highest of all European countries, as seen in Fig. 9.

Local incentives provide consumers with added economic benefits, valued by users to be worth 1500 Euros per year (median 1100 Euros), on average. Half comes

[2]Insurance costs have been cheaper for BEVs. One of the major insurance companies indicated in June 2018 that this would change (Gjensidige 2018). Insurance cost is therefore not included in the estimates. The costs of servicing, other than oil change, are in general not visible to users at the time of purchase and vary in different parts of the country. The cost should be lower for BEVs, but may not be, as the servicing price for many brands is set independently by the dealers. Servicing cost has therefore not been included.

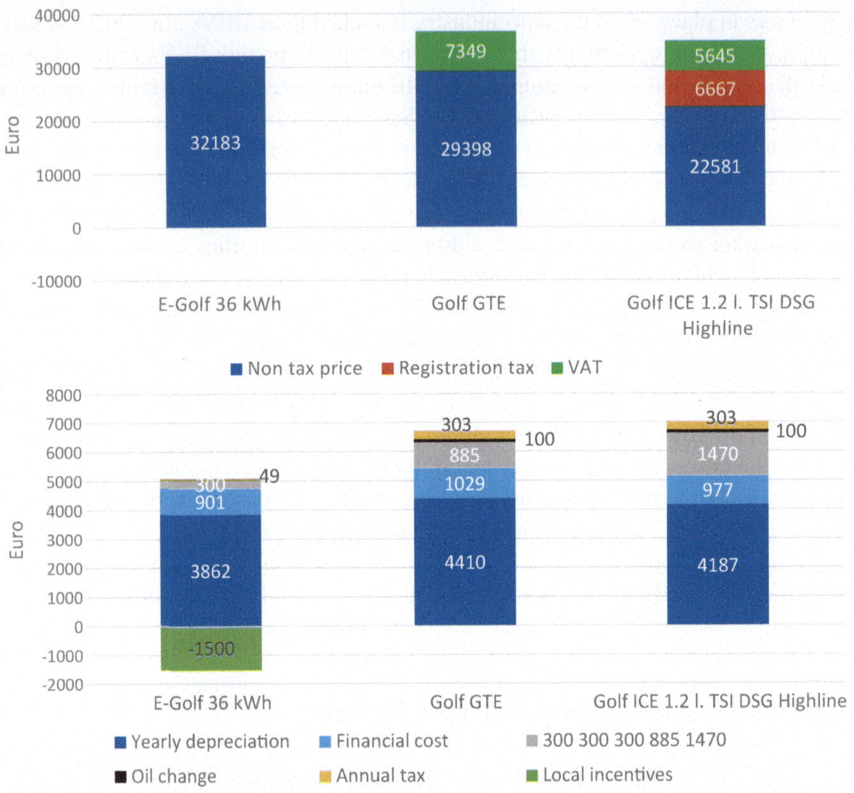

Fig. 8 BEV purchase price (left) and cost of ownership (right) in Euros compared to ICEVs and PHEVs. Cost of ownership does not contain insurance cost (assumed to be the same) and service cost (not transparent to buyers) Figenbaum (2018)

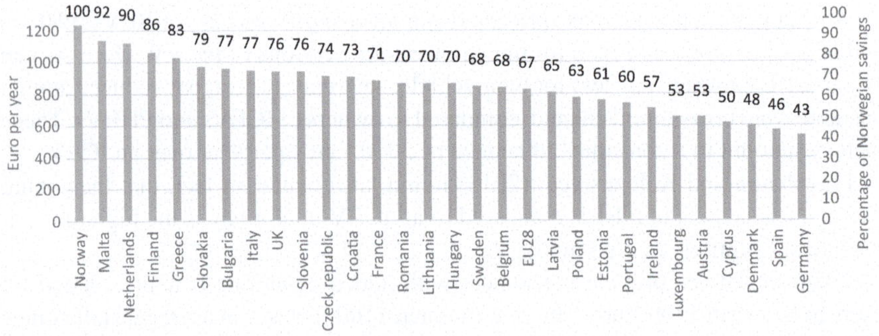

Fig. 9 Estimated annual savings in energy cost of driving a BEV 15,000 km compared to driving an ICEV. Cost saving in Euros and compared to Norway (100%). Adapted from Figenbaum and Kolbenstvedt (2015)

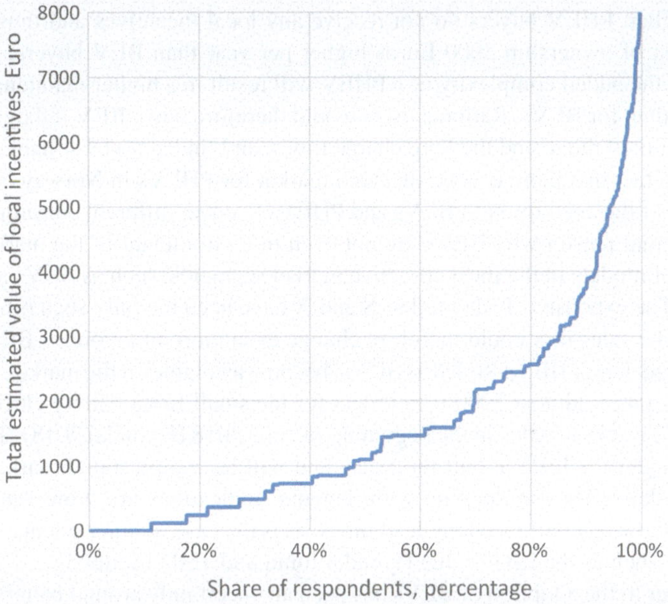

Fig. 10 Individual user estimates of the value of local incentives, i.e. exemptions from road tolls and parking charges, reduced rates on ferries, and the value of time saving using bus lanes. Average total value per province (Figenbaum and Kolbenstvedt 2016; Figenbaum 2018)

from free toll roads, almost a third from the value of the time savings in bus lanes, and the rest mainly from free parking, with a small contribution from free ferries (Figenbaum and Kolbenstvedt 2016). The spread between users and regions is, however, large, as seen in Fig. 10. Some consumers can benefit from several incentives, whereas others may have none or only one incentive available to them where they live. The ability to offer a varied package of incentives is therefore one of the success factors of BEVs in Norway, as the incentives reach more consumers. The toll road incentive has, for instance, stimulated the geographical diffusion of BEVs across Norway seen in Fig. 7. These factors made it economically sound for multi-vehicle households to replace one vehicle with a BEV even before 2013, when BEVs were more expensive to buy than comparable ICEVs, and consumers were worried about the second hand value of BEVs (Figenbaum et al. 2014). Critical mass could thus be achieved earlier.

It is therefore not surprising that low user cost is seen as the most important reason for consumers to buy a BEV, both by users and non-users (Figenbaum and Kolbenstvedt 2016). The toll road exemption is seen as the most important local incentive.

Compact sized PHEVs get a large enough rebate for low CO_2 emission to nullify the other registration tax elements. PHEVs operate partially on electricity, leading to reduced energy cost, as compared to an ICEV. The overall result is a small reduction in the overall cost of ownership compared with ICEVs, although their purchase price can be higher. PHEV buyers do not receive any local incentives and thus have an annual cost of ownership 3000 Euros higher per year than BEV buyers. It is also likely that the added complexity of a PHEV will result in a higher maintenance cost over time than for BEVs. Rational users would therefore buy a BEV if they can cope with the limited range and the long charge times, and find a model suitable for their needs. The fact that there is nevertheless a market for PHEVs in Norway is a strong indication of the need for both BEVs and PHEVs to cover different consumer needs.

A potential reason why BEVs are not even more dominant in the market could be a lack of models in the most attractive market segments, such as SUVs and 4WD vehicles. The expensive Tesla models S and X have been the only such offerings up to 2018. The sales mix could therefore change even more in favour of BEVs when longer range, larger BEVs, styled as SUVs, become available on the market. Hyundai had, for instance, almost 7000 firm orders for the small Kona battery electric SUV after only 2 weeks of sales in the beginning of June 2018 (Hyundai 2018). The Kona has the range of a Tesla at half the price, and will have a profound market impact if vehicle deliveries can keep up with demand. Indications are, however, that the buyers of the Kona will face long lead times, as is the case of other popular BEVs in mid-2018, such as the E-Golf, the Hyundai Ioniq and Tesla Model 3.

A barrier to the adoption of BEVs has been an uncertainty around battery life and the resale value of the vehicles. Both barriers have more or less been removed during the seven years since the first modern BEVs became available from the automakers in 2011. People that bought BEVs between 2011 and 2013 suffered a high value loss, mainly because the new vehicle sales price of the models on the market were substantially reduced (Figenbaum and Kolbenstvedt 2015). After 2013, the depreciation of BEVs has been more or less the same as for ICEVs (Figenbaum and Kolbenstvedt 2015, 2016, Dine Penger 2016). The share of consumers seeing second hand value as a disadvantage of BEVs was substantially reduced from 2014 and 2016, and is no longer a large barrier to adoption (Figenbaum et al. 2014; Figenbaum and Kolbenstvedt 2016).

Experiences and Opportunities with BEVs

BEV owners are on average 7–8 years younger than other vehicle buyers, have a higher employment rate, higher education, more children, live in suburban and urban areas, and have longer commutes than ICEV and PHEV owners (Figenbaum and Kolbenstvedt 2016), as seen in Table 3.

BEV owners are more inclined to own more than one vehicle, 79% are multi vehicle households versus 52% of ICEV owners. They are apparently wealthier, but when controlling for vehicle age and other sociodemographic factors, there were no significant differences in the household income level between BEV, PHEV and ICEV owners (Figenbaum and Kolbenstvedt 2016). With both parents typically working, the BEV owning families face constant time pressure getting to and from work, and transporting children to and from kindergartens, schools and activities. BEVs

Table 3 Socio-demographics of vehicle owner groups (Figenbaum and Kolbenstvedt 2016)

	BEV	PHEV	ICEV
Employed or self-employed (%)	91	77	67
Retired/benefit recipient/student (%)	9	23	33
Primary, secondary, high school (1–13th grade) (%)	22	25	33
Higher education up to 4 years (%)	38	38	37
Higher education in excess of 4 years (%)	40	37	29
Gender, male share of respondents (%)	80	83	78
Average age (year)	47	55	56
Average number of persons in household	3.2	2.6	2.5
Share of households with children (%)	56	32	27
Multivehicle households (%)	79	54	52
Average distance to work (km)	24	17	17
Prior transportation to work: vehicle based-driver/passenger (%)	87	85	74
Gross household income: 0–600,000 NOK (%)	12	15	30
600,001–1,000,000 NOK (%)	37	41	44
>1,000,000 NOK (%)	51	44	26

perform these local transportation tasks more time efficiently than public transport, and at a much lower cost than an ICEV, given the Norwegian incentives.

Comparable groups of BEV, ICEV and PHEV owners, i.e. owners that are employed full time, have children, and own 2011 and newer year models, all drive their vehicles about 15,700 km per year on average (Figenbaum 2018). The overall tendency is the same for single vehicle and multi-vehicle households (Figenbaum and Kolbenstvedt 2016). BEVs are, however, used more frequently for local trip types than ICEVs (Figenbaum and Kolbenstvedt 2016), indicating that in multi-vehicle households a redistribution of trips between the household vehicles occurs when a BEV replaces an ICEV, so that the BEV offsets some of the trips with the remaining ICEV. The reason is obviously the reduced user cost associated with electric driving, but also that a BEV's driving characteristics are ideal for local trips.

The practical user experiences in Norway have demonstrated that BEVs with a nominal range of 160–250 km can work well in the daily transportation activity of multi-vehicle households (Figenbaum and Kolbenstvedt 2016). That is the case even though users experience a 25% range reduction in real summer traffic, and up to 50% during the rough Norwegian winters (Figenbaum and Kolbenstvedt 2016), as seen in Fig. 11. This seems to counter claims that range must increase for BEVs to reach the mass market.

A lesson from Norway is, thus, that if BEV prices come down to a level that makes them affordable and cheaper to own and use than ICEVs, then multi-vehicle households can take them efficiently into use. Other conditions for adoption are that consumers need to be aware of BEVs existence and have somewhere to charge them. Clean electricity, high potential for home charging, competitive purchase price and

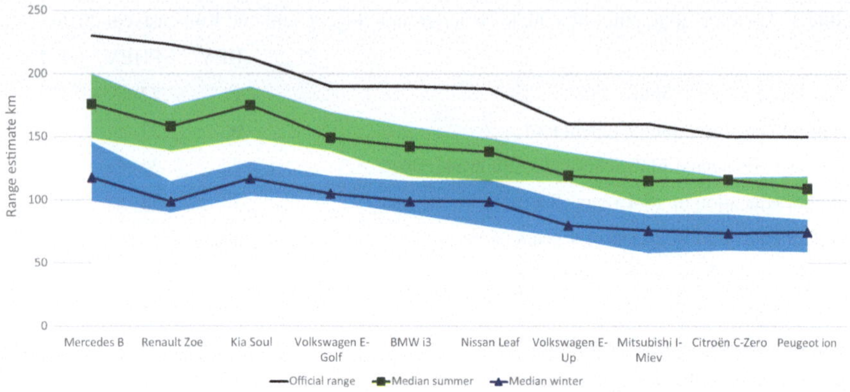

Fig. 11 BEV users' assessed real-world range in summer and winter. Median and 20–80% percentile of users, green summer and blue winter (Figenbaum and Kolbenstvedt 2016)

low running costs are other conditions that have led to the popularity of BEVs in Norway. For single vehicle households the results are less conclusive.

In most other countries the running cost savings of using BEVs instead of ICEVs are, as seen in Fig. 9, less than in Norway, and the electricity will contain varying percentages of fossil fuels. Consumers in other countries will therefore see fewer cost and environmental advantages of BEVs than Norwegians do.

Multi vehicle households cope well with BEVs' range and charge time limitations. In 2016, 6% of BEV owners said that they have had to abort a trip due to a problem on the go and 27% had avoided doing a trip due to the range being too short (Figenbaum and Kolbenstvedt 2016). The rest had never experienced these issues with their BEVs. About half of these incidents were, according to the respondents, related to inferior, faulty or lacking charging infrastructure (Figenbaum and Kolbenstvedt 2016). When these issues occur, consumers plan their travels better, drive more economically, fast charge on the way or find other means of transport (Figenbaum and Kolbenstvedt 2016). The latter is easy for multi-vehicle households, as they merely swap their vehicles according to needs. Single BEV owning households would have to borrow or rent a vehicle, or use public transport. Between the two surveys in 2014 and 2016, the share that would find other means of transport had decreased, and the share that would make the trip with their vehicle anyway had increased (Figenbaum and Kolbenstvedt 2016). More people stated in 2016 they would be driving economically and fast charge along the way to be able to do a trip longer than the range of their vehicle permitted. This answer is most likely influenced by the longer range of newer models and by the fact that a rudimentary network of fast chargers was installed in many regions between 2014 and 2016. These chargers have enabled long distance driving, and have solved many daily range problems (Figenbaum 2018).

Although current BEVs work well for Norwegian households, BEV owners do want greater range to be able to cover more of their driving with the BEV. An all year range of 300 km, i.e. on par with the Tesla Model S, could open up 50–80% of the total market (Figenbaum and Kolbenstvedt 2016). A range of 200 km could open up half of that potential. But these potentials can only be reached when, in addition, more models in different sizes become available (Figenbaum and Kolbenstvedt 2015; Figenbaum 2018; Fearnley et al. 2015). Vehicle owners in general (currently ICEV owners) say that a greater range is needed than that which BEV owners say. Either the difference is due to self-selection, i.e. that those who can use BEVs with a shorter range most efficiently have been the ones that bought BEVs, or due to a disparity in the knowledge of how BEVs work in practice, i.e. that ICEV owners overestimate the range that is actually needed.

Vehicle manufacturers have stated that the infrastructure needs to be improved before they can sell BEVs in significant numbers. Yet, users in Norway rarely use the public charging infrastructure. 75% of all Norwegian households in municipalities with more than 20,000 inhabitants say they can park a vehicle on their own land, while another 14% have dedicated access to parking less than 100 m from their home (Hjorthol et al. 2014). A private parking space can most likely be used for charging a BEV, especially when on own land, and is a prerequisite for owning a BEV (Figenbaum 2018). Of the BEV owners in the 2016 user survey, 94% charge their vehicle at home. 59% do it daily, another 21% 3–5 times per week, the remaining 14% on a weekly basis (Figenbaum and Kolbenstvedt 2016). Public charging is rarely used; only 11–12% use it on a weekly basis, another 38% only monthly, while 24% never use them (Figenbaum and Kolbenstvedt 2016). About a third are satisfied with the quality, position and reliability of fast chargers, about a third are not (Figenbaum and Kolbenstvedt 2016).

The energy supply to BEVs is not a big issue. If all passenger vehicles in Norway were BEVs they would only consume about 7 TWh per year, which corresponds to a 6% increase (Figenbaum et al. 2015) in the total normal electricity consumption in Norway. The power requirement for simultaneous charging of large number of BEVs could cause more challenges. The typical time to plug in at home is between 4 and 6 PM when owners come home from work (Figenbaum and Kolbenstvedt 2016), which means that the bulk of the home charging coincides with the peak power demand (NVE 2016) in the Norwegian grid. A large increase in BEV usage may thus potentially lead to a need to strengthen local grids to support home charging. Most current users use their vehicle to commute to work, and plug them in when arriving back home. When BEVs reach a much broader adoption in the general population, people who do not drive to work daily will also buy BEVs, and the charging start up time will be much more spread out (NVE 2016). The Norwegian Water Resources and Energy Directorate (NVE 2016) found therefore few issues with the high power transmission system related to charging of BEVs, but some issues can occur in the low power distribution system. NVE found however that less than 10% of the low power distribution transformers and 5% of cables will be overloaded if the average power for simultaneous BEV charging can be kept below 2 kW per household (NVE 2016). Upgrading cables that are buried underground can be very expensive, whereas early retirement of easily accessible transformers is much less costly. NVE also stated that the future peak power need from a large population of BEVs is likely to occur at night in a 2030 timeframe, when the distribution network has more spare capacity.

The reason is the deployment of smart electricity meters across Norway by 2020 that will allow for differentiated prices day and night.

Fast charger stations in Norway typically consist of a minimum of two 50 kW DC CCS/Chademo fast chargers[3] and two semi-fast AC 22 kW type 2 sockets. The fast charge stations are typically connected to the grid through an upgraded transformer with excess capacity allowing for later expansion.[4] Any costs incurred to upgrade the grid are thus included in the installation cost of the station. Fast charge operators experience that different grid operators have different costs of grid connection and power availability. They therefore sometimes change the position of fast charge stations to reduce grid and power costs.

The average BEV owner use fast chargers slightly more than once per month (Figenbaum and Kolbenstvedt 2016), while Tesla owner do it twice as often. The usage frequency stated by the users increased somewhat between 2014 and 2016 (Figenbaum and Kolbenstvedt 2016). The users mainly plan to use fast chargers to complete a trip, and most commonly for non-regular trips (Figenbaum and Kolbenstvedt 2016). So far, the bulk of fast charging occurs in the outskirts of cities where most BEVs are located.[5] These fast chargers support daily traffic, long commutes and professional users.

Long distance driving with a non-Tesla BEV is rarely done (Figenbaum and Kolbenstvedt 2016). Non-Tesla BEV owners state that short range is the dominant reason not to use their BEVs on long distances, but also that the vehicle is too small, or that charging infrastructure is lacking along the way or at the destination (Figenbaum and Kolbenstvedt 2016). Most BEV owners also have an ICEV they can use for these trips. It is likely that there will be more long distance driving with BEVs when a longer range becomes available in cheaper and smaller vehicles, such as the Hyundai Kona.

89% of current users say they will repurchase a BEV (Figenbaum and Kolbenstvedt 2016), further supporting the conclusion that BEVs work for them in practice. This answer should also be seen in relation to the large incentives that BEV owners receive, including local driving privileges.

Plugging into the Electric Future

Norway has ambitious climate policy targets, of which the most important ones are (1) a 40% reduction of Greenhouse gas emissions in 2030 compared to 1990, (2) carbon neutrality by 2030 (emissions reductions abroad to make up the difference), and (3) to become a low emission society by 2050 (Ministry of the Environment 2018). BEVs have become an essential part of meeting those targets.

---

[3]From different manufacturers for redundancy in case of system issues with one manufacturer/model.

[4]According to information from the main operators.

[5]See Footnote 4.

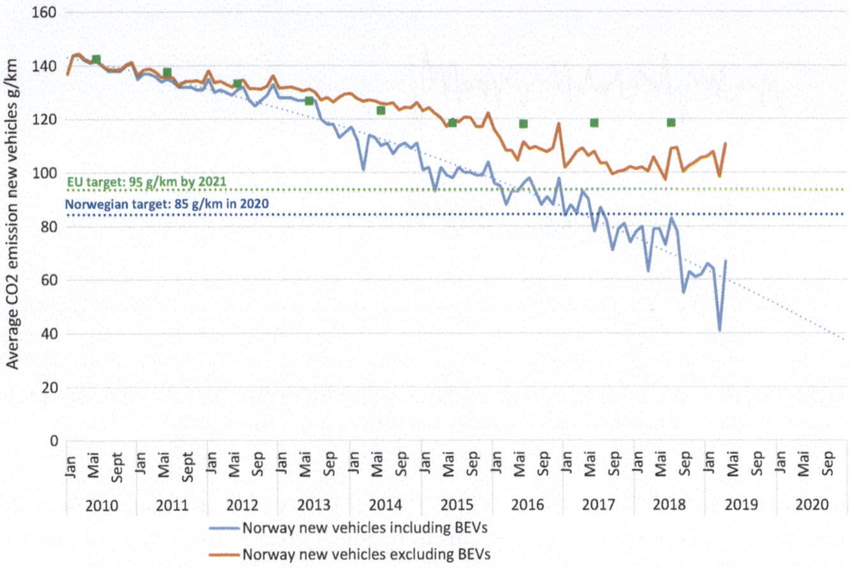

Fig. 12 Norwegian progress towards 85 g CO_2/km by 2020 target, and the EU progress towards their 95 g CO_2/km target for 2020. Monthly average CO_2-emissions from all new vehicles and from new ICEV (including PHEVs) vehicles in Norway. Average annual CO_2-emission of all new vehicles in EU (Figenbaum 2018)

A transportation sector sub-target has been to get the average new car CO_2 emission down to 85 g/km by 2020. That target was, as seen in Fig. 12, reached already in 2017. The EU progress towards 95 g CO_2/km by 2020 is not as convincing. The reduction has more or less stagnated over the last years, partly due to the falling diesel vehicle market shares in the aftermath of the diesel scandals that have hit the sector since 2015.

The question in Norway is now whether BEVs could cover all types of driving needs, as the government set an aspirational target in 2017 (which the Parliament supported) that only zero emission vehicles (i.e. BEVs or FCEVs) should be sold after 2025 (Figenbaum 2018). Although 89% of current BEV owners say they will buy a BEV again, reaching 100% of new vehicle sales by 2025 will still be a daunting task given that more sceptical and more demanding customers must find BEVs (or FCEVs) attractive, i.e. matching their needs at a cost they can afford. It will definitely require the continuation of effective policies and large incentives. The European Economic Area agreement between Norway, Lichtenstein, Iceland and the European Union makes it impossible for Norway to ban sales of ICEVs, as long as they are legally sold within the EU. The growing BEV development in the auto industry since 2015 should, however, make the target less demanding to reach. More models in all segments and with longer ranges will come on the market (Figenbaum 2018). The

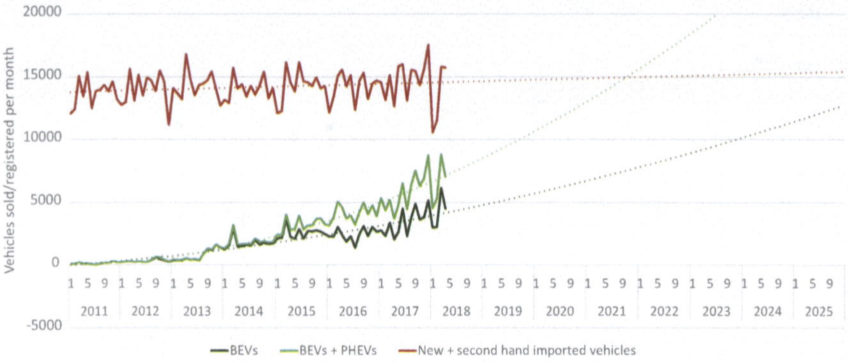

Fig. 13 Progress towards the national target of only selling BEVs from 2025. Monthly sales of all new and second hand imported vehicles, BEVs and PHEVs (Figenbaum 2018)

progress towards the target is shown in Fig. 13. If PHEVs are also counted towards the target, the potential to reach it seems to be much greater than when only BEVs count.

The main advantages of BEVs that vehicle buyers of all types acknowledge, are their low user costs, their benign environmental characteristics, and the home charging ability (Figenbaum and Kolbenstvedt 2016). BEV owners also like their comfort and driving characteristics, whereas ICEV owners are in general more sceptical towards the technology, and less inclined to see the advantages of BEVs (Figenbaum and Kolbenstvedt 2016). Longer range, more charging infrastructure and the preservation of incentives are seen by all vehicle owning groups as being important for future BEV market growth (Figenbaum and Kolbenstvedt 2016).

When long range BEVs become a larger share of the vehicle fleet and more people start using them for long distance driving, charging infrastructure and charge time will likely emerge as bigger barriers than range and battery size (Figenbaum 2018). On peak travel days it might not be possible to avoid long queue times for charging, especially on roads where the traffic volume on peak travel days varies substantially from the everyday traffic (Figenbaum 2018). As an example, Fig. 14 shows light duty vehicle activity through selected toll roads over a year. Travel peaks occur during typical vacation times. The RV7 road is both a national road between Oslo and Bergen, and a road used by owners of mountain cabins during vacation times. This road has huge variations in traffic between weekdays, the weeks themselves, and vacation times. By comparison, the European main road E6 south of Oslo (leading to Gothenburg in Sweden) has much less variation in traffic over the year and between weekdays. The challenges of establishing charging infrastructure will thus be totally different for these two different roads. Many more long-range BEV models will become available in the coming years. If users with the longest driving needs buy the vehicles with the longest range, then the infrastructure challenges will be reduced.

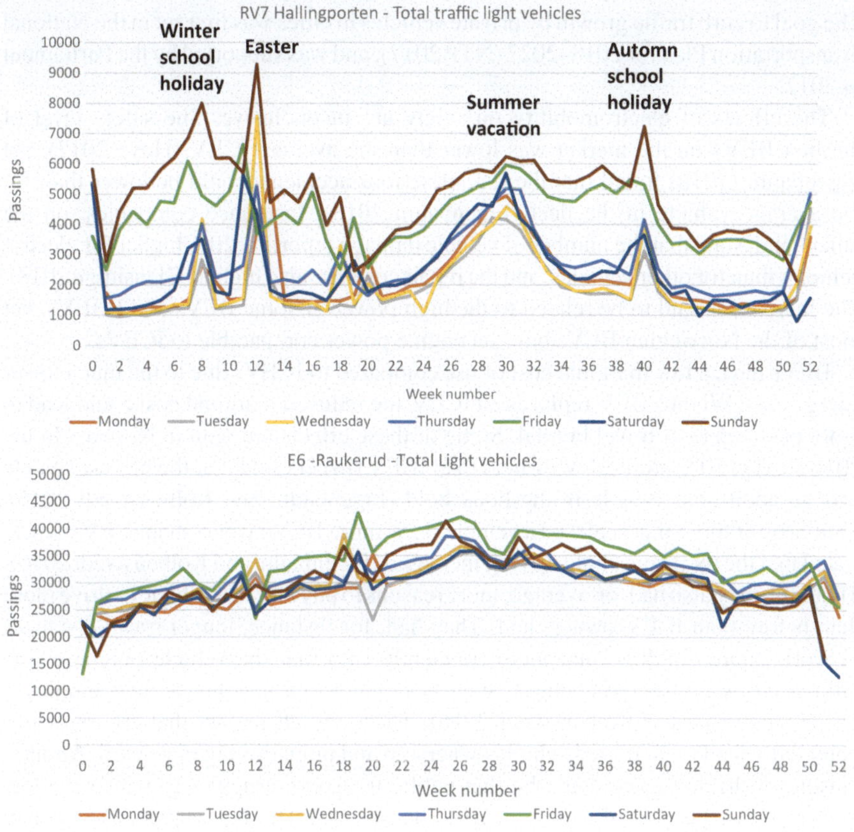

Fig. 14 Light duty vehicle transportation volumes by week and weekday over the year 2016, vehicles passing through toll gates on the main roads RV7 west of Oslo and the E6 South-east of Oslo (Figenbaum 2018)

Relation to Other Societal Targets

The Norwegian government has several societal targets for the transportation sector. Electromobility is one of the most important measures to reach the national goals of reducing greenhouse gas emissions and local pollution. Other targets that may influence the transition to electromobility are:

1. The Vision Zero for traffic fatalities
2. Traffic growth in cities shall be taken by walking, cycling and public transport.

Vision Zero, which can be summarized as: "A vision of no road fatalities or road accidents causing lifelong injury for the long-term", was established during the work with the National Transportation Plan in 2002 (Ministry of Transport and Communications 2002). It guides the transportation sector's work on traffic safety.

The goal to curb traffic growth by private vehicles in cities was first set in the National Transportation Plan for 2018–2027 (NTP 2017), and was supported by the Parliament in 2017.

The effects of electromobility on safety are inconclusive. The safety level of the first BEVs on the market was lower than the average ICEV (Høye 2017), yet Figenbaum (2018) found that the risk of serious accidents could be lower than for the average vehicle in the fleet (Figenbaum 2018). Insurance companies, on the other hand, state that the number of vehicle damage reports is 20% higher for electric vehicles than for other vehicles, and the repair costs are also higher (Gjensidige 2018). The increase is said to be related to the high power of some BEVs and PHEVs, yet most of the Norwegian BEVs have an engine power comparable to ICEVs.

BEVs have a low marginal cost of use compared to ICEVs due to the much lower energy cost. When a BEV replaces an ICEV, the reduced marginal cost could lead to more passenger car travel in total. Some of these effects are seen in Norway. In the 2016 survey, 10% more BEV owners than ICEV owners said that the newest vehicle was an additional vehicle in the household (Figenbaum and Kolbenstvedt 2016). Similarly, of those that replaced a vehicle, 12% more BEV owners than ICEV owners stated that the total household driving increased (Figenbaum and Kolbenstvedt 2016). That said, they also had, on average, more reasons to buy an extra vehicle or drive more than before than ICEV owners had. They had, for instance, longer travel distances to work, more children, and more frequently said that the vehicle purchase was influenced by household changes, such as moving to a new house, new employer etc. (Figenbaum and Kolbenstvedt 2016). These are all factors that are known to potentially lead to increased vehicle ownership and more driving in general. Another reason for driving more could be that public transport struggles to match the low marginal cost per km of driving a BEV in Norway. The Norwegian BEV policy could thus risk being in conflict with the transportation sector goal of limiting traffic growth and congestion in cities.

Historically, the passenger vehicle taxes and the fuel taxes have provided Norwegian governments with substantial revenues. This revenue has enabled road construction and compensated for most of these vehicles' external costs. The rapid increase in BEV sales has started to substantially erode these revenues. An ICEV uses 2–4 times the energy per km as a BEV measured from battery/tank to the wheel. The energy tax income per km is therefore much lower from BEVs than from ICEVs, given the taxes in Norway (Fig. 3). The contribution towards covering the external costs of driving will thus be much lower. BEVs do generate lower externalities than ICEVs due to their zero tailpipe emissions and quiet operation. BEVs contribute, however, equally to congestion, traffic accidents and the wearing of roads as any other vehicle does. The income loss in the new vehicle taxes is already being observed. For 2017 the loss was estimated to be about 440 million Euros, of which VAT exemption was the main cost at 340 million Euros (National budget 2018). Another 70 million Euros was due to the registration tax exemption and 30 million Euros due to the reduced annual tax. Some of the loss can be recovered by increasing the taxation of ICEVs, which would also increase the pressure to switch over to BEVs. The much slower growth in the BEV share of the total fleet than in new vehicle sales will keep up the income from fuel taxes much longer than the purchase taxes.

The growing BEV share on the road influences the income of toll road companies. These companies have borrowed money to finance the construction of a new road, a bridge or a tunnel, which is repaid with the collected tolls over a predetermined collection period. The toll rates and the collection period are based on a prognosis for the future traffic when the toll road opens. In most cases these prognoses turn out to be conservative, with traffic often growing faster than expected, more than offsetting the impact of increased BEV shares. If the toll road income becomes too low, the toll road company can increase tolling rates by up to 20%, and the tolling period for up to 5 years (Figenbaum 2017). No toll road companies have yet had to raise their rates due to increased BEV shares. Some cities use income from toll roads to improve and support public transport. These schemes may faster run into financial issues if the number of paying vehicles goes down.

The same financial issues will occur for national ferry services where BEVs pay a lower rate than ICEVs. The reduced income is, however, immediately seen in the day to day profitability of the ferry operation. These ferry services are based on tenders. Most of the current services were tendered before the main growth in the share of BEVs started. Some of the ferry operators thus end up in an economically challenging situation when the BEV share increases. Ferry services controlled by the provinces could however apply the same rates to BEVs and ICEVs.

Extreme Outlier or a Lab to Learn from?

Figenbaum and Kolbenstvedt (2015) propose a circle of decisive factors, shown in Fig. 15, that needs to be in place in order to achieve an efficient BEV adoption among consumers. Consumers must have access to quality information to be able to build up

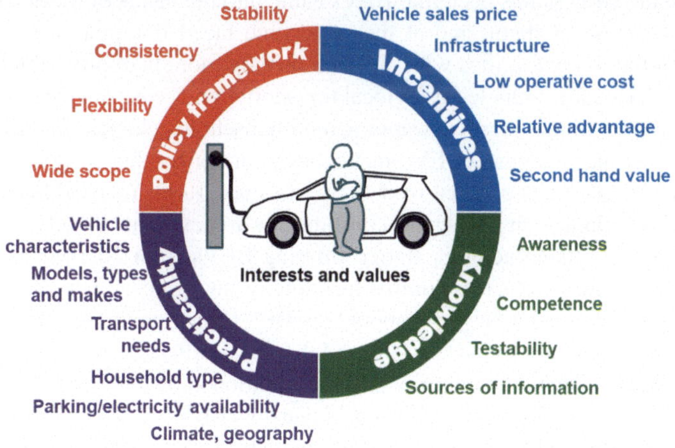

Fig. 15 The circle of factors that affects BEV adoption among consumers (Figenbaum and Kolbenstvedt 2015)

knowledge about BEVs. BEVs must meet basic needs of different consumer types, i.e. have a range that supports driving needs and meet requirements for safety, and be available in different sizes and body styles. Consumers must have space to park and charge their vehicle, at home but also at destinations and during trips. A stable policy framework will reassure businesses and consumers about the future viability of the technology. Incentives can be used to reduce disadvantages and high initial costs of BEVs, and to provide consumers with unique user advantages that help generate the initial markets. By addressing these factors, main consumer concerns with BEVs have been alleviated in Norway.

The Norwegian BEV policy seems extreme compared to other countries. It has produced extreme results in terms of sales volumes and annual electric km driven. The growth in BEV sales has occurred because consumers have experienced substantial economic and practical advantages of owning and operating a BEV compared to an ICEV. These benefits have been achieved through tax exemptions made possible because Norway had very high vehicle taxes and fuel taxes at the outset, and by providing consumers with local driving privileges. These policies and incentives have been stable over a very long time period, in some cases since the 1990s. While effective in increasing BEV market shares, these tax exemptions lead to a substantial loss of income for the government when sales take off. These issues may deter other countries from introducing similar policy packages. Other countries that want to expand the diffusion of BEVs should therefore aim at lowering the cost of BEVs to consumers through sustainable incentives such as, for instance, a self-financed bonus/malus system. In these systems a small malus, i.e. a tax, on polluting vehicles is used to provide a big bonus for buyers of BEVs. User privileges such as access to bus lanes should be temporary so that they can be scaled back when the BEV fleet increases.

The case with which BEVs have been integrated into the Norwegian vehicle fleet, shows that BEVs can work for consumers in practice under tough climate conditions. The key to the success has been that BEVs came into the hands of those who could make the best use of them, and at the same time faced few practical challenges when using them, i.e. families with children owning more than one vehicle. These households have large vehicle-based local transportation needs, and can easily solve range and charge time issues by swapping their household vehicles. The installation of fast chargers has increased BEV practicability further for this group.

The BEV market has evolved over the years since 1990. With 3000 consumers owning BEVs already in 2009, the consumer awareness was already high when the major vehicle manufacturers started offering BEVs from 2010. In most other countries BEVs are so far a less familiar technology among consumers, and very few existing owners can tell their peers about how BEVs function in everyday traffic. A survey of the population in Norway showed that the 5% fleet share was sufficient to lead to a situation where the majority of the population knows someone with a BEV.

Virtually all of the Norwegian consumers that own BEVs can charge their vehicles at home. Countries with less private parking available should focus on ways to make charging available where potential BEV buyers live as consumers are unlikely to buy a BEV they cannot recharge at home. Home charging is also seen as one of the

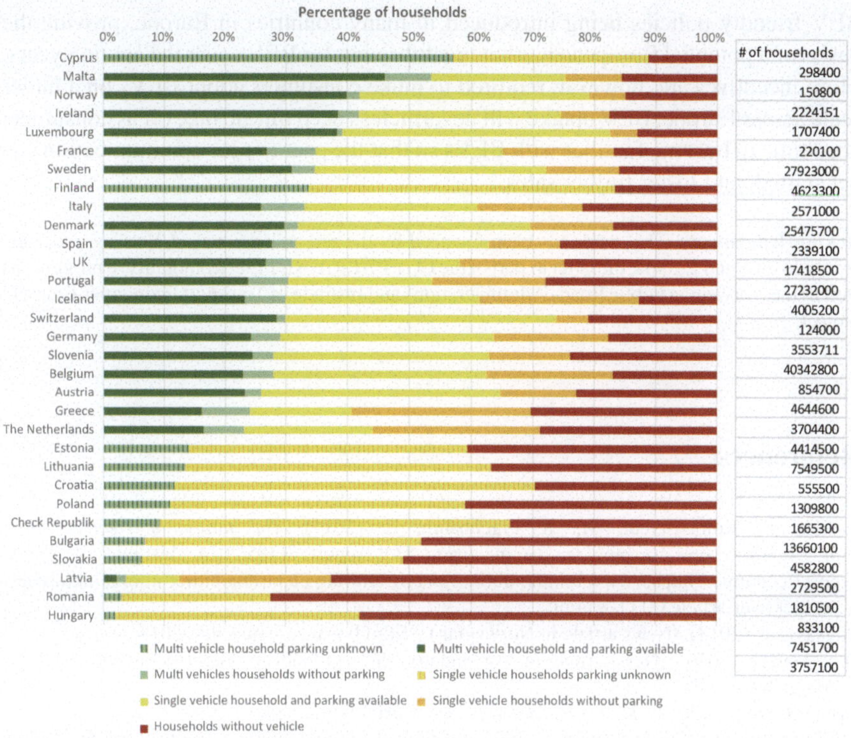

Fig. 16 Households in Europe, vehicle and parking space availability and total number of households. Adapted from Figenbaum and Kolbenstvedt (2015)

most important consumer benefits of owning a BEV (Figenbaum and Kolbenstvedt 2016). There are 45 million European (European Union + Norway, Iceland and Switzerland) households that own more than one vehicle and have access to a private parking space (Figenbaum and Kolbenstvedt 2015), as seen in Fig. 16. Most of these could likely take a BEV successfully into use as one of their household vehicles if the price was competitive. Another 55 million European households have access to parking at home, and own one vehicle. For these single vehicle households, the barriers are cost, range and charge time, and the lack of charging infrastructure that can supports long distance trips. Range is a much bigger barrier for single vehicle households and the improvements in range seen for BEVs coming on the market from 2018 (Figenbaum 2018) will be essential to increase their BEV adoption.

The regional differences in Europe are however large. Eastern European countries have much lower vehicle ownership levels than in Western and Northern Europe, and much smaller shares of multi vehicle households (Figenbaum and Kolbenstvedt 2015). The availability of parking is not known for these countries.

The huge focus on BEVs in the automotive industry, the technological improvements, the build out of charging infrastructure, the large potential market, and more

BEV friendly policies being introduced in many countries in Europe, provide the basis for a potential European market breakthrough for BEVs over the coming years. More incentives are, however, required to make consumers adopt BEVs on a larger scale across Europe. BEV uptake will nevertheless need time to take off as consumers need time to become familiar with BEVs so that the peer to peer communication can get started, and foster further sales.

Acknowledgements This work has been financed by the research council of Norway under the grants 267848 and 225564, the latter as part of the EU FP7/ERA-NET Electromobility + program, as well as the Norwegian Public Roads Administrations and Institute of Transport Economics Emiroad project.

References

Assum, T., Kolbenstvedt, M., & Figenbaum, E. (2014). *The future of electromobility in Norway—some stakeholder perspectives*. TØI report. 1385/2014. Institute of Transport Economics. https://www.toi.no/publikasjoner/framtidig-elektromobilitet-i-norge-aktorenes-perspektiver-article33111-8.html.

Dagsavisen. (2018). (News article in Norwegian). "Elbil bonanza". June 4th, 2018.

Dine Penger. (2016). (News article in Norwegian). «Stort prisfall for elektrisk storselger». Dine Penger Nr. 6 2016.

Drivkraft Norge. (2018). *Fuel price statistics*. www.drivkraftnorge.no.

Elbilbarometer. (2018). Data from a survey of vehicle buying intentions carried out by Opinion AS in January 2018 for the EV Association and Nordic Energy Research. Presentation at Nordic EV Summit February 1–2, 2018.

Eurostat. (2018a). *Mean and median income by household type—EU-SILC survey*. http://appsso.eurostat.ec.europa.eu/nui/show.do?dataset=ilc_di04&lang=en.

Eurostat. (2018b). *Comparative price levels of consumer goods and services*. http://ec.europa.eu/eurostat/statistics-explained/index.php/Comparative_price_levels_of_consumer_goods_and_services.

Eurostat. (2018c). *Passenger vehicles in the EU*. Downloaded August 17, 2018. http://ec.europa.eu/eurostat/statistics-explained/index.php/Passenger_cars_in_the_EU.

Fearnley, N., Pfaffenbichler, P., & Figenbaum, E. (2015). *E-vehicle policies and incentives—Assessment and recommendations*. TØI Report 1421/2015. Oslo, Norway: Institute of Transport Economics.

Fetene, G. M., Kaplan, S., Mabit, S. L., Jensen, A. F., & Prato, C. G. (2017). Harnessing big data for estimating the energy consumption and driving range of electric vehicles. *Transportation Research Part D, 54*(2017), 1–11.

Figenbaum, E. (2018). *Electromobility status in Norway—Mastering long distances—The last hurdle to mass adoption*. TØI report 1627/2018. Oslo, Norway: Institute of Transport Economics. https://www.toi.no/publications/electromobility-status-in-norway-mastering-long-distances-the-last-hurdle-to-mass-adoption-article34903-29.html.

Figenbaum, E. (2017). Perspectives on Norway's supercharged electric vehicle policy. *Environmental Innovations and Societal Transitions, 25*(2017), 14–34. https://doi.org/10.1016/j.eist.2016.11.002.

Figenbaum, E., & Kolbenstvedt, M. (2016). *Learning from Norwegian battery electric and plug-in hybrid vehicle users. Results from a survey of vehicle owners*. TØI report 1492/2016. Oslo, Norway: Institute of Transport Economics. https://www.toi.no/publications/learning-from-norwegian-battery-electric-and-plug-in-hybrid-vehicle-users-results-from-a-survey-of-vehicle-owners-article33869-29.html.

Figenbaum, E., & Kolbenstvedt, M. (2015). *Competitive electric town transport. Main results from COMPETT-an Electromobility+project*. TØI-report 1422/2015. Oslo, Norway: Institute of Transport Economics. https://www.toi.no/publications/competitive-electric-town-transport-main-results-from-compett-and-electromobility-project-article33368-29.html.

Figenbaum, E., & Weber, C. (2016). *Experimental testing of plug-in hybrid vehicles*. TØI report 1539/2016. Oslo, Norway: Institute of Transport Economics. https://www.toi.no/publications/experimental-testing-of-plug-in-hybrid-vehicles-co2-emission-energy-consumption-and-local-pollution-article34298-29.html.

Figenbaum, E., Assum, T., & Kolbenstvedt, M. (2015). Electromobility in Norway—Experiences and opportunities. *Research in Transportation Economics, 50*, 29–38. https://doi.org/10.1016/j.retrec.2015.06.004.

Figenbaum, E., Kolbenstvedt, M., & Elvebakk, B. (2014). *Electric vehicles—Environmental, economic and practical aspects. As seen by current and potential users*. TØI, Report1329/2014. Oslo, Norway: Institute of Transport Economics. https://www.toi.no/publications/electric-vehicles-environmental-economic-and-practical-aspects-as-seen-by-current-and-potential-users-article32644-29.html.

Galvin, R. (2017). Energy consumption effects of speed and acceleration in electric vehicles: Laboratory case studies and implications for drivers and policy makers. *Transportation Research Part D, 53*, 234–248.

Geels, F. W. (2012). A socio-technical analysis of low-carbon transitions. Introducing the multi-level perspective into transport studies. *Journal of Transport Geography, 24*, 471–482.

Gjensidige. (2018). (press release in Norwegian). «Miljøbiler gir dårligere trafikksikkerhet» . Pressemelding. Gjensidige. June 4th, 2018.

Haakana, A., Laurikko, J., Granström, R., & Hagman, R. (2013, December). *Assessing range and performance of electric vehicles in Nordic driving conditions—Project Final Report*. Norden—Energy and Transport. Nordisk Energiforskning.

Hjorthol, R., Engebretsen, Ø., & Uteng, T. P. (2014). (in Norwegian). «Den nasjonale reisevaneundersøkelsen 2013/14 – nøkkelrapport». TØI rapport 1383/2014. Oslo, Norway: Institute of Transport Economics. https://www.toi.no/publikasjoner/den-nasjonale-reisevaneundersokelsen-2013-14-nokkelrapport-article32973-8.html.

Hyundai. (2018). (press release in Norwegian). «Rekordsalg av Hyundai Kona electric: 6969 orders in two weeks. http://www.mynewsdesk.com/no/hyundai-motor-norway-as/pressreleases/rekordsalg-av-hyundai-kona-electric-6-969-ordre-paa-to-uker-2546447.

Høye, A. (2017). (in Norwegian). «Trafikksikkerhetseffekter av bilenes kollisjonssikkerhet, vekt og kompatibilitet». TØI rapport 1580/2017. Institute of Transport Economics. https://www.toi.no/publikasjoner/trafikksikkerhetseffekter-av-bilenes-kollisjonssikkerhet-vekt-og-kompatibilitet-article34632-8.html.

Kolbenstvedt, M., & Assum, T. (2018). *The sales of electric vehicles—The role of the salesepeople and the customers' assessments*. TØI report 1639/2018. Oslo, Norway: Institute of Transport Economics.

Ministry of Climate and Environment. (2018). *Norway's Seventh National communication. Under the framework convention of climate change*. https://www.regjeringen.no/contentassets/52d65a62e2474bafa21f4476380cffda/t-1563e.pdf.

Ministry of Transport and Communications. (2002). *Road safety in Norway. Strategy 2002–2011*. https://www.regjeringen.no/globalassets/upload/kilde/sd/bro/2002/0001/ddd/pdfv/160260-road_safety.pdf.

National budget. (2018). (in Norwegian). Meld St. 1 (2017–2018). «Nasjonalbudsjettet 2018». https://www.statsbudsjettet.no/upload/Statsbudsjett_2018/dokumenter/pdf/stm.pdf.

NPRA. (2018). Dataset from the Norwegian vehicle register, fleet of BEVs and PHEVs per 01.01.2018. Provided by the Norwegian Public Roads Administration.

NTP. (2017). (in Norwegian). Innst. 460 S. (2016–2017). «Innstilling fra transport- og kommunikasjonskomiteen om Nasjonal transport plan 2018–2029 – BERIKTIGET». Stortinget 2017. https://www.stortinget.no/globalassets/pdf/innstillinger/stortinget/2016-2017/inns-201617-460s.pdf.

NVE. (2016). (in Norwegian). "Hva betyr elbiler for strømnettet? NVE rapport 74/2016. ISBN: 978-82-410-1527-4.

SSB. (2018a). *Energy price statistics*. Statistics Norway.

SSB. (2018b). Table 10453: «Husholdninger med utvalgte varige forbruksvarer (prosent) 1976–2012)». Statistics Norway.

SSB. (2018c). Statistics Norway. 08940: Greenhouse gases, by source, energy product and pollutant 1990–2017. https://www.ssb.no/en/statbank/table/08940.

Plug-in (Hybrid) Electric Vehicle Adoption in the Netherlands: Lessons Learned

Rick Wolbertus and Robert van den Hoed

Abstract The Netherlands is considered one of the frontrunners in the field of electric mobility, both in number of plug-in electric vehicles sold as in the number of publicly available charging stations. This chapter analyses the policy measures that led to the growth in the number of plug-in hybrid electric vehicles. It also provides room for a critique of the policies, with a special focus on the actual reduction of emissions. Consequently the number of public charging stations also rose. The chapter provides an analysis of how these charging stations are used and in which way this is related to sales policies. The chapter provides several lessons learned on how to shape policy to get a group of early adopters to drive electric.

Keywords Policy · Public Charging Infrastructure · Plug-in Hybrid · Emissions

Introduction

The story of plug-in electric vehicle (PEV) adoption in the Netherlands is known for its high adoption rates and the relative high share of plug-in hybrid vehicles (PHEVs). In parallel the development of public charging infrastructure resulted in one of the highest public charger to vehicle ratios in the world (International Energy Agency 2017). This chapter focuses on the drivers behind the successful growth of the PEV market in the Netherlands and goes deeper into how the fiscal climate has lured consumers into buying a large number of PEVs and PHEVs in particular. It reflects on the pros and cons of the Dutch policies on stimulating PEV adoption and public charging infrastructure development.

In the past ten years the Dutch government has had a set of ambitious goals and subsidy measures to stimulate PEVs as part of an effort to reduce emissions from transport. In 2016 ambitions were set to achieve 10% of all new vehicles to have an electric drivetrain by 2020, with subsequent increases towards 50% by 2025 (of

R. Wolbertus (✉)
TU Delft, Delft, The Netherlands
e-mail: r.wolbertus@tudelft.nl

R. van den Hoed
Amsterdam University of Applied Sciences, Amsterdam, The Netherlands

M. Contestabile et al. (eds.), *Who's Driving Electric Cars*, Lecture Notes in Mobility,
https://doi.org/10.1007/978-3-030-38382-4_7

which 30% full electric) and 100% electric by 2030 (Formula E-Team 2016). From 2012 generous fiscal incentives were put in place to stimulate PEV sales, leading to a spur in sales especially for PHEVs. Up to 2017, when incentives for PHEVs were substantially reduced, the Netherlands was among the countries with the highest adoption rates of PEVs in the world.

This chapter studies two defining elements of the Dutch case namely (i) the impact of fiscal incentives on consumer decisions, particularly in relation to the high share of PHEVs; and (ii) how the dense public charging network has facilitated PEV drivers. Regarding PEV penetration, fiscal incentives that have been in place up to 2018 and the resulting PEV sales in this period are discussed. Particularly the fiscal measures that stimulated PHEVs in early years and the discontinuation of which lead to a major drop in PHEV sales in later years are reviewed. These serve as an illustration of how substantial shifts lead to major changes in adoption decisions. Apart from sheer sales, we will explore actual emission mitigation effects by PHEVs and the difference between real life usage of PHEVs and NEDC cycles. Regarding charging infrastructure, the roll-out of public charging stations in the Netherlands is discussed. It sheds light on the problem of on-street parking in urban areas for PEVs and provides some possible solutions for public charging based on Dutch practices. In the final part we reflect on the achievements as well as drawbacks of Dutch policies to stimulate consumers to switch to electric, and provide recommendations for policymakers as well as future PEV-research to lower barriers for consumers through incentives and provision of charging infrastructure.

The remainder of this chapter is structured as follows: in the next section we set the scene by discussing the fiscal incentives that affect consumer adoption of PEVs and the policies regarding charging infrastructure. In the following section we take a closer look at how effective the policies have been at reducing emissions. In order to do so we review the factors that drive the share of electric kilometres of PHEV drivers. We then have a look at how consumers use the public charging infrastructure and identify three key metrics. The final section concludes the chapter by providing policy recommendations on how to further influence consumer adoption rates and make use of the charging infrastructure in a cost effective way.

Dutch Context on E-mobility

Structured planning on how to stimulate electric mobility in the Netherlands dates back to 2011 when the first plan of action for E-mobility was presented (Netherlands Enterprise Agency 2011). With ambitions to realize 1 million PEVs in 2025, a set of governmental instruments were put in place. These are (i) financial incentives for purchasing and/or leasing PEVs, (ii) support to the rollout of charging infrastructure and (iii) demonstration programs for particular target groups including commercial and commuter traffic, logistics, taxis and government vehicles. Lease car drivers were particularly supported with tax measures, given their relatively high mileage and kilometres driven in urban areas.

In recent years the ambitions for PEV sales have increased to the previously mentioned targets of 50% of all car sales to be electric, 30% of which fully electric, by 2025, with an intermediate goal of 10% of vehicle sales having a plug by 2020 (Formula E-Team 2016). The types of support instruments used did not change over the years, although financial incentives were significantly altered. Regarding public charging points the supporting role of municipalities has been a major factor. In particular the four main cities Amsterdam, Rotterdam, The Hague and Utrecht as well as the Metropolitan region around Amsterdam (covering 80 municipalities) developed one of the densest charging networks worldwide. The combination of financial incentives and development of charging infrastructure both nationwide and in cities provided an environment where the market for electric vehicles surged and made the Netherlands one of the frontrunners of electric mobility worldwide.

Purchase Incentive Scheme

The Dutch government has come up with a broad incentive scheme to stimulate the sales and lease of electric vehicles. The scheme comprises of four main measures including reductions in purchase tax, annual vehicle tax, and two measures aimed at lease business drivers. These measures are discussed below.

Purchase Tax

Direct purchase incentives are in place for PEVs through a purchase tax that is based on CO_2 emissions. Purchase taxes on new vehicles are calculated by adding two components: a fixed amount based on CO_2 emission bands and an additional amount directly proportional to the CO_2 emitted. Both are estimated using NEDC test cycles. Differences in taxation across different vehicle models can be substantial: zero-emission vehicles only pay €365 (2018 levels) while vehicles that emit more than 162 g of CO_2 per km pay at least €12,593 and €458 more for every gram of CO_2 above this threshold.

Annual Vehicle Tax

Zero-emission vehicles are exempt from annual vehicle taxes. Annual taxes are determined based on fuel type (gasoline or diesel), the weight of the vehicle and the area where the vehicle is registered. For mid-size passenger cars these taxes are in the range of €800–€1500 per year.

Addition for the Private Use of a Company Car Tax

In the Netherlands nearly 50% of all new sold cars are leased (Vereniging Nederlandse Autoleasemaatschappijen 2018). Despite being a large portion of new vehicle sales, they constitute less than 10% of the entire vehicle stock. Most (88%) of the leased cars are company leased and are used for business as well as private purposes. The Dutch government considers any company car that is used privately for more than 500 km per year as additional income over which taxes have to be paid. This tax is known as the *addition for the private use of a company car tax,* mostly referred to in Dutch as "*bijtelling*" ("addition").

The "addition" to the income level, over which income taxes need to be paid, is calculated on the basis of the retail price of the new car. Depending on the CO_2 emissions of the vehicle, 0–25% of the new car value is added to the annual taxable income. The "addition" tax has been used by policymakers in recent years to steer company car users towards lower CO_2 emitting vehicles. An overview of the changes in this tax since 2012 is shown in Table 1. The percentage of addition tax is set for 5 years from the moment the car is registered.[1] Notable changes can be seen in 2013 where both Battery Electric Vehicles (BEVs) and PHEVs were strongly favoured through 0% "addition" tax, and increases in "addition" tax in subsequent years particularly for the 0–50 g category, making PHEVs increasingly less favourable.

Table 1 Addition tax for leased vehicles per CO_2 emission category from 2013 to 2019

Year	0 g CO_2 emission (only BEV)	0–50 g CO_2 emission (PHEVs)	>50 g CO_2 emission (non EVs)
Pre-2012 (%)	0	14	14–25
2012 (%)	0	0	14–25
2013 (%)	0	0	14–25
2014 (%)	4	7	14–25
2015 (%)	4	7	14–25
2016 (%)	4	15	14–25
2017 (%)	4	22	22
2018 (%)	4	22	22
2019	4% < €50,000 22% > €50,000	22%	22%

[1]For vehicles with more 50 g of CO_2 emissions the addition tax varied between 14 and 22% depending on fuel type and CO_2 emissions. This was simplified to 22% in 2018 for all vehicles. In the text we have limited ourselves by focussing on EVs. A complete overview, going back to 2011 can be found on the website of the Dutch tax authority: https://www.belastingdienst.nl/bibliotheek/handboeken/html/boeken/HL/thema_s-vervoer_en_reiskosten.html#HL-21.3.4.

Environmental Investment Deduction (EID)

Beside a reduction in addition tax the Dutch government has also provided businesses with the opportunity to get a tax reduction on the yearly depreciation costs. PEVs have been placed on the Environmental Investment Deduction (MIA) list which allows a certain percentage of the investment to be deducted from the company profit tax (25%). This reduction is 36% from the depreciation costs, which are capped at €50,000. The MIA particularly favours entrepreneurs and freelancers with own businesses, given their opportunity to deduce company car depreciation costs as environmental investment.

Impacts of Different Incentives

The above incentives target different vehicle owner and powertrain types in different ways. The first two incentives (purchase tax and annual vehicle tax) mainly benefit private car owners. Although the *vehicle purchase tax* could build up to a significant amount, it only compensates for a part of the higher PEV prices generally observed in the market. The *annual vehicle tax* only provides a relatively small additional incentive.

The "addition" measure is directly aimed at the company car/lease market and provides a much more powerful incentive for lease drivers than the purchase taxes for private users. Over the lifetime of the lease contract the addition tax largely compensates the price premium for PEVs thereby enabling to drive electric at similar costs as gasoline cars. Given that in early years (2013) the "addition" for PHEVs was at the same level as BEVs (0%), PHEVs were largely favoured thanks to the absence of range limitations together with the relatively low price premium compared to BEVs.

All in all the set of incentives put in place provides a scheme that particularly favours company cars and small business owners. This market was deliberately targeted in the plan of action (Netherlands Enterprise Agency 2011) but partly lead to unforeseen consequences, particularly the surge of PHEVs, making the Netherlands the country with the highest market share of PHEVs. The private car market enjoyed far fewer incentives. This approach was taken amongst others because the Dutch tax system does not allow for a further deduction in price for private PEV buyers besides the purchase tax. A direct incentive for private buyers turned out to be complicated. Given the characteristics of the Dutch vehicle market, the focus on company cars was quite rational: close to 50% of all new vehicles sold are company lease cars, while the average purchase price for company cars is much higher than for the private market. Especially when the tax benefits were introduced in early 2013, the PEV market mainly had models available in the higher price segment. Targeting this segment was therefore most effective and opportune.

Electric Vehicle Sales

How did these financial instruments play out on the sales of PEVs? An overview of the PEV sales from 2012 to 2018 is given in Fig. 1. PEV sales took off in 2012 and first spiked at the end of 2013 ewhen the "addition" tax for PHEVs was raised from 0 to 7%. In the last month of 2013 many drivers ordered a PHEV to profit from the lower tax for another 5 years. Additionally there was a small bump in BEV sales as the addition tax was raised from 0 to 4%. Altogther PEVs had a market share over 15% of newly sold vehicles in December of 2013. Even higher market shares followed in the December months of 2015 and 2016. This was also facilitated by the rising amount of available PEV models on the market, particularly PHEVs such as the Mitsubishi Outlander, which sold nearly 10,000 vehicles in those two years. By the end of 2016, just before taxes for PHEVs were to be raised to 15%, more than 20% of all new vehicles sales were PHEVs and BEVs. A last spike in PHEV sales occurred in late 2016 when taxes were raised to a level that was equal to gasoline and diesel cars. PHEV sales have since fallen flat and the number of vehicles on the road has slightly decreased due to exports.

BEV sales, although enjoying much of the same or better benefits as PHEVs, remained at a slow but steady pace until late 2016. After the subsidies for PHEVs were cut at the end of 2016 lease drivers turned to BEVs. BEV sales have been on a steady rise since early 2017, accounting for 3–4% of all sales in the first months of 2018.

The fiscal incentives have played a very powerful role in the sales of PEVs in the Netherlands. The sheer size of the incentive has made PHEVs and to a lesser extent

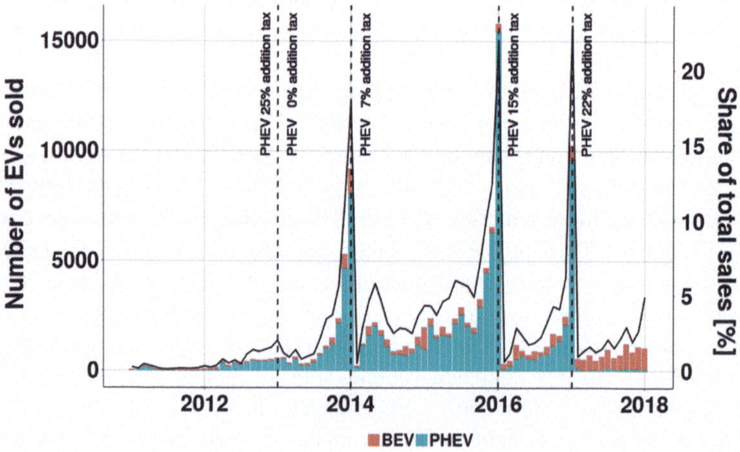

Fig. 1 EV sales in the Netherlands since 2012, separated in BEV and PHEV (coloured bars). Share of total vehicle sales per month (black line), based on car registration data of the Dutch road authority (RDW 2018)

BEVs an economic alternative to regular fossil fuel powered vehicles. The end-of-year spikes as well as the drop in PHEV sales from 2017 onwards illustrate just how influential these incentives have been. Beside the fiscal measures, it is worthwhile mentioning other supporting drivers for PEV sales in the Netherlands. Particularly in inner cities where parking pressures are high, having a PEV provides preferential access to parking spots equipped with chargers, especially as in the Netherlands there is no fine for long connections to charging stations. Combined with the economics of leasing PEVs, this has provided an additional motivation for PEV purchases. Similarly a range of companies (e.g. banks, consultancies) with strong corporate social responsibility policies have provided restrictions to leasing polluting cars and have encouraged their employees to lease PEVs. Several cities in the Netherlands have provided a positive climate for purchasing PEVs through supporting measures such as free parking, free charging and facilitating placement of charging infrastructure close to prospective PEV drivers' homes. Nevertheless, the fiscal incentives to make these vehicles come in the same price range as conventional internal combustion engine cars have been instrumental to their success. A last important enabling factor has been the development of public charging infrastructure, which we will discuss now.

Roll-Out of Charging Infrastructure

An important part of the Dutch policy in stimulating PEVs has been the development of charging infrastructure for public usage. Given that charging and related range anxiety is generally seen as one of the main barriers for electric mobility, and that in the Netherlands an estimated 65% of households do not have a dedicated parking space with charging facilities, enabling public charging infrastructure has been one of the priorities in Dutch PEV policy.

At national level, financial support was given to a program set up in 2009 by joint grid operators (ELaadNL) to develop a public charging network of 10,000 charging points nationwide. This was complemented by municipal initiatives to develop public chargers through public tenders in the 4 major cities in the Netherlands. By 2018 the Netherlands had one of the most dense public charging infrastructures worldwide with almost 17,500 public chargers installed (Netherlands Enterprise Agency 2018), which corresponds to nearly one public charger every 7 electric vehicles. The dense charging infrastructure is generally seen as one of the success factors in overcoming the first hurdles for purchasing electric vehicles (Sierzchula et al. 2014).

In the early stages of the roll-out of charging infrastructure, the main focus was on placing charging stations in strategic locations such as city centres. However as PEV adoption started to take off also among those that relied on on-street parking facilities, the focus shifted to a more demand driven roll-out (Helmus et al. 2018). PEV drivers could request a charging station to be placed near their home while these charging stations remained publicly accessible. When few electric vehicles were on the road this also meant that those drivers also created a private parking spot for themselves as the accompanying parking area was exclusively accessible to electric vehicles. In areas with high parking pressure this served as an additional incentive for potential buyers.

Due to the demand-driven roll-out strategy, the ratio between the number of electric vehicles and public chargers has remained relatively stable and one of the lowest of the world (European Alternative Fuel Observatory 2018). Figure 2 shows an overview of the number of public chargers in the 4 major municipalities and the PEV to charger ratio for this public charging infrastructure. The latter is calculated based on the number of PEV drivers actually using the infrastructure, based PEV charging card data for the level 2 public charging stations in the four major municipalities. Apart from the steady growth in the number of charging stations, the figure shows how higher PEV sales at the end of years 2015 and 2016 lead to peaks in the PEV to charger ratio in the same months. It is only near the end of 2016, when the highest number of PEVs was sold, that the ratio of PEVs to charging stations increased from around 5.5–7. The ratio provides a metric for the level of public charging infrastructure required to support PEV adoption, also in consideration of the fact that a great deal of PEVs are known to only make use of the public chargers few times over the years, and otherwise rely on private charging infrastructure.

The Dutch government has also been pushing to provide interoperable standards that allow everyone to get access to public charging stations with a single charging card. The so-called Open Charge Point Protocol (OCPP) provided access for all PEV drivers across all charging stations, lowering the barrier for charging station usage. In several countries competing mobility service providers and limited interoperability hamper widespread access to charging stations leading to lower utilization.

Fig. 2 Number of public charging stations (black) and the ratio "number of unique users" to "public charging stations" (red) in the four major cities and the metropolitan region Amsterdam in the Netherlands (Amsterdam University of Applied Sciences 2018)

Effectiveness of PEV Adoption Support Policies

With relatively high PEV sales, the Netherlands became known as one of the countries that most supported PEV adoption, particularly PHEVs. The question is to what extent the policies have been effective in mitigating vehicle tailpipe emissions. In the Netherlands, the debate mainly focused on the extent to which PHEV drivers actually charge and drive on electricity. Attention was also devoted to the effectiveness of the public charging infrastructure provided and its actual utilisation. This section presents findings of early studies on charging behaviour on PHEVs in the Netherlands and explores insights into the utilization of public charging stations and related concerns for policymakers.

Plug-in Hybrid Vehicles Utilizaton

Comparison with Drive Cycle Assessments

The real use of PHEVs in the Netherlands and share of electric kilometres driven has been a point of national debate. Expectations of emission reductions brought by PHEVs were largely based on the results of NEDC cycle tests. However, the NEDC cycle is not representative of real world driving. Moreover, PHEV users may fail to drive on electricity due to a number of reasons, ranging from limitations in available charging facilities to resistance to charging the vehicle on a regular basis. The question around real use of PHEVs is important for policymakers seeking to establish effective incentive schemes that realize air quality improvements and CO_2 emission reductions.

A first study looking into this topic comes from Ligterink et al. (2013). They found that PHEV users in the Netherlands more often used gasoline than assumed in the NEDC test cycle. Their findings resulted in a more structural monitoring of PHEV usage in the Netherlands, research that was continued until 2016 (Ligterink and Smokers 2016). Monitoring of fuel usage and charging habits was based on fuel card data, that registered both the use of charging stations (public and private) as well as that of regular gasoline pumps. The study includes 4 years of charging and refuelling data (2013–2016). The study compared for different vehicle models on the Dutch market the average percentage of electric kilometres as expected from NEDC cycles with actual percentage of kilometres on electricity for the whole group (average). The results published in 2016 (see Table 2) show that on average the number of kilometres driven on electricity for PHEVs is far below the test cycle results. Particularly PHEVs with relatively small battery sizes tend to drive significantly fewer electric kilometres than assumed in NEDC cycles, in most cases less than half of that. Only the BMW i3 has comparable electric mileage to the NEDC cycles, which is likely due to the large battery pack in these vehicles. Striking is that the share of electric kilometres is

Table 2 Share of electric driven kilometres per vehicle type

Vehicle	NEDC cycle % electric (%)	Share of electric driven kilometres (yearly average)			
		2013 (%)	2014 (%)	2015 (%)	Q1 2016 (%)
Opel Ampera	77	48	49	46	46
Chevrolet Volt	77	47	48	46	48
Toyota Prius-Plug-in Hybrid	50	17	18	17	18
Volvo V60 PHEV	67	26	25	24	27
Mitsubishi Outlander PHEV	68	29	34	31	33
Audi e-tron	67			31	31
VW Golf GTE	66			28	30
Ford C-Max plug-in hybrid	64			34	28
Mercedes-Benz c350e	55			23	25
VW Passat GTE	67			29	30
BMW i3 range extended	86		89	87	85
Average	NEDC cycle	68	67	67	66
	Practice	35	33	30	31

From Ligterink and Smokers (2016)

stable across the different years, indicating that drivers have a stable charging routine where one might have hypothesized shifting norms over time.

Ligterink and Smokers continued their efforts to monitor the PHEV usage on a yearly basis and included more vehicle models once data became available. It was not until 2016 that their results caused a public debate about major subsidy amounts being spent on stimulating sales of PHEVs while only limitedly contributing to reduced emissions. As a result the addition tax came increasingly under scrutiny. Eventually the advantage for PHEVs was abolished. Sales of PHEVs collapsed accordingly, after spiking one last time at the end of 2016.

These results were rather remarkable as studies in other countries found that the percentage of electric kilometres driven by PHEVs was rather high (Ge et al. 2017; Idaho National Laboratory 2015; Zoepf et al. 2013). Studies in the US indicate that Chevrolet Volt drivers averaged 75% of all miles driven electric and those with workplace charging even 83%. In the Netherlands the Chevrolet Volt and Opel Ampera drivers only averaged 49%.

The main reason that may explain the differences between the Dutch and US studies is that the Dutch PHEVs were leased, where in the US the PHEVs were purchased by private owners. Ligterink and Smokers (2016) used fuel card data provided by lease companies to calculate PHEV utilization, thereby limiting the study to lease fleets [note that this is a significant portion of the PHEV market; given that around 90% of all PEVs are leased in the Netherlands (RVO.nl 2016)]. There may be differences in motivation between private users and company car users in purchasing PHEVs. Given the high cost premium for private users, this group may have been more motivated to realize fuel cost savings. In a lease construction, many drivers are free to use their fuel card, leading to limited or no incentive to reduce on fuel costs. The drivers personally do not have an incentive to save on fuel costs and therefore are less motivated to recharge their car. The additional possible inconvenience, i.e. seeking for an available charging point is not offset by possible cost savings. Hence business drivers in the Netherlands could profit from lower addition tax but did not have to care about fuel savings.

Charging Behaviour of PHEVs

PHEVs in the Netherlands tend to drive less electric kilometres than assumed in the NEDC cycle, but charging data in four major Dutch cities shows how PHEVs are responsible for a significant portion of total PEV charging sessions and in absolute terms do contribute to emission reductions. A study by Van den Hoed et al. (2016) sought to quantify the contribution of PHEVs to total PEV charging sessions and kilowatt-hours charged relative to BEVs in a urban context. Data from charging sessions on public infrastructure were gathered and analysed, where sessions were attributed to BEVs, PHEVs and an unknown category in all those cases where the charging data did not provide sufficient information to determine whether a PHEV or BEV had been charging.

Building on this earlier study, we have analysed the utilization of the charging infrastructure based on a set of more than 5 million charging transactions involving more than 4000 charging points until early year 2018. The same dataset as in Wolbertus and Van den Hoed (2016) has been used but data gathering has continued until March 2018. Data on the charging sessions in the cities of Amsterdam, Rotterdam, The Hague, Utrecht and 80 municipalities part of the so-called Metropolitan Region Amsterdam has been gathered. For each of the charging sessions data as shown in Table 3 has been collected. The database is referred to as the Charging Infrastructure Efficiency Forecast (CHIEF) (Amsterdam University of Applied Sciences 2018).

The analysis compared the number of PHEVs on the road and their comparative charging infrastructure usage. The results showed that PHEVs used public charging station only slightly less than would be expected by the number of vehicles on the road. Between 2014 and 2017 on average 85% of all EVs on Dutch roads were PHEVs. Figure 3 shows that in the same period on average a share of 75% of all charging sessions on public infrastructure networks in the four major cities were performed by PHEVs. Only in 2018 the share of transactions by PHEVs has reduced to 55–60% mainly due to the shift in the PEV fleet composition towards BEV. This suggests that the share of PHEV drivers that never charge may be limited and at least is not the norm. A significant share of PHEV drivers charge frequently, although

Table 3 Data variables, examples and descriptions

Variable	Example	Description
User code	60DF4D78	Code of charging card
Charging point operator	Essent	Owner of the charging point
Location_skey	456	Unique location key per charging station
City	Amsterdam	City in which charging point is placed
Address	Prinsengracht 767	Address charging point is placed
Start connection date time	24-04-2015 13:56:00	Start date and time of charging session
End connection date time	24-04-2015 17:14:00	End date and time of charging session
Charging time	1:45:00	Time the car is charging
Volume	6.73	Amount charged (kWh)

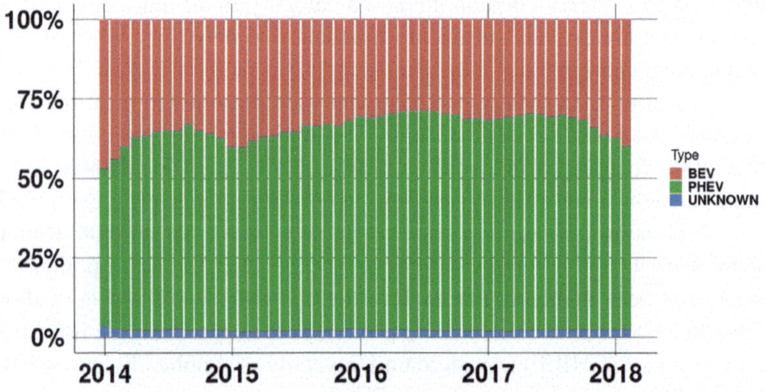

Fig. 3 Share of charging sessions by type of electric vehicle on public charging stations in the four major cities and the metropolitan region Amsterdam in the Netherlands (Amsterdam University of Applied Sciences 2018)

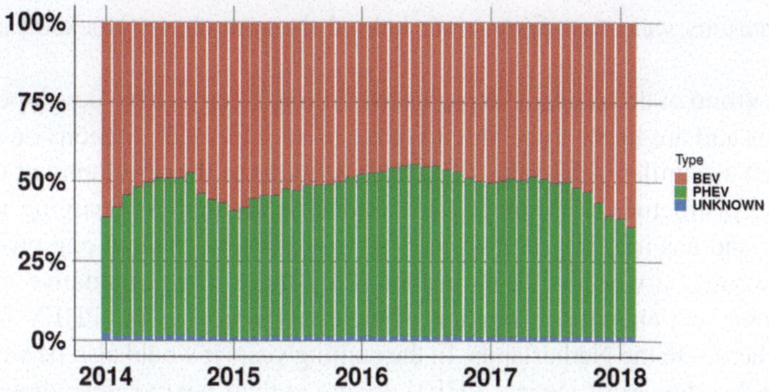

Fig. 4 Number of kWh charged by type of electric vehicle in the four major cities and metropolitan region Amsterdam in the Netherlands (Amsterdam University of Applied Sciences 2018)

some may skip a charging session if a convenient charging location is not accessible. This idea was re-enforced by additional analysis of Ligterink and Smokers (2016), showing that a small portion of users hardly ever charged but that the majority of PHEV drivers charged regularly. Nearly 25% of all PHEV drivers charged more than once a day.

Not surprisingly in terms of average kilowatt hours charged per session PHEVs lag behind BEVs, given the larger size of batteries of the latter. Figure 4 shows the contribution of electricity charged by PHEV and BEVs. Until early 2018 PHEVs have been responsible for more than half of the total electricity charged by PEVs in the four major cities, and as such have made a considerable contribution to emission reductions. PHEVs may in practice drive less electric than expected based on the NEDC drive cycle, but the sheer size of the PHEV fleet in the Netherlands has in fact contributed to a significant share in electric kilometres driven in urban environments. A further step would be to compare the actual emission reductions with invested subsidies between BEVs and PHEVs.

Secondary Effects

Apart from the direct contribution of PHEVs to electric kilometres driven, possible secondary effects of PHEV sales in the Netherlands can be reflected upon. For one, PHEVs have contributed to PEVs (vehicles with a plug) becoming a legitimate alternative to gasoline vehicles for a fairly large public. The accumulated fleet of 120,000 PEVs still only reflects 1.5% of the total Dutch vehicle market but is beyond a small niche market and not limited to an elite group but in relatively mainstream lease markets. Yet, PEV drivers are still a predominantly middle aged male group with above average education and income levels. BEV drivers are however slightly more technology and environmentally orientated while PHEV drivers often indicated that financial reasons were the main driver behind their purchase (Hoekstra and Refa 2017).

A large group of drivers have experienced driving electric, have developed charging routines and are likely to be able to make more informed decisions on pros and cons of PEVs. Similarly the sales of PHEVs stimulated the development of public charging infrastructure to a point that it is one of the densest charging networks worldwide and has in a way helped to overcome the first chicken-egg problem for PEVs. It would be worthwhile to further study these more qualitative effects of PHEVs to make a nuanced evaluation of the effectiveness of the PHEV favouring subsidy schemes in the Netherlands. In the coming years it would also be very useful to monitor the choices of current PHEV drivers in their next vehicle choice. Many of the lease contracts have a lifespan of 4–5 years which faces these drivers with the decision to go full electric or not.

Charging Infrastructure Utilization

Beside large investments in the number of vehicles on the road through government tax breaks, there has also been a significant investment in public charging infrastructure in the Netherlands. The extent to which the charging infrastructure provided is actually used is important, because not only do policymakers need to legitimise the investments made but also the additional pressure on public space and parking opportunities that developing the charging infrastructure has caused. Moreover, understanding utilisation of public charging infrastructure provides insights into the business opportunities for companies seeking to build and operate public charging points. The question is how much is the charging infrastructure used, and to what extent can public investment in charging infrastructure be justified.

We have particularly focused on: (i) utilization rate, (ii) charge/connection ratio, and (iii) connection times. These are discussed in turn in the following sections. Where possible a comparison is made with relevant other cities and/or countries.

Utilization Rate

Utilization rate is defined as the amount of time charging points are connected with PEVs divided by the total available time. 100% indicates that all charging points are fully occupied; 0% indicates that all charging points are always accessible. For most municipalities in the Netherlands charging infrastructure has been built aiming to strike a difficult balance between providing sufficient accessibility to charging points and not having over-capacity of charging infrastructure. This is in order not to give up parking spots for under-utilized charging spots in neighborhoods with high parking pressure.

Fig. 5 Historic development of average utilization rate across several cities (January 2014–March 2018) (Amsterdam University of Applied Sciences 2018)

Figure 5 shows the development of utilization rates in the four major cities and metropolitan region. The figure shows that the extent to which the charging infrastructure is occupied varies from around 20–30% in The Hague, Rotterdam and MRA, to 35–40% in Amsterdam and Utrecht). It also shows seasonal effects, with lower utilization rates in summer vacation periods. In all regions the utilization rates show an upward trend; indicating that the utilization of charging infrastructure is steadily increasing by several percent-points per year.

Occupancy of 30–40% may seem low. However policymakers do not strive for 100% as this would limit accessibility for new PEV drivers. In fact as a rule of thumb policymakers in Dutch cities generally regard a utilization rate of 50% as high and a legitimate percentage at which to consider adding charging points in that neighborhood. Charging station occupancy is relatively high compared to other countries as many of the charging stations were installed for so-called primary use. This means that the charger mainly functions as a home or office charger for those that have to park their car in public places. In the Netherlands the majority of home owners (estimates are in the range of 60–80%) rely on on-street parking facilities. A smaller number of charging stations were installed for so-called secondary use. These charging stations are meant to facilitate visitors in the area. However as both types of chargers are public, they can be used interchangeably for both purposes.

The importance of providing primary chargers is supported by a stated choice experiment involving 149 respondents who relied on on-street parking, that indicated that lack of accessible charging opportunities near home is one of the biggest barriers to PEV adoption (Wolbertus et al. 2018). The stated willingness-to-pay to have a 'dedicated' charging station in a public area is €2248–€2557, in the range of the cost of installing a charging station at home. It is considered very important for prospective PEV owners that rely on-street parking to be able to charge their car near their home. High utilization rates may thus not only frustrate current PEV drivers in being able to charge, but also demotivate prospective PEV drivers to purchase one.

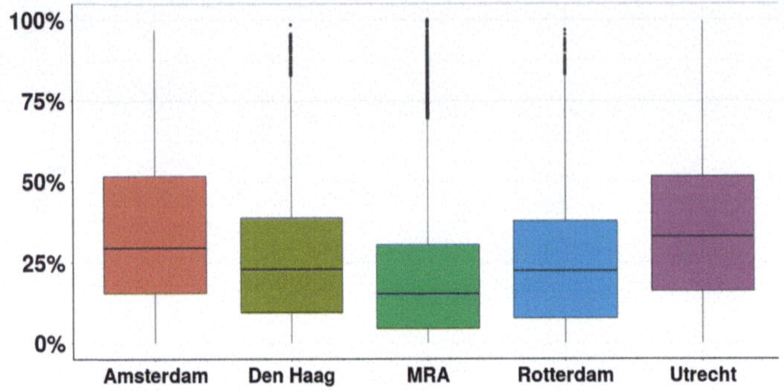

Fig. 6 Boxplot of utilization rates across charging stations in The Netherlands in March 2018 (Amsterdam University of Applied Sciences 2018)

These averages give a good picture of how much charging infrastructure is used in Dutch cities as a whole. However, they do not account for differences across locations within the cities. Figure 6 shows the distribution of charging station occupation within the cities over the course of a month (March 2018). While the median occupancy rate is somewhere between 15 and 30% across the cities, some charging stations are nearly always occupied (the dots indicate the outliers) while some remain empty. Local authorities track the occupancy of single charging stations and those in their surroundings and if these are too high in a certain area this can be a reason to place additional stations nearby.

Apart from locational differences, utilization is also influenced considerably by time of day. Figure 7 provides an overview of utilization rates of two exemplary charging stations over 24 h, averaging out all days in March 2018. This includes both weekdays as weekends. The figure illustrates how charging stations may differ in daily charging profiles. While both charging stations have an average utilization rate of about 50% their usage profile is completely different. As a result utilization rates vary geographically depending on different urban environments and shares of residential versus office buildings.

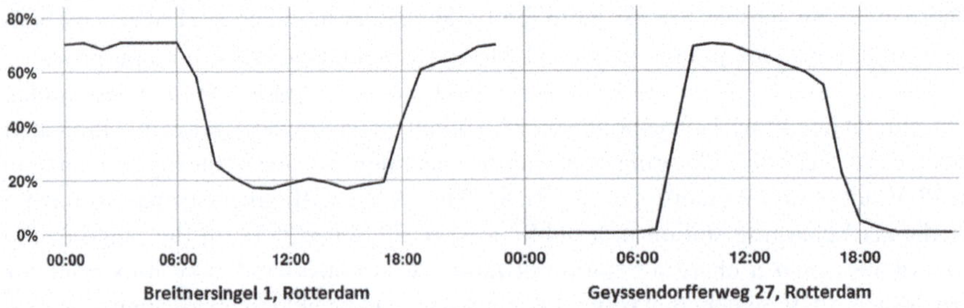

Fig. 7 Utilization rates over the course of a day (average over March 2018, Rotterdam) (Amsterdam University of Applied Sciences 2018)

Utilization rates may thus differ from station to station, as well as from hour to hour. As a result there may be charging stations that are relatively under-utilized, with a negative impact on the business case and possible scrutiny from local inhabitants that valuable parking space is lost on charging stations that are seldom used. On the other hand there are charging stations that tend to be over-utilized, with a positive effect for the business case but possible frustrating effects for PEV drivers that are dependent on charging facilities but find all charging stations in use.

Reducing both over- as well as under-utilization are major concerns for policymakers. Measures that policymakers can take to avoid or reduce *under-utilization* include applying window times at charging stations, i.e.: making charging spots available to non-PEVs after a certain time, e.g. 10 pm (Wolbertus and van den Hoed 2017b). Another option is to attract additional user groups with complementary charging profiles to the same charging infrastructure. An example is that of attracting electric car sharing schemes and electric taxis to charging places otherwise only used by residential building PEV owners (Helmus and van den Hoed 2015).

Common measures to reduce *over-utilization* of charging points in Dutch cities include placing new charging points or charging hubs that extend the number of charging points at one location when certain thresholds in utilization are reached. Particularly the strategy to place them on-demand means that the charging station will be used by at least the PEV driver who requested it. A new public charging station is only built if the PEV user requesting it does not have access to off-street home or workplace charging. This means that usage on a near daily basis per charging station is almost guaranteed. Helmus et al. (2018) have demonstrated that on-demand charging stations lead to more transactions with higher kWh charged than ‘strategically’ placed secondary charging points that should provide opportunity charging. These secondary points do have a higher number of unique users which may be explained by the logic that “strategic” poles are placed at points of interest such as shopping malls and are used by those who visit them. The demand-driven strategy is a typical element of the Dutch rollout strategy.

The roll-out of infrastructure has created charging stations that are used for home and office charging but are also available to visitors at times when they are not occupied by their primary users. At more than 800 vehicles, the city of Amsterdam has one of the largest electric taxi fleets in Europe which relies on the dense charging infrastructure. Free floating electric car sharing schemes, such as Car2Go and Hyundai Car sharing, have flourished in the city, while in other PEV friendly cities such San Francisco and San Diego comparable schemes had to be shut down due to lack of charging infrastructure. In other cities such as Madrid the shared electric vehicles rely on centralized charging hubs, while in Amsterdam the PEVs can be parked and charged across the city.

Connection Times

A second topic that has been subject of debate in the Netherlands is the connection times of PEVs at public charging points. Given that there is no penalty for leaving a car at a charging station once fully charged, PEV drivers are likely to leave their car connected much longer than strictly needed.

Figure 8 provides an overview of connection times of all transaction over 2017 in the four cities and MRA. Slightly over 25% of all charging sessions are shorter than 4 h, probably in the majority of cases related to secondary charging use. Striking is that more than 40% of the sessions are between 8 and 16 h, which most likely relates to overnight parking/charging of inhabitants owning a PEV. Close to 5% of all sessions are longer than 24 h and less than 1% are longer than 72 h. The latter is important to note as the issue of 'long-parking' became a point of debate in the national media and there was widespread perception that this was the norm. However evidence shows that long sessions are an exception.

Nevertheless, long sessions are a burden on the availability of charging infrastructure (Wolbertus and van den Hoed 2017a). Whereas only 5% of all sessions are over 24 h, they are responsible for more than 25% of the total occupancy of the infrastructure. Providing incentives to disconnect charged PEVs and making them available to other PEV drivers is important for improving both accessibility and the business case of charging points.

Something to consider here is that a great deal of PEVs start charging in the evening, so disconnecting PEVs from the chargers as soon as the battery is fully charged means doing this at nighttime. Not only is this inconvenient for PEV drivers but it would also result in limited increase of utilization of the charging point. As can be seen from start-connection times in the database (Amsterdam University of Applied Sciences 2018), demand for charging reduces sharply after 10 pm. Hence stimulating PEV drivers to disconnect their cars when charging overnight does not make much sense. Instead such measures should target charging sessions with connection times longer than 16 h. Options available for incentivizing drivers to disconnect their PEV from the charger once the session is complete include providing time incentives (Wolbertus and Gerzon 2018) and the use of apps that allow nudging fellow PEV drivers to disconnect their car. This system is currently in use in the Netherlands and works on a voluntary basis for PEV users, possibly complemented by an incentive program.

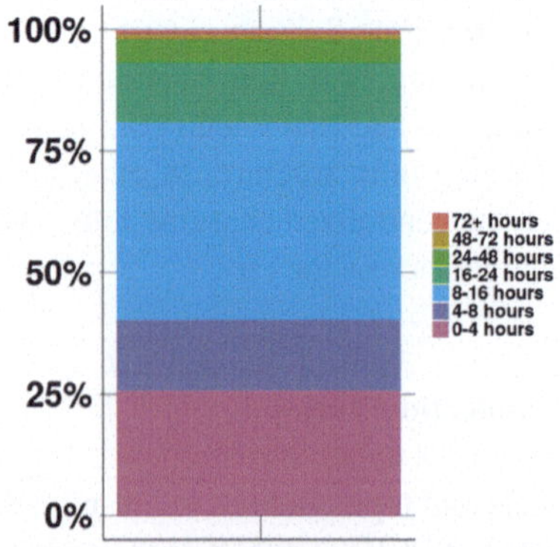

Fig. 8 Distribution of connection times in different time bins (based on 2017 data in G4/MRA) (Amsterdam University of Applied Sciences 2018)

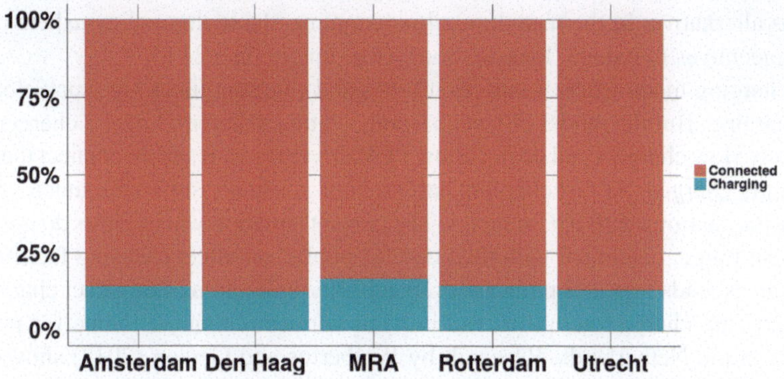

Fig. 9 Charge time ratio for G4/MRA (Amsterdam University of Applied Sciences 2018)

Charge Time Ratio

In the previous section we showed how connection times at Dutch charging points are relatively high. In this section we show how actual charging only lasts a small share of the time the PEV is connected to the charger. Figure 9 shows the charge time ratio, defined as charging time divided by connection time, for the G4/MRA. Where average connection times are 9–10 h, actual charging times are close to 2–3 h. On average the charge time ratio is between 0.15 and 0.22, indicating that more than 75–85% of the time EVs are connected they are not charging. As such they are not using the charging infrastructure, which theoretically would have been available to other PEV users.

Low charge time ratios can largely be explained by the fact that charging infrastructure in residential areas is mainly used overnight, when re-parking a car is not an option. However destination charging also shows a relatively lower charge time ratio that one would have expected. Most problematic are longer sessions (24+ hours) which are likely to have charge time ratios of less than 10%.

Combining this analysis with utilization rates leads to the following picture. Utilization rates indicate that on average 30–40% of the time charging points are occupied. Charge time ratios indicate that 15–25% of this connection times are used for charging. Combined this implies that actual charging on average only takes place 5–10% of the total time on current infrastructure, thereby raising questions regarding inefficiency of capacity utilization and low business case opportunities for public chargers. One might argue that incentives should be used to convince PEV drivers to re-park their car once fully charged (excluding night times), thereby enabling higher utilization. Alternatively one might argue that this lower utilization may just be the price for facilitating PEV adoption in city centers. Given that there are no alternatives for most PEV drivers without an own parking facility in city centers at this point, the lower utilization may be justifiable for policymakers in order to facilitate clean mobility in their cities. In the medium to long term rapid chargers may become a serious alternative. In the meantime, it is recommendable that policymakers put in place incentives to reduce 'long sessions' (e.g. longer than 24 h).

A charging mode where connection times and charging times are much closer is fast charging. But the model of fast charging is quite different. Level 2 chargers are relatively slow chargers and do not allow PEV drivers to wait and disconnect their car once fully charged. At fast chargers half an hour is usually sufficient, much similar to gasoline stations with ICE vehicles. The current situation where PEVs do not have sufficient range, combined with the need for public, on-street charging facilities in urban areas leads to a suboptimal overall capacity utilization of on street chargers.

Improving charge time ratios and utilization rates is a major theme for policymakers in the Netherlands. Research by Wolbertus and Gerzon (2018) shows that in general time-based tariff structures are supported by most PEV drivers, as long as they do not interfere with their primary charging needs. As such, incentives can play a role in increasing utilization rates and are likely to improve the business case and lessen the need for additional rollout of charging infrastructure. However, it will come at a cost for PEV drivers, either financially (e.g. through tariff structures), in terms of more limited accessibility (due to higher utilization rates) or in terms of the inconvenience of disconnecting their cars once charging is complete.

Conclusions and Recommendations

The electric mobility history in the Netherlands provides interesting material to reflect on the interplay between regulation, uptake of PEVs by consumers and the utilization of infrastructure by PEV drivers. This has allowed us to study how consumers react to policy measures and how they use public charging infrastructure. Particular in the Dutch case is the significant incentive scheme for stimulating PEV sales in combination with the strong development of public charging infrastructure. These factors have played an important role in the breakthrough of PEVs, but also lead to questions concerning selective stimulation of particular technologies (PHEVs) and target groups (business lease) as well as discussions on the investments in a public charging network.

It can be concluded that the regulatory context and financial incentives have played a determining role in the Dutch uptake of PEVs in general and PHEVs in particular. Where in early years there was little to no differentiation in the level of financial incentives between PHEVs and BEVs, PHEVs were largely favored by consumers due to lack of range anxiety and generally lower purchase prices compared to BEVs. The shift in available subsidies for PHEVs between 2013 and 2018 and subsequent dramatic reductions in PHEV sales illustrate the determining role the subsidy schemes played in driving adoption. The shifting support to PHEVs can hardly be characterized as a stable, long term regulatory context and it has been scrutinized for both over—(2013–2015) as well as under-subsidizing (2017–2018) PHEVs. The need for a more balanced subsidy scheme, factoring in different price premiums of PHEVs versus BEVs as well as taking into account actual driving and charging behavior is one of the lessons learnt from the Dutch case.

The policy choice to focus on business lease drivers has had both up- and downsides. This target group had been chosen due to their high mileage, relative low age of lease vehicles and relative high average price of lease vehicles. Financial incentives have been very effective to stimulate this group to move towards electric.

PHEVs have been the preferred choice for many lease drivers in the Netherlands. Data from charging and fueling cards have shown that PHEVs tend to drive much fewer electric kilometers than the NEDC cycle would suggest. It shows how particularly PHEVs with small batteries fall short of NEDC expectations. There are limited incentives for PHEV drivers to charge as many PHEVs are company leased, in which case the driver does not pay for the refuelling cost. Despite these limitations, charging data also showed how PHEVs were responsible for more than half of all kWhs charged on the public charging infrastructure and have played a substantial role in reducing emissions of air pollutants from vehicles in urban areas. A more balanced approach in stimulating PHEVs should be considered, in which the size of the battery is taken into account, together with measures for stimulating actual charging by PHEVs such as providing the users with free charging credits. Lastly, where company cars have been a successful policy focus, it has proven to be much harder to make the desired shift to stimulate private users to purchase PEVs. Chapter 6 on Norway has provided evidence on the level and type of incentives that may be required in order to turn a large part of the latter group of car users into PEV adopters. However given the substantial differences between the two countries, also in terms of vehicle and fuel taxation, adopting a similar approach in the Netherlands appears problematic.

A second theme of this chapter concerns the public charging infrastructure in the Netherlands. Most prominent concerns for policymakers include (i) under-utilization (limited use, low business case, scrutiny of local community), (ii) over-utilization (access limitations for PEV drivers, failed charging sessions), and (iii) low charge time ratios (long connections with short charging times). Municipalities in the Netherlands seek to strike a difficult balance between providing sufficient accessibility to charging points without providing over-capacity, because the latter implies ineffective rollout of expensive charging stations while giving up scarce parking spots.

Data shows that charging infrastructure utilization depends on location and time characteristics. Neighborhoods with high income tend to attract more PEV users, leading to high charging demand. Furthermore residential areas tend to lead to charging profiles with demand peaks at night versus office neighborhoods with peaks during the day. The demand-driven strategy applied in many Dutch cities to select which locations require new charging stations has proved to be effective in increasing utilization rates in the early phases of rollout. Also attracting different target groups (e.g. electric car sharing scheme, electric taxis) has increased utilization rates and complementary charging profiles. This target group approach provides policymakers with options for improving the utilization of charging infrastructure. Taxis and car sharing systems have shown to have distinctly different charging profiles than regular residents and can thereby fill in the gaps in available chargers.

Given the concerns about effective use of public charging infrastructure, policymakers have to balance over- and under-utilization of charging infrastructure, while devising ways to increase charge time ratios. Measures such as window times (making charging spots available to non-EVs after a certain time), attract new target groups (e.g. taxis, sharing concepts), set up tariff structures as economic incentives and stimulate more social incentives (such as the social charging app) should be further investigated to come to a structured approach to optimizing use of charging infrastructure.

References

Amsterdam University of Applied Sciences. (2018). *Charging infrastructure efficiency forecast database (CHIEF)*. Amsterdam.

European Alternative Fuel Observatory. (2018). *Electric vehicle charging infrastructure: Total number of PEV charging positions*. Retrieved July 1, 2018, from http://www.eafo.eu/electric-vehicle-charging-infrastructure.

Formula E-Team. (2016). *Electric transport green deal 2016–2020*. The Hague. Retrieved from https://www.greendeals.nl/wp-content/uploads/2016/04/Green-Deal-Electric-Transport-2016-2020.pdf.

Ge, Y., Mackenzie, D., & Keith, D. R. (2017). Gas anxiety and the charging choices of plug-in hybrid electric vehicle drivers. *Transportation Research Part D*. https://doi.org/10.1016/j.trd.2017.08.021.

Helmus, J. R., Spoelstra, J. C., Refa, N., Lees, M., & Van den Hoed, R. (2018). Assessment of public charging infrastructure push and pull rollout strategies: The case of the Netherlands. *Energy Policy, 121*, 35–47. http://doi.org/10.1016/j.enpol.2018.06.011.

Helmus, J., & van den Hoed, R. (2015). Unraveling user type characteristics: Towards a taxonomy for charging infrastructure. In *EVS28 International Electric Vehicle Symposium and Exhibition*, pp. 1–16.

Hoekstra, A., & Refa, N. (2017). Characteristics of Dutch EV drivers. In *Electric Vehicle Symposium 30* (pp. 1–13). Stuttgart.

Idaho National Laboratory. (2015). *Plugged In: How Americans charge their electric vehicles* (pp. 1–24). Retrieved from https://avt.inl.gov/sites/default/files/pdf/arra/SummaryReport.pdf.

International Energy Agency. (2017). *Global EV Outlook 2017: Two million and counting*. IEA Publications.

Ligterink, N. E., & Smokers, R. T. M. (2016). *Monitoring van plug-in hybride voertuigen (PHEVs) april 2012 t/m maart 2016*. Delft.

Ligterink, N. E., Smokers, R. T. M., & Bolech, M. (2013). *Fuel-electricity mix and efficiency in Dutch plug-in and range-extender vehicles on the road* (pp. 1–4). http://doi.org/10.1109/EVS.2013.6914979.

Netherlands Enterprise Agency. (2011). *Electric mobility gets up to speed: 2011–2015 Action Plan*. The Hague.

Netherlands Enterprise Agency. (2018). *Statistics electric vehicles in the Netherlands*. Retrieved from https://www.rvo.nl/sites/default/files/2017/10/Elektricvehicles-Special-Analysisover2016.ENG.pdf.

RDW. (2018). Basisregistratie Voertuigen. Zoetermeer. Retrieved from https://www.rdw.nl/over-rdw/dienstverlening/open-data.

RVO.nl. (2016). *Special: Analyse over 2015*. The Hague. Retrieved from http://www.rvo.nl/sites/default/files/2016/01/SpecialAnalyseover2015.pdf.

Sierzchula, W., Bakker, S., Maat, K., & Wee, B. V. (2014). The influence of financial incentives and other socio-economic factors on electric vehicle adoption. *Energy Policy, 68,* 183–194. https://doi.org/10.1016/j.enpol.2014.01.043.

Van den Hoed, R., Helmus, J., & Kooij, M. (2016). *Laadgedrag van PlugIn Hybride Elektrische Voertuigen op publieke laadpunten*. Amsterdam. Retrieved from http://www.idolaad.nl/gedeelde-content/publicaties/publicaties-algemeen/laadgedrag-van-plugin-hybride-ev.html.

Vereniging Nederlandse Autoleasemaatschappijen. (2018). *Autoleasemarkt in cijfers 2017*. Retrieved from https://www.vna-lease.nl/website/iedereen/feiten-cijfers/infographics-2017.

Wolbertus, R., & Gerzon, B. (2018). Improving electric vehicle charging station efficiency through pricing. *Journal of Advanced Transportation, 2018*, 1–11. http://doi.org/doi.org/10.1155/2018/4831951.

Wolbertus, R., Kroesen, M., van den Hoed, R., & Chorus, C. G. (2018). Policy effects on charging behaviour of Electric Vehicle owners and on purchase intentions of prospective owners: Natural and stated choice experiments. *Transportation Research Part D: Transport and Environment, 62*, 283–297. http://doi.org/10.1016/j.trd.2018.03.012.

Wolbertus, R., & van den Hoed, R. (2016). Benchmarking charging infrastructure utilization. In *EVS29 Symposium* (pp. 1–15). Montreal, Quebec, Canada.

Wolbertus, R., & van den Hoed, R. (2017a). Charging station hogging: A data-driven analysis. In *Electric Vehicle Symposium 30* (pp. 1–14). Stuttgart.

Wolbertus, R., & van den Hoed, R. (2017b). Managing parking pressure concerns related to charging stations for electric vehicles: Data analysis on the case of daytime charging in The Hague. In *European Battery, Hybrid & Fuel Cell Electric Vehicle Congress* (pp. 1–8). Geneva, Switzerland.

Zoepf, S., MacKenzie, D., Keith, D., & Chernicoff, W. (2013). Charging choices and fuel displacement in a large-scale plug-in hybrid electric vehicle demonstration. *Transportation Research Record: Journal of the Transportation Research Board, 2385*, 1–10.

What Can Driving Patterns Reveal About the Suitability of PEVs in Sweden? Analysis and Policy Implications

Frances Sprei

Abstract Sweden has ambitious climate goals. For the transport sector the goals is a 70% reduction of emissions from domestic transport by 2030 relative to 2010 levels. Even if Sweden has not set any specific goals for PEVs, electrification of the transport sector is one of the major strategies to achieve the climate goals. In this chapter we analyze the suitability of PEVs with a special emphasis on BEVs, in Sweden. The main approach has been to study driving patterns and how these relate to the limited range of a BEV. Special emphasis has been set on two-car households since these are potential early adopters. Data on driving patterns is complemented with interviews of households that have trialed a BEV for at least 3 months. We find that the longer driving distances in Sweden might make it more challenging for BEVs to be adopted. On the plus side the economic conditions in Sweden with low electricity prices, higher fuel taxes and subsidies make the BEV economically viable from a TCO perspective. The combination of these factors might explain the relatively high shares of sales of PEVs in Sweden (5.3% sales share in 2017) and why these are dominated by PHEVs. The chapter also looks at the lessons learned from the failed introduction of flex-fuel vehicles and finds that is important to maintain the economic advantages of PEVs through, e.g., higher fuel taxes and that PEVs should have a relative advantage for the user that goes beyond their environmental performance.

Keywords Driving patterns · Two-car households · Flex-fuel vehicles · Climate policy

Introduction

As of the first of January 2018, Sweden has a Climate Act that forces its government to pursue a climate policy based on the goals adopted by the parliament. Every year the government has to present a climate report in the budget proposal and every four

F. Sprei (✉)
Chalmers University of Technology, Gothenburg, Sweden
e-mail: frances.sprei@chalmers.se

M. Contestabile et al. (eds.), *Who's Driving Electric Cars*, Lecture Notes in Mobility,
https://doi.org/10.1007/978-3-030-38382-4_8

years it has to produce a climate policy action plan presenting how the climate goals are to be achieved (Government Offices of Sweden 2018a).

The Swedish climate goals state that by 2045 there should be no net emissions of greenhouse gases (GHG) and after that they should be negative. No net emissions means that flexibility measures, such as reducing emissions outside of Sweden and increased uptake of carbon in forests, can be used. However, emissions from activities on Swedish territory have to be cut by at least 85% compared to 1990 levels. Specifically for the transport sector the goal is 70% reduction of emissions from domestic transport by 2030 relative to 2010 levels. Domestic air transport is not included since this is covered by the EU European Trading Scheme of carbon emissions (Government Offices of Sweden 2018b). There are no specific targets when it comes to plug-in electric vehicle (PEV) sales and many of the policies and incentives set in place to achieve the climate goals have at least on paper been technology neutral (see next Section for an overview of relevant policies).

Looking at PEV sales shares, Sweden is one of the top countries in the world, with 5.3% of the new sold cars in 2017. This is only second to Norway and Iceland. As of September 2018, there were 62,000 PEVs on the road in Sweden. Plug-in electric vehicles (PHEVs) dominate over battery electric vehicles (BEVs) (73% PHEV) and they were mainly sold as company vehicles (Circle 2018). The dominance of company vehicles is not surprising. New vehicle technology in Sweden has often first been adopted in company vehicles since the taxation schemes for these vehicles have made expensive vehicles more economically viable. Company cars constitute roughly 60% of the new cars sold in Sweden.

The Swedish PEV owners live mainly in detached houses, are predominantly males and well-educated, and 85% own at least one more car (Granström et al. 2017). Granström et al. (2017) find that BEV owners are more satisfied with their car than PHEV owners.[1] Based on one day travel diaries, Langbroek et al. (2017b) find that PEV users in the Greater Stockholm area perform significantly more trips by car than conventional car owners and choose cars for a larger share of their travel distance. Wikström et al. (2014) focuses on commercial fleets and finds that over time the driving and usage between charging occasions increases as the accumulated experience of BEVs grows.

The Swedish Energy Agency has performed surveys to map attitudes, knowledge and willingness to purchase regarding PEVs among Swedish drivers. The survey has been distributed in 2013, 2015 and 2017 with a sample size of around 2000 respondents each year. They find that all three aspects have increased in a positive direction and that in 2017 70% of respondents are either positive or very positive towards PEVs. 45% of the respondents also state that they would consider purchasing a PEV. They also find a positive correlation between knowledge about PEVs and willingness to purchase. Other surveys find that policies such as free parking and access to bus lanes (currently not implemented in Sweden) would increase willingness to purchase a PEV (Langbroek et al. 2016). Langbroek et al. (2017a) find that knowledge and the

[1] Unfortunately the authors don't give any explanation for the discrepancy.

relative advantage of PEVs over conventional vehicles are more important than environmental concerns when trying to boost PEV sales, while Rezvani et al. (2017) find that consumers that perceive strong social norms in relation to pro-environmental behavior are more prone to adopt PEVs.

The studies presented in this Chapter take a different approach by starting from driving patterns rather than attitudes. As Langbroek et al. (2016) conclude, driving patterns and how well these are suited to a BEV may have an impact on the willingness to purchase. They will also affect the economic viability of the vehicle, since the amount of driving will determine if the higher investment cost can be recovered thanks to the lower running costs. In one of the studies the quantitative driving data is complemented with interviews to better understand the experience of using a BEV. While the driving data collected has been used to assess PHEVs as well, in this Chapter, we focus on BEVs since their adoption is more challenging from a consumer perspective. Moreover, if Sweden is to reach the 2045 climate goals, a larger market adoption of BEVs will most probably be necessary.[2]

The Chapter starts with a closer look at the policies to promote PEVs that have been and are in place in Sweden. Thereafter the methodological approach and data used in the studies are presented. While this book is focused on PEVs one Section is dedicated to flex-fuel vehicles, i.e., vehicles that can be driven on a mixture of ethanol and gasoline, since their development gives insights and lessons that can be useful for the adoption of PEVs as well. Subsequently the results from studies related to driving patterns and BEV are presented, with a special focus on two-car households. The chapter ends with conclusions and policies implications.

Swedish Incentives Related to PEVs

Achieving a shift in technology in the transportation sector has been high on the policy agenda in Sweden even before the Climate act. The previous government had set the goal for the Swedish domestic transport sector to be fossil fuel independent by 2030 which was a step towards reaching zero net emissions by 2050. In the investigation that was set up on how to reach these goals a fossil fuel independent fleet was defined as one that is primarily driven by biofuels or electricity. Besides fuel taxes there have been a number of incentives in place with the aim to shift the vehicle fleet. This section looks at the most important ones.

The 'super green vehicle rebate' (*supermiljöbilspremien*) (Transportstyrelsen 2018) was put in place on January first 2012 and provided an incentive of 40,000 SEK[3] for the purchase of a vehicle with emissions below 50 g CO_2/km, thus mainly BEVs and PHEVs. After 2016 the incentive was reduced to 20,000 SEK for PHEVs but remained the same for BEVs. Many companies in Sweden provide a car to employees as a benefit. For these the incentive was, after 2016, 35% of the price difference

[2]PHEVs with biofuels might be another option.

[3]One SEK is roughly 0.1 EURO and 0.11 USD.

between the BEV and an equivalent internal combustion engine (ICE) vehicle—with a maximum incentive of 40,000 SEK. For PHEVs the share was 17.5% and the maximum 20,000 SEK. The majority of BEVs and PHEVs receive the maximum amount but not all. For a full list see Transpoststyrelsen (2018). The number of paid out 'super green rebates' increased from 500 in 2012 to 18,416 in 2017. The large majority of these, over 90%, have been to company vehicles.

The incentive was set to last until June 2018 and has been replaced by a bonus-malus system. A bonus-malus, also called a feebate, gives an incentive (bonus/rebate) to some vehicles, in this case those with tailpipe CO_2 emissions below 60 g/km, while imposing extra fees (malus) to other vehicles, those with emission above 95 g/km. The extra fees imposed on new vehicles are to be paid for three years and are 82 SEK/g CO_2 for the emissions between 95 and 140 g/km, and 107 SEK/g for emissions above 140 g/km. To this a base fee of 360 SEK is added. For diesel vehicles there are extra fees, both a constant one of 250 SEK and an extra fuel fee of 13.52 times the CO_2 emissions. This implies that a gasoline vehicle with CO_2 emissions of 125 g/km will have to pay a fee of 2820 SEK per year for the first three years (thereafter the yearly circulation tax will be 668 SEK per year). If instead the vehicle is a diesel vehicle the fee would be 4760 SEK per year for the first three years.

The bonus is calculated based on CO_2 emissions and the price of the vehicle, with a cap at 60,000 SEK. The maximum amount is given to zero-emission vehicles (basically BEVs but could include fuel cell vehicles). The bonus is then reduced by 833 SEK per g CO_2. For example, the bonus for a PHEV that costs 345,000 SEK and has emissions of 38 g CO_2/km would be 28,346 SEK.

The bonus malus is also applicable to company cars and in this case the bonus is calculated slightly differently. Since the company vehicles are a benefit for the employee, they are subjected to a fringe benefit tax. The fringe benefit tax is based on the sales price of the vehicle. For PEVs there is a reduction that results in a lowering of the tax that has to be paid on it. The size of this incentive will vary with the price of the PEV, the price of the equivalent ICE vehicle and the general income level of the employee.

In general, the policies set up in Sweden aim at being technology neutral and Sweden doesn't have a specific goal for the number of PEVs sold. The incentives in place are based on criteria instead. For example, the incentive for 'super green vehicles' (*supermiljöbilspremie)* did not specify the technology; instead it specified that the tailpipe emissions should be below 50 g CO_2/km. Also, the pump law mandating fueling stations above a certain size to provide an alternative fuel did not specify what fuel should be provided. However, even if many of these policies are technology neutral on paper, they will favor one technology, generally the one closest to being price competitive. This might not always be the best solution in the long run and might make it harder for technological solutions that are less price competitive at the moment but that have a higher mitigation potential in the long run to emerge (Azar and Sandén 2011).

Methods and Data Used

Our analysis is mainly based on three databases of car movement data collected by GPS. The first one consists of data from 429 randomly selected conventional vehicles from the national vehicle registry. The average observation period is 58 days and the data was collected between 2010 and 2012 (Karlsson 2013; Kullingsjo and Karlsson 2013). The second dataset contains 64 two-car households where both cars' driving patterns are measured. The average observation period is 74 days (Karlsson 2017). The third set consists of a subset of 25 vehicles from the second dataset where one of the conventional vehicles has been replaced with a BEV (Volkswagen e-Golf). The average observation time is 106 days. The third dataset is also complemented with interviews with the households prior to and after the BEV trial (Jakobsson 2016). For comparison reasons, three car movement datasets from other countries are also analyzed. The datasets are from Germany (MOP 2010), Seattle, US, (Council 2008) and Winnipeg, Canada (Smith et al. 2011).

The data are used for statistical and techno-economic analysis such as estimation of distributions of daily driving distance, number of days with driving above a certain range, economic viability based on battery range and driving distances, and Total Cost of Ownership (TCO) calculations.

One central measure that is computed to assess the BEVs is Days Requiring Adaptation (DRA). These are the days in which the daily distance driven exceeds the range of the BEV. They are calculated based on the assumption that the vehicles are only recharged at night and have a full range in the morning. The adaptation needed can take various forms such as additional charging, both slow and fast, abstained trips, trips taken with other modes, renting or borrowing of another vehicle, etc. In the TCO calculations these days imply an extra cost for the consumer.

We focus on two-car households since these have been identified as suitable for BEV adoption. This builds on two assumptions. The first assumption is that these households have different cars for different purposes; where one car is used for longer trips, towing and transporting more people and goods; while the other car is used for shorter everyday trips. This second car would be more suitable to be replaced by a BEV. We address this assumption by analyzing the driving patterns of the first and second car in households using the first data set. What we don't manage to capture in this analysis is the purpose of the trips and issues such as extra space needed. The second assumption is that households may be able to shift trips between the cars to circumvent the range limitations of the BEV. This assumption is analyzed using the second data set and estimating the potential for this type of flexible usage of the vehicles. The third data set allows us to observe how households actually use a BEV and to what extent they change their driving patterns.

Since the analysis is based on actual driving data, it gives an accurate picture of how vehicles are used in households on a daily basis. The data is collected over a relatively long period (at least two months) for each vehicle and measurements are spread over the year to allow better representation of seasonal variation. Still, not a whole year is covered for any individual vehicle so at this level the accuracy is lower. Traditional travel surveys follow individuals and not vehicles and since a household may have more than one driver, the surveys might underestimate the actual usage of the vehicle. The drawback of following the vehicle is that we don't have any information about the drivers and how they shift the vehicle(s) between them. Another weakness of our analysis is that we miss the perceptions and attitudes from the users. This is somehow addressed in the third study which is complemented with interviews. However, in this case the sample size is too small to be representative. There is thus a trade-off between sample size and richness of information.

The technological development of PEVs is moving at a rapid pace and thus studies are fairly quickly outdated. Battery ranges that were the standard in 2016 are quickly replaced with longer ranges. Still there is a merit to understanding the role and adoption of even shorter range BEVs since there is an economic and environmental advantage to keeping the batteries small. There might also be a market of shorter-range BEV both for used cars and in less expensive models even in the future.

The analysis of flex-fuel vehicles is of a different nature and relies on more aggregated and secondary data. A statistical analysis of sales data, incentives, fuel prices, and number of fueling stations is performed. There is also an analysis of the reporting of the media on ethanol and flex-fuel vehicles.

Lessons Learned from Flex-Fuel Vehicles

In its attempt to decouple from fossil fuels Sweden has previously been on the path to shifting its vehicle parc. In 2008 the sales of flex-fuel vehicles, i.e., vehicles that can be driven on a mixture of gasoline and ethanol, reached almost 25% of new car sales. However, after that they plummeted and since 2012 hardly any flex-fuel vehicles were sold. So the question is: what happened and what lessons can be learned?

The advantage and disadvantage of flex-fuel vehicles is that they are not that different from conventional vehicles. This implies that the adoption threshold shouldn't be too large for the users. The additional cost of the vehicle is also low, around an extra 100 USD (Jackson 2007). However, the observed market price difference between a flex-fuel vehicle and its equivalent conventional vehicle was up to 1000 USD,[4] presumably due to higher mark-ups. The disadvantage of the flex-fuel vehicles is that they do not provide any additional utility to the consumers, except lower environmental impact, and thus there is no additional motivation to purchase these vehicles. Sales were driven by incentives (Sprei 2018) and to a certain extent a growing interest in the environmental performance of vehicles (Sprei and Wickelgren 2011).

The fuel for flex-fuel vehicles, E85 (a fuel with 85% ethanol and 25% gasoline), was made available through a pump law that mandated fueling station above a certain volume to provide an alternative fuel. The type of alternative fuel was not specified,

[4]For some cases identifying the equivalent conventional vehicle is not straightforward since engine power might differ.

i.e., on paper the requirement was technology neutral, but given that E85 was the technically easiest option with the lowest costs, most fueling stations chose to provide E85 as the alternative fuel. This led to a rapid increase in fueling stations providing E85 and by 2012 E85 could be found in two-thirds of the fueling stations in Sweden. The regulation on reducing CO_2 emissions from personal cars in the EU allowed automakers to lower the reported tailpipe emissions of flex-fuel vehicles sold in countries with a high density of fueling stations providing E85. This implied that automakers had an interest in providing and selling flex-fuel vehicles in Sweden. The model availability thus increased and in 2008 flex-fuel vehicles were available in 74 different models. Notably the two major Swedish automakers, Volvo and Saab were providing flex-fuel options in their most sold models such as the Volvo V70. Saab also utilized the higher octane-level of ethanol to increase the available power of the engine.

Sprei (2018) looked at the discontinued diffusion of the flex-fuel vehicles and found that there are three main reasons for it. First, when gasoline prices decreased due to falling oil prices, the economic advantage of the flex-fuel vehicles was no longer there, despite the tax exemptions from both the CO_2 tax and energy tax on fuels. The relatively high price of ethanol made it less competitive despite the ample subsidies. Second, the incentive structure for flex-fuel vehicles changed from a point of sales rebate of 10,000 SEK to a five-year exemption from the annual circulation tax. This implied both a reduction in the actual monetary value of the incentive and a lowering of the effect since it was spread over time (Greene et al. 2005). And third, diesel vehicles with tailpipe emissions below 120 g CO_2/km, also labelled as green vehicles, received subsidies and started to outcompete the flex-fuel vehicles. Figure 1 illustrates this by showing the sales of Volvo station wagon cars (the most popular models sold in Sweden). As can be seen, when the "green" diesels were introduced the demand shifted from flex-fuel vehicles to diesels instead.

One more aspect that might have influenced sales, but for which the effect has been harder to quantify, is the media reporting of ethanol as a fuel. From presenting it as a "green" fuel (*miljöbränsle*) with promises of lower emission, energy security and local jobs, to a more nuanced picture, tending to negative. The wording shifted to being an "alternative" fuel (*alternativbränsle*) instead of "green" questioning the environmental benefits and also pointing out its possible contribution to increased food prices (Sprei 2018).

The case of flex-fuel vehicles in Sweden shows that if the alternative vehicles don't have a clear advantage for the user compared to the incumbent it will be hard for them to compete in the long run. This is specifically illustrated in the case of the 'green' diesel vehicles substituting the flex-fuel vehicles. This is well in line with Rogers Diffusion of innovation theory where one of the key aspects of the innovation is its relative advantage (Rogers 2003). The relative advantage of flex-fuel vehicles was their possible environmental performance and lower refueling costs. When both these were questioned and the incumbent technology improved its performance it was hard for the flex-fuel vehicles to compete. It is also an illustration of the importance of model availability and especially within the most popular models, which in the case of Sweden are station wagons. In 2007 if you wanted a 'green' vehicle the

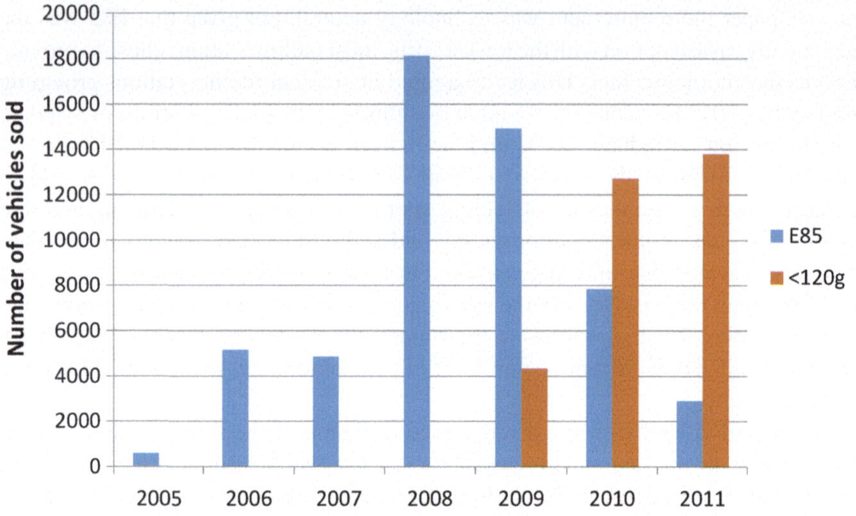

Fig. 1 Sales of Volvo's station-wagons, i.e., V50, V60, V70. The blue is the sum of flex-fuel models and red the sum of models with CO_2 emissions below 120 g/km (not including flex-fuel vehicles). *Source* (Sprei 2018). *Data source* Bilsweden

choice was either between smaller gasoline vehicles or flex-fuel station wagons, but by 2009 station wagon diesels with lower CO_2 emissions started to appear and as can be seen in Fig. 1 they were a direct competition to the flex-fuel vehicles.

Is there a risk that PEVs will follow the same path? If we start by looking at their relative advantage this depends on what attributes are taken into consideration. Range is often mentioned as a disadvantage for BEVs but as range increases this might be less of a problem. Also if PHEVs are considered then range is no longer a problem. On the other hand, the silent drive, more responsive and smoother driving, not needing to refuel at fueling stations might be seen as relative advantages that go beyond environmental impacts and lower running costs, even if at the moment the higher purchase price is a barrier. However, to reach larger sales PEVs have to be available in popular models such as station wagons. For this reason PHEVs will probably continue to play an important role in Sweden, not only because they don't have range limitations but because they are available in bigger vehicle models that are more popular in Sweden.

How Well Suited are BEVs for Swedish Consumers and Their Driving Patterns?

Driving Patterns

As mentioned, one of the disadvantages of BEVs is their limited range. So one question is: how big of a problem is this for Swedish vehicle owners? It is sometimes argued that Swedes need bigger vehicles and vehicles that can handle longer trips to country houses and skiing resort (Sprei and Wickelgren 2011). GPS measurements of Swedish conventional vehicles in Western Sweden show that given a full recharge at night and no adaptation only 20% of the vehicles would manage all their driving if they had a range of 130 km, which has been the range of BEVs such as the e-Golf and early Nissan Leafs (see Fig. 2 and (Jakobsson et al. 2016b)). If instead we look at ranges closer to 300 km, 65% of the vehicles would manage all their driving on a fully charged battery in the morning.

If we presume that their drivers resort to some kind of adaption on average once per month, the number of vehicles that can cover all their driving increases to 40% for 130 km range and 85% for 300 km range. And if the adaptation happens once a week then these numbers increase even more and reach 80% and 100% respectively for 130 and 300 km range.

So how does Sweden compare to other countries? To analyze this, we compare the Swedish driving data with three other datasets (see Methods and data section).

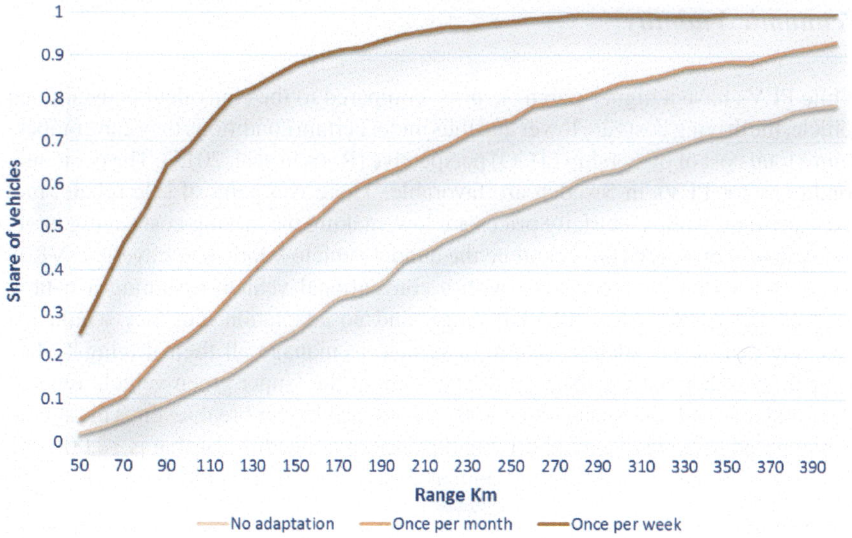

Fig. 2 Share of measured vehicles in Western Sweden that can manage all their driving after a full night's recharge with no adaptation; adaptation once per month and adaptation once per week. *Source* Jakobsson et al. (2016b)

The estimated share of vehicles that manage all their driving given a certain range will depend on the distribution chosen, e.g., the log-normal tends to give a more conservative estimate compared to the Weibull distribution (Plötz et al. 2017). Fitting the driving data to a Weibull distribution we find that 21% (±3.9%) of the Swedish vehicles would manage all their driving with a range of 150 km with only overnight charging, compared to 53% (±1.2%) of the German users. Even compared to other two datasets, the share of users that can manage all their driving is lower. For Seattle the share is fairly similar, 28% (±4.3%), while for Winnipeg, a smaller Canadian city, the share is much higher, 61% (±6%).[5] The low share of vehicles that manage all their driving with a range of 150 km can be explained by Sweden being a long country and that Swedish vehicles do travel longer distances, making BEVs less attractive. The driving patterns of Swedish vehicles could also be one of the explanations why PHEVs so far have been more popular than BEVs.

The numbers presented rely on the assumption that BEVs will always provide their nominal range. This is far from being sure as it is today. Driving style and conditions will have an effect on the actual range. In Sweden the cold winters will reduce the available range for different reasons. First, winter tires have more friction and thus will reduce range. Second, more energy is needed for in-cabin heating, and third, the cold directly affects battery performance. Interviews with households that tried a BEV for 3-4 months experience a 25% reduction in range and measurement show a reduction up to 50% (Karlsson et al. 2017). This implies that actual range might have to be even larger to accommodate all the driving.

Economic Viability

While PEVs have a higher purchase price compared to the equivalent conventional vehicle, the driving costs are lower and thus under certain conditions they can compete from a total cost of ownership (TCO) perspective (Roosen et al. 2015). The economic conditions for PEVs in Sweden are favorable. There is a point of sale rebate, fuel taxes are high, while electricity prices are low, making the driving cost saving large.

Jakobsson et al. (2016b) calculate the annual vehicle kilometers travelled (VKT) needed for a PEV to break even with a conventional vehicle assuming a battery price of 333 EUR/kWh, a 120 km range, and no adaptation, i.e. they would not have any extra costs such as renting a vehicle to manage all their driving.[6] They compare Sweden and Germany with and without the 'super green vehicle rebate'. They find that with the rebate about 8600 km are needed for break-even. This should be compared to average annual driving distance of a Swedish car that is 12,000 km. However, without the subsidy the annual distance needed would be much larger,

[5]It should be mentioned that the driving patterns for the measured vehicles in Winnipeg are not directly comparable to the other cases considered, given it is a smaller city in a Canada where the majority drive within the urban area and not over longer distances.

[6]The calculation assumes an investment horizon of 8 years and a discount rate of 5%.

14,800 km. In Germany the break-even distance is much longer, i.e. 28,900 km, mainly due to the higher electricity prices.

Given that battery prices are rapidly declining (Nykvist et al. 2019) and that the subsidy has been increasing, break-even can now be reached with even lower VKT. These calculations are fairly simple and presume that there is no additional cost due to adaptation. Extra costs that users might encounter are higher fees for fast charging, cost of renting a vehicle for a long range trip or membership in car sharing services. On the other hand, the calculations also do not take into consideration other costs that might be lower for BEV such as service and maintenance.

The conclusions are that despite the higher investment costs, PEVs face relatively favorable economic conditions in Sweden which might explain the high sales shares from a global perspective. However, the high purchase price is still a barrier since most vehicle buyers are not always aware of running cost of vehicles (Turrentine and Kurani 2007). Not to mention other barriers such as the availability of models including features such as towing capability.

Looking Closer at Two-Car Households

One way of handling the limited range of a BEV is to have a second car in the household. In this case one vehicle could be used for the long-distance trips thus reducing the need for other forms of adaptation such as fast-charging or renting a vehicle. Both in Sweden and in Norway most PEVs have been adopted in households with more than one car. In some cases the PEV has been added as an extra car, increasing the number of cars in the household. If we look at all the households in Sweden, roughly 25% have two or more cars (and 23% have no cars at all). The share is also increasing since this category of household is the one with the largest growth rate (Trafikanalys 2015). Two-car households also often live in detached houses facilitating overnight charging. They are thus an interesting group to look closer at as possible early adopters of BEVs.

Comparing the First and Second Car of a Household

When looking at two-car households the first step is to analyze which car is best to replace with a BEV. Generally in a two-car household there will be a car that is driven the most, which we will call the 'first' car and a 'second' car that is driven less. Studying the driving patterns, we find that the second car is more suitable to replace. In fact the share of second vehicles that can manage all their driving on electricity is 20 percentage points higher compared to first cars and single car households. The electric range needed to cover all the driving of a second car is also smaller being only 220 km compared to 390 km for an average car (Jakobsson et al. 2016b). 220 km is below the range of BEVs that have been introduced to the market in 2017 implying that the range should, at least theoretically, not be a barrier for adoption of BEVs in these households.

If the second car is driven less then there is a risk that it will not be as economically viable as the first car since it will be harder for it to recover the higher investments costs. Jakobsson et al. (2016b) however show that if you take into consideration the costs of adaptation, i.e. the extra costs for those days when not all the driving can be managed, a larger share of second cars will be economically viable compared to first cars. They base their calculations on a BEV with 120 km range and battery price of 333 Euro/KWh and for a range of different DRA. The assumptions on life time and discount rate are the same as previously mentioned, i.e. an investment horizon of 8 years and a 5% discount rate. BEVs as second cars will also be economically viable at shorter ranges compared to single car households and first cars. The analysis shows that roughly 25% of the second cars would be economically viable from a TCO perspective at 100 km range. It should be noted that the battery prices assumed in these calculations are conservative and thus with decreasing battery prices one can expect that the share of second cars that will be economically viable will be larger. On the other hand, as previously mentioned, consumers may not be fully aware of TCOs and may have an implicit discount rate that is much higher than 5%.

The Role of Flexibility

The calculations in the previous section are based on replacing the driving of one of the vehicles with a BEV without shifting any trips between the vehicles in a household. Having a BEV in a two-car household however may give the opportunity to optimize the driving between the vehicles to reduce the number of unfulfilled trips but also to maximize the number of electric vehicle kilometers travelled and thus increase the savings during the user phase of the BEV. Karlsson (2017) uses the data set with GPS data from both cars in 64 two-car households to estimate how the driving could be optimized in order to maximize the electric vehicle kilometers travelled. In his model, the drivers in the household can shift vehicles between themselves during the day in order to maximize the electric driving. He also assumes that the BEV is only recharged at home during the night. For battery ranges above 100 km the average potential BEV driving in two-car households can increase between 55 and 95% if the BEV use is maximized compared to substituting only the second car's driving. Karlsson (2017) also calculates the monetary value of this flexibility to an average of 6700 USD. This added value implies that a BEV is economically viable for all the measured households from a TCO perspective.

How Do Two-Car Households Use a BEV?

We have seen that BEVs are well suitable for multi-car households. However, the analysis is based on modeling the driving of conventional vehicles and not actual BEV driving. So the question is how do households actually use a BEV? To answer this a subgroup of 25 of the 64 households measured in Karlsson (2017) was selected for an intervention where one of their conventional vehicles was replaced by a BEV. The households chose which vehicle to replace themselves. The BEV they were offered was an e-Golf with a range of 130 km based on the US Environmental Protection Agency rating, which is a much more realistic estimation of range compared to the European test driving cycle. For all 25 households we thus had two measurement periods: one with two conventional vehicles (comparison period) and one with a BEV and a conventional vehicle (trial period). Interviews were carried out before and after the trial period to be able to capture how the households experienced the BEV but also their perceived adaptation. For a more in-depth description of the data and trial please see (Jakobsson et al. 2016a, 2018)

The first question we looked at was: did the households manage all their driving during the trial period? The answer is yes, with the exception of one occasion for one of the households. For this occasion the household stated that the BEV was only a contributing factor. There were however cases in which the households had to do major re-planning, borrow or rent a car. These were few and overall the household didn't experience any major problems with having a BEV with 130 km range. Despite the fact that they basically managed all their driving and that they perceived the range limitation as a minor problem, most of the households would want a longer range were they to purchase a BEV. Ten of the households tried the BEV during winter time and overall these were more negative about the range limitation since they experienced that the cold had a severe impact on the available range. Besides the actual range, the uncertainty in range was also an issue. Many felt that the information on remaining range in the vehicles was unreliable. Part of this was due to the large decrease in available range at the start since the BEV was marked with the range rating of 170–190 km from the European test driving cycle. In reality the range was closer to 130 km.

The limited amount of days and trips that needed adaptation may be a sign that calculations of DRA for multi-car household may be overestimating the problem and that in reality these are fewer and easier to handle. At the same time, given the small sample size this could be also specific for these households. When calculating the DRA during the comparison period we find that these 25 are below the average found in Jakobsson et al. (2016b). The households also chose to replace the car with the lowest number of DRAs. The households made their choice without any input from us, showing that the observed households are well aware of which car is used for longer distance trips and thus less suitable to be replaced by a BEV.

We find that on average the BEV is driven more for daily driving distances between 40 and 90 km compared to the replaced car in the comparison period. Thus, relative to the car it replaced, the BEV reduces both the amount of long daily driving distances (those above 90 km) and the short daily driving distances (below 30 km). The reduction of the long daily driving distances might be due to range anxiety and that the households have not to a large extent charged outside of the home. The decrease in shorter daily driving distances might be an example of the intent to utilize the BEV more than the replaced car, and thus letting the conventional car take the shortest daily driving.

Behind the average numbers there is a lot of heterogeneity among the individual households. We find that some households have to a large extent kept the same driving distances as the replaced car, in some the BEV has much shorter daily driving distances while in others the daily driving distance of the BEV is longer. There is also a spread of the fractional difference in driving share between the BEV and the replaced car, ranging from a decrease of 44% to an increase of 159%, while the majority falls between no change and an increase of 10%. For the household with the 159% increase, the daily driving pattern of the BEV and the replaced car were fairly similar. The increase can rather be explained by a reduction in the household's total driving between the two periods due to more long-distance trips for outdoor activities during the comparison period.

In the interviews we asked how they had planned the choice of vehicles for the different trips. Both vehicles were used for commuting since this was one of the selection criteria for the participating households. But they could assign the vehicles depending on the length of the commute. For some household this implied that the BEV was used for the longest commute, however, not always. In one case, the driver with the longest commute did neither trust the range of the vehicle to last for the return trip, nor that the charging station at work would always be available and working, thus the driver with the shortest commute used the BEV instead. 18 of the 25 households stated that they used the BEV more than the car it replaced. Some did this by planning in the morning the usage of vehicles to maximize the BEV usage, while others used the BEV more in the evening when this was possible. The first type of behavior is similar to the flexibility simulated in Karlsson (2017). In three of the households the cars were strictly divided by driver, i.e. there was no mixing and attempt to increase the driving of the BEV relative to the car it replaced. This behavior is similar to the one calculated in Jakobsson et al. (2016b) looking separately at the first and second car in the households.

Overall the households had a good impression of the BEV and had a more positive view of electric vehicles after the trial period than before. They appreciated the silent drive of the vehicle, the acceleration capacity, the smooth driving and the fact that it was environmentally friendly. When asked if they were willing to purchase or lease a BEV in the future, six of the households replied that they would not consider this. The main reasons were the usual barriers for BEV adoption such as range limitation, high investment costs and lack of charging infrastructure. Some households would instead consider a PHEV, while others said that once the prices would decrease a BEV would be an attractive option. Leasing was an option for seven of the households, mainly to avoid the risk of battery deterioration and its effect on the depreciation of the vehicle. One of the households did sign up for a lease of a Nissan Leaf by the end of trial period, one started queuing for the Tesla Model 3, one would consider purchasing a BEV within the coming year, and one household was disappointed that they had signed a lease of a conventional vehicle right before the trial period. They said that had it been after the period they would have chosen a BEV instead.

Concluding we can see that despite the overall positive impression of the BEV, and that the households managed all their driving, the limited range is still a barrier as is the high investment cost. These barriers hinder the majority of the trial households to actually purchase or lease a BEV at the moment. However, given the experienced technological development, lower prices and longer ranges are to be expected which will facilitate the further adoption of PEVs given the overall positive experience from the trial households. One challenge will be to spread the knowledge about PEVs and the possibility for households to try and experience a BEV.

Conclusions

We have analyzed the suitability of PEVs, with a special emphasis on BEVs, in Sweden. The main approach has been to study driving patterns and how these relate to the limited range of a BEV. We have looked more in detail into two-car households since these are a potential early adopter group.

We find that the longer driving distances in Sweden might make it more challenging for BEVs to be adopted. On the plus side the economic conditions in Sweden with low electricity prices, higher fuel taxes and subsidies make the BEV economically viable from a TCO perspective in many cases. The combination of these factors might explain the relatively high shares of sales of PEVs and why these are dominated by PHEVs. The failed introduction of flex-fuel vehicles indicates that it is important to maintain these economic advantages through, e.g., higher fuel taxes.

Two-car households are increasing in Sweden and they are a well-suited group for replacing at least one of the vehicles with a BEV. Overall the driving patterns of the second car are better suited for the BEV and if the driving of the BEV is maximized then, even with conservative values for batteries costs, the BEV could be economically viable.

Two-car households that got one of their conventional cars replaced with a BEV had an overall positive impression of the BEV. Besides the lower environmental impact, they enjoyed the quiet and smooth drive as well as the acceleration capacity. This implies that there is a better chance that the diffusion process won't be interrupted as was the case with flex-fuel vehicles. Still, barriers such as limited range and higher investment costs hinder many of them from actually wanting to purchase a BEV in the near future. Model availability, especially in popular segments such as station wagons is also a critical factor.

From a policy perspective information to the households, and especially to those with more than one vehicle is important. More awareness is needed of the suitability of BEVs despite the lower range and that TCOs can be favorable. Increasing the possibility of actually experiencing and testing a PEV may also increase willingness to purchase one. Since the higher purchase price is a major barrier, subsidies continue to play an important role. While a longer range might not be desirable for environmental reasons and not always necessary for the user either, at least in two-car households, ranges of at least 300 km might be needed for users to feel comfortable. This holds especially true in Sweden where cold winters will lower the available range.

References

Azar, C., & Sandén, B. A. (2011). The elusive quest for technology-neutral policies. *Environmental Innovation and Societal Transitions, 1*(1), 135–139.

Circle, Power (2018). *Laddbara bilar i Sverige—oktober* 2018, 2018-11-12.

Council, Puget Sound Regional (2008). *Traffic choices study*, (Technical report).

Granström, R., et al. (2017). Användarnas beteende och syn på laddbara bilar, *Rapport från projketet SELF-I. TRUM-Rapport,* 1.

Greene, D. L., et al. (2005). Feebates, rebates and gas-guzzler taxes: A study of incentives for increased fuel economy. *Energy Policy, 33*(6), 757–775.

Jackson, M. (2007). Ethanol Biofuels Implementation Plan. (CEC-ARB Workshop on Developing a State Plan to Increase the Use of Alternative Transportation Fuels).

Jakobsson, N. (2016). On battery electric vehicles: Driving patterns, multi-car households and infrastructure. *Technical report/Department of Energy and Environment,* (2016: 1).

Jakobsson, N., Karlsson, S., Sprei, F. (2016a). How are driving patterns adjusted to the use of a battery electric vehicle in two-car households. *Electric Vehicle Symposium 29* (Montreal).

Jakobsson, N., et al. (2016b). Are multi-car households better suited for battery electric vehicles?—Driving patterns and economics in Sweden and Germany. *Transportation Research Part C: Emerging Technologies, 65,* 1–15.

Jakobsson, N., Karlsson, S., Sprei, F. (2018). How do users adapt to a battery electric vehicle in a two-car household?. *Under Review.*

Karlsson, S. (2013). *The Swedish car movement data project*. PRT Report, (2013: 1).

Karlsson, S. (2017). What are the value and implications of two-car households for the electric car? *Transportation Research Part C: Emerging Technologies, 81,* 1–17.

Karlsson, S., Jakobsson, N., & Sprei, F. (2017). *The variation of BEV energy use and range wit climate and driving behaviour in real driving in Sweden.* Geneva: EEVC, European Electric Vehicle Congress.

Kullingsjo, L. H., Karlsson, S. (2013). GPS measurement of Swedish car movements for assessment of possible electrification. *EVS27 Symposium*, Barcelona, Spain.

Langbroek, J. H., Franklin, J. P., Susilo, Y. O. (2017a). *Change towards electric vehicle use in Greater Stockholm.*

Langbroek, J. H. M., Franklin, J. P., & Susilo, Y. O. (2017b). Electric vehicle users and their travel patterns in Greater Stockholm. *Transportation Research Part D: Transport and Environment, 52,* 98–111.

Langbroek, J. H., Franklin, J. P., & Susilo, Y. O. (2016). The effect of policy incentives on electric vehicle adoption. *Energy Policy, 94,* 94–103.

MOP, Mobilitätspanel (2010). Deutschland 1994–2010, Technical report, Project conducted by the Institute for Transport Studies at the Karlsruhe Institute of Technology (KIT). Available at: www.clearingstelleverkehr.de.

Nykvist, B., Sprei, F., & Nilsson, M. (2019). Assessing the progress toward lower priced long range battery electric vehicles. *Energy Policy, 124,* 144–155.

Plötz, P., Jakobsson, N., & Sprei, F. (2017). On the distribution of individual daily driving distances. *Transportation Research Part B: Methodological, 101,* 213–227.

Rezvani, Z., Jansson, J., Bengtsson, M. (2017). Consumer motivations for sustainable consumption: The interaction of gain, normative and hedonic motivations on electric vehicle adoption. *Business Strategy and the Environment.*

Rogers, E. M. (2003). *Diffusion of innovations* (5th ed.). New York: Free Press.

Roosen, J., Marneffe, W., & Vereeck, L. (2015). A review of comparative vehicle cost analysis. *Transport Reviews, 35*(6), 720–748.

Smith, R., et al. (2011). Characterization of urban commuter driving profiles to optimize battery size in light-duty plug-in electric vehicles. *Transportation Research Part D: Transport and Environment, 16*(3), 218–224.

Sprei, F. (2018). Discontinued diffusion of alternative-fueled vehicles—The case of flex-fuel vehicles in Sweden. *International Journal of Sustainable Transportation, 12*(1), 19–28.
Sprei, F., & Wickelgren, M. (2011). Requirements for change in new car buying practices—Observations from Sweden. *Energy Efficiency, 4*(2), 193–207.
Sweden, Government Offices of (2018a). *The climate policy framework*. 28-11-2018.
Sweden, Government Offices of (2018b). *As of today, Sweden has a new Climate Act*!. 2018-11-28.
Trafikanalys (2015). Peak car is sikte? Statistik och analys över Sveriges personsbilsflotta och dess användning', (PM 2015: 14; Stockholm: Trafikanalys).
Transportstyrelsen (2018a). Frågor och svar om supermiljöbilspremie. 11-29-2018.
Transpoststyrelsen (2018b). 2018-06-29.
Turrentine, T. S., & Kurani, K. S. (2007). Car buyers and fuel economy? *Energy Policy, 35*(2), 1213–1223.
Wikström, M., Hansson, L., & Alvfors, P. (2014). Socio-technical experiences from electric vehicle utilisation in commercial fleets. *Applied Energy, 123,* 82–93.

Beyond Consumer Innovators: Adoption of Plug-in Vehicles by Mainstream Consumers in the U.K.

Stephen Skippon and George Beard

Abstract This chapter discusses key methodological issues involved in researching potential Plug-in Electric Vehicle (PEV) uptake among mainstream consumers: all those who are not Innovators in terms of Rogers' (Diffusion of innovations. Free Press, New York, 2003) diffusion model. In the U.K. in 2019 mainstream consumers do not have their own PEVs and have little or no direct experience of using them. Issues considered in this chapter include the problems of psychological distance, the Hawthorne effect, the "Bed of Procrustes", and the role of symbolic meaning and self-congruity in influencing uptake. Two studies that have worked with mainstream consumer samples, one qualitative and one quantitative, are described in depth to illustrate how these issues can be addressed.

Introduction

The U.K. has a current (2019) population of some 66.5 million people according to United Nations estimates, amounting to 0.87% of the global population. Around 81.2% of that population live in urban areas. In 2019, the U.K. remains the world's fifth largest economy, with a gross domestic product estimated to be around 3.2 trillion dollars (Focus Economics 2019). At the end of March 2019 there were 32.7 million cars and 4.2 million Light Goods Vehicles (mainly vans) in the U.K. (Department for Transport 2019a). Among these at the end of March 2019 there were 190,939 plug-in electric cars and 8344 plug-in electric Light Goods Vehicles in the U.K. (Department for Transport 2019b), with around 125 models available (Next Greencar 2019). The U.K. government's *Road to Zero* strategy sets out the intention to end the sale of new conventional gasoline and diesel fuelled cars and vans by 2040, with the further intention that by 2050 almost every car and van in the U.K. will be an Ultra-Low Emission Vehicle (ULEV).

S. Skippon (✉)
Lemonade Consulting Ltd., 2 Byron Close, Deeside CH53TE, UK
e-mail: stephen.skippon@outlook.com

S. Skippon · G. Beard
TRL Ltd., Crowthorne, UK

M. Contestabile et al. (eds.), *Who's Driving Electric Cars*, Lecture Notes in Mobility,
https://doi.org/10.1007/978-3-030-38382-4_9

Threats from Car Traffic, a 2007 book edited by Tommy Garling and Linda Steg, highlighted two threats posed by road transport that will be familiar to readers of this book: first, it contributes to urban air pollution, and second, it is a major contributor to carbon dioxide emissions that are drivers of global climate change. The U.K. government has set the target of a complete transition away from conventional petrol and diesel engines in new cars by 2040. Plug-in Electric Vehicles (PEVs) are generally considered the main option to replace them. It is anticipated that over this timescale there will be substantial technical developments in PEV technology, for instance reducing the cost of the vehicles and the time it takes to recharge their batteries, and increasing their all-electric range (AER). However the transition to a PEV car parc will involve more than just a technological revolution. It will be a *socio-technical* revolution (Geels 2005; Geels et al. 2012), dependent as much on how individuals and society respond to PEVs as on the technology of the new vehicles themselves. The pace and extent of the transition will depend on when (or whether) people chose them for their personal travel. Intimately intertwined with these choices, the development of new social practices such as charging at home or at public charging stations will need to become norms, and others such as visits to petrol stations will consequently fade from relevance. This means that we cannot simply use the speed of PEV technical development to predict the pace of transition to a PEV-dominated car parc. We must understand the conditions under which U.K. consumers will adopt them in preference to Internal Combustion Engine vehicles (ICEVs).

The Importance of Mainstream Consumer PEV Adoption

The research community has learnt a great deal about the potential for consumer adoption of PEVs from studies with 'Innovators', those members of the public who have been the first to start using these still relatively new types of vehicles. In the U.K., Innovators have responded enthusiastically to PEVs: perceiving them as desirable and fun to drive, adapting quickly to features such as regenerative braking, reporting preferences for charging at home compared with refuelling at petrol stations, and expressing a willingness to persuade others of their merits, (e.g. Bunce et al. 2014; Carroll et al. 2013). Studying these responses has helped researchers, policymakers, and the automotive sector to understand what needs to be done to facilitate the early stages of PEV adoption by consumers. However if we want to look beyond those first stages and understand what is needed to encourage mass-market adoption of PEVs, we need to look to the wider car-user population.

There is no a priori case for assuming that Innovators' responses to PEVs will be representative of the responses of members of the wider car-user population. Indeed the opposite is the case. The U.K.'s Energy Technologies Institute (ETI) commissioned a segmentation study with a large sample ($n = 2729$) of U.K. car users (Anable et al. 2014) that showed that attitudes towards PEVs differed markedly between consumer segments. The attitudes of what the study termed the "Plug-in Pioneers" segment (2% of the sample) were very unrepresentative of those of the other segments. While Plug-in Pioneers had very favourable responses to PEVs, other, larger segments had very unfavourable attitudes: for instance, the 'Image-Conscious Rejectors' (18% of the sample) had very unfavourable responses, while 'Uninspired Followers' would consider a PEV only if a majority of others already had them. The attitudinal heterogeneity identified in this and subsequent studies implies that to understand the potential for mass adoption of PEVs, we need to research the attitudes of what we shall refer to as *mainstream* consumers. In attempting to do that we encounter multiple challenges; in this chapter we shall explore what these are, some ways to address them, and some findings from recent research with mainstream consumers in the U.K. that has done so.

Who Are Mainstream Consumers?

We shall define mainstream consumers in relation to Rogers' (2003) diffusion model, which proposes that the time to adoption of an innovation by members of a population of potential users is normally distributed, as shown in Fig. 1. Rogers defined a set of population segments in relation to the standard deviation of this distribution, and

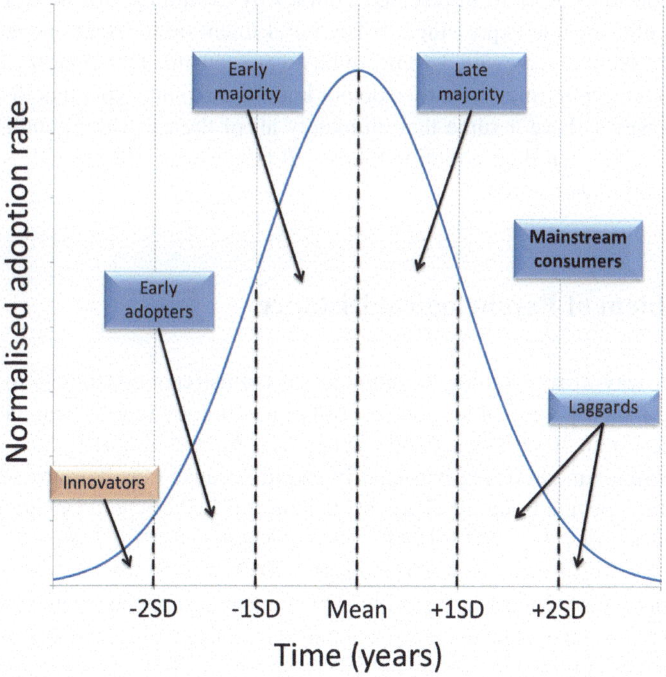

Fig. 1 The distribution of times to adoption of an innovation according Rogers' (2003) diffusion model, and Rogers' segments, defined in relation to the standard deviation of the distribution

his segment labels have become a staple in discussion of the adoption of many innovations, including PEVs. Those consumers who adopt more than two standard deviations before the mean adoption time are 'Innovators'; those adopting between two and one standard deviation before the mean adoption time are 'Early Adopters'; and so on for segments labelled 'Early Majority', 'Late Majority', and 'Laggards'.

In this model Innovators represent the first approximately 2.5% of the eventual adopter population. PEVs represent around 2% of the present U.K. car parc (EAFO 2018), so all of the present users of these vehicles are defined as Innovators in terms of the diffusion model (we note in passing that the term "Early adopter" is often used for these people, but in the diffusion model this refers to the segment that adopts later than the Innovators).

We define mainstream consumers as all those who adopt PEVs later than two standard deviations before the mean time to adoption: the Early Adopter, Early Majority, Late Majority, and Laggard segments, that is everyone except the Innovators. There are two reasons for this. The first is theoretical. In the diffusion model, most of the population adopts an innovation as the diffusion of knowledge about it and positive attitudes towards it extend to more and more people. That diffusion has to start somewhere: it starts with Innovators, who themselves are not influenced by diffusion but rather are its initial agents. In this sense, mainstream consumers are those who adopt PEVs as a consequence of diffusion processes.[1] The second is pragmatic: it enables us to exclude Innovators from research samples by excluding anyone who already has a PEV plus anyone expressing a behavioural intention to adopt one in the near future. One potential flaw with defining mainstream consumers in relation to the distribution of adoption times is that we do not know in advance what that distribution will be. However if we assume that ultimately all of the car-using population will adopt a PEV, and that their adoption times will be normally distributed, we have a simple operational criterion.

The Problem of Psychological Distance

A common way of researching the attitudes of mainstream consumers to PEVs is to conduct a survey with an appropriate sampling strategy (e.g. Axsen and Kurani 2013; Carley et al. 2013; Delang and Cheng 2012; Shin et al. 2012). Surveys using stated preference methods are particularly useful because they help to identify the ways in which people trade off different attributes of PEVs when considering them (e.g. Dimitropoulos et al. 2013; Hackbarth and Madlener 2013; Jensen et al. 2013; Lebeau et al. 2012; Lieven et al. 2011; Ziegler 2012).

Focussing on the U.K., a survey with 2729 mainstream car users with recent experience of buying a new or nearly new car identified a variety of factors that influenced their willingness to consider having a PEV (Anable et al. 2014). Lower running costs

[1] We note that there is evidence for other social influences as well as diffusion operating in relation to adoption of PEVs. Axsen et al. (2013) explored these in a U.K. context.

were identified as the main factor in favour of PEV adoption, but stacked against those were high price compared with ICEVs, limited availability of vehicle categories (e.g. small hatchback vs. executive saloon vs. Sports Utility Vehicle), limited availability in terms of brands, limited public charging infrastructure, short range, long charging times, and insufficient incentives. Schuitema et al. (2013), using data from the same survey, investigated the role of instrumental, hedonic, and symbolic attributes in the intention to adopt PEVs. They found that instrumental attributes were important mainly because they were associated with other attributes such as the pleasure of driving (a hedonic attribute) and the projection of personal identity through ownership and use of a PEV (a symbolic attribute). They also found that those with a "pro-environmental" identity were more likely to have positive perceptions of PEV attributes, while positive perceptions of PEV attributes were only weakly associated with having a "car-authority" identity.

However when the topic of interest is what Hoeffler (2003) referred to as "really new" products like PEVs, surveys of mainstream consumers have an important weakness. In the early market for a really new product category, the majority of participants in a survey study that uses a population sample will have no direct experience of the category to draw on when expressing attitudes towards it or making choices between it and more familiar categories. This can lead to some specific biases.

Lack of direct experience is one form of the phenomenon of *psychological distance*, a central concept in Construal Level Theory (CLT) (Eyal et al. 2009; Liberman et al. 2007; Trope and Liberman 2003). An object is psychologically distant when it is detached from someone's direct experience, but is psychologically near when a person can sense it directly. Psychological distance can involve separation in time (the object is construed as belonging in the past or future), social separation (an object can only be sensed by someone else rather than oneself), separation in space (an object is located far away), or separated by hypotheticality (if it is seen as having a low probability of being experienced directly). Eyal et al. (2009) further proposed that novelty would have similar effects to these dimensions of psychological distance. CLT proposes a generalised association between psychological distance and the level of abstraction at which an object is construed. A psychologically distant object is represented by higher-level, more abstract construals; a psychologically close object is represented by lower-level, more concrete construals.

The appeal or value of an object in CLT depends on a weighted sum of the value associated with high-level construals of that object plus the value associated with low-level construals. Psychological distance increases the weight attached to high-level construals, and psychological closeness increases the weight given to lower-level construals. There is compelling evidence for these effects in a variety of circumstances (reviewed by Eyal et al. 2009).

Higher-level construals relate to the value of the end-state that an object or an action such as choosing that object can deliver; and to the higher-level personal goals that the object can contribute to fulfilling (for an account of the hierarchical nature of personal goals, see Carver and Scheier 1998). In less technical terms, higher-level construals refer to *why* a person might choose and use that object: its *desirability*. On the other hand lower-level construals relate to the means used to attain an end-state, or *how* a person might choose and use an object: its *feasibility*. Liberman and Trope (1998) showed that when outcomes were desirable but difficult to obtain, their attractiveness was higher when they were psychologically distant, and conversely when outcomes were less desirable but easily obtained, their attractiveness was higher when they were psychologically close.

Further, when considering the pros and cons of a choice object (its desirable attributes compared to its undesirable attributes), Eyal et al. (2004) showed that pros are higher-level construals than cons, and therefore weighted more highly in evaluations of psychologically distant objects.

Returning to the study of adoption of PEVs, Construal Level Theory suggests that studies of potential adoption by mainstream consumers where participants are psychologically distant from PEVs will have certain biases. Participants' evaluations of PEVs will tend to be weighted more towards desirability considerations than feasibility considerations, and more towards the pros than the cons of PEVs. CLT also tells us how to reduce the influence of such biases: we should reduce the psychological distance of mainstream consumer participants from PEVs.

The Problem of the Hawthorne Effect

One obvious way to reduce psychological distance is to find research participants who are already psychologically close to PEVs, for instance Innovators who already have them, and those who are actively considering acquiring one or are willing to consider having one. This approach was adopted by the U.K. Technology Strategy Board's *Ultra Low Carbon Vehicle (ULCV) Demonstrator Trial* that was aimed at providing a real-world assessment of PEV technology and accelerating the introduction of PEVs into the U.K. market. Carroll et al. (2013) summarised key findings from this trial, which involved 349 vehicles from a variety of manufacturers, and established a good understanding of the motivations and usage behaviours of PEV Innovators in the U.K. market. Useful as studies of this kind are, there are two methodological challenges that limit how far their findings can be extrapolated beyond Innovators.

First, as discussed above, the heterogeneity of the car user population means that information derived from Innovators cannot be extrapolated to mainstream consumers. An early attempt to address this issue was described by Skippon and Garwood (2011). Their study involved providing 56 employees of the electricity supplier E.ON with a brief direct experience of driving a Mitsubishi i-MiEV BEV, followed by an attitudinal questionnaire. The participants did not have PEVs themselves and their work was not connected with them. The sample therefore comprised mainstream consumers rather than Innovators, though it had other sample biases, particularly a higher average educational level than the general U.K. car-user population. This study was the first to use the attribution-vignette method to characterise the symbolic meaning of BEVs, a topic we shall discuss later.

Second, Carroll et al.'s (2013) study, along with the Skippon and Garwood study and many PEV trials, was potentially confounded by the Hawthorne effect. This takes its name from studies carried out at the Hawthorne Works in Cicero, Illinois, from 1924 to 1932 on the effects of changes to workplace conditions such as lighting and break times, on the productivity of the workforce. In a re-analysis of these studies, Landsberger (1958) suggested that the measured productivity gains could be accounted for by increased motivation of the workers, *knowing that attention was being paid to them*. Tellingly, productivity was seen to increase whenever a variable was changed, even when this change involved reversion to an earlier condition. Landsberger's specific interpretation of the Hawthorne data has since been challenged, but the effect he uncovered is now widely recognised: participants often modify their behaviour in response to being observed, i.e. to the knowledge that they are participating in research, independently of any effects of variables that are manipulated in the study.

The possibility of a Hawthorne effect creates a potential ambiguity in the results of trials where all participants receive the same experimental condition: any outcome measures (whether behavioural or attitudinal) cannot be unambiguously attributed to the experimental condition, since they could also have been caused by the Hawthorne effect.

The Hawthorne effect can be controlled for readily using a between-participants experimental design in which participants are randomly allocated to two groups, one that experiences the experimental condition (e.g. experience of using a PEV) and another that experiences a control condition (e.g. experience of using an ICEV). Since both groups are aware that they are participating in the study, both of them experience the Hawthorne effect but only the experimental group experiences the experimental condition, so the effects are disambiguated. Controlled designs like this are common in many fields of research, but regrettably not so in transport research (Graham-Rowe et al. 2011). There have been relatively few controlled studies of responses to PEVs to date.

The Influence of Symbolism and Self-congruity

It seems reasonable to assume that likelihood to adopt PEVs will be related to motivations for car acquisition and use. A starting point for understanding such motivations is the variety of theories of the social psychology of material possessions (Dittmar 1992; Dittmar and Brown 2008; Dolfsma 2008; Fournier 1991; Keller 1993; Ligas 2000; Park et al. 1986). Steg (2005), for instance, used Dittmar's social constructionist theory in a study of car use. Steg's version of Dittmar's theory had three dimensions of evaluation: *instrumental*, *affective* and *symbolic*.

Instrumental or functional motivations for car use include such factors as cost, journey time, convenience, effort, carrying capacity, etc. (Steg 2005). Individuals differ in the extent to which they experience instrumental benefits from owning or having access to a car with a particular combination of instrumental attributes. For some, instrumental benefits of car use are so great, and/or so difficult to achieve through other means, that they describe themselves as "car dependent" (Goodwin

1997; Jeekel 2003). Vehicles clearly offer instrumental benefits, and instrumental motivations play a significant role in most vehicle purchase choices. In fact many studies have assumed that instrumental motivations dominate vehicle choice. Models that assume choices are made on the basis of maximizing 'utility', usually defined in instrumental terms (Puttaswamaiah 2002). Choo and Moktarian (2002) analysed eleven vehicle choice models and found that the focus in all of them was on monetary costs and instrumentally-defined attributes.

However there are two other categories of motivations that have been argued to be major influences on vehicle choices. We shall focus here on symbolic motivations [a discussion of theoretical reasons for scepticism about the distinctiveness of affective motivations is given by Skippon (2014)].

It is well established that consumers acquire and use material goods in the pursuit of symbolic as well as instrumental motives (Belk 1985; Dittmar 1992; Dittmar and Brown 2008; Fournier 1991; Keller 1993; Ligas 2000; McCracken 1988; Miller 2009, 2010; Park et al. 1986). Symbolic motives concern signalling something about the owner/user to other people, such as personal identity (Dittmar 1992; McCracken 1988; Miller 2009; Sirgy 1982a, b, 1985; Tucker 1957), status (Griskevicius et al. 2007; Nelissen and Meijers 2011; Sundie et al. 2010; Veblen 1899), or social category membership (Dittmar 1992; McCracken 1988). Possession or use of material goods conveys such signals because goods carry symbolic meaning—i.e. they stand for something else. To be effective as means to fulfil symbolic goals, their meanings must be shared by the possessor/user and by their audience—those people to whom something is signalled. These socially or culturally defined meanings are then used by the individual to express something about themselves (Dittmar 1992; McCracken 1988; Mick 1986; Miller 2010).

A product has symbolic value to the extent that its possession or use supports the pursuit of a symbolic goal: i.e. it holds symbolic meanings that can act as effective signals of something a person is motivated to signal. Symbolic meanings confer the symbolic value that products have for the pursuit of symbolic goals. Thus, the key to understanding the symbolic value of vehicle types is to identify their symbolic meanings. In self-congruity theory (Sirgy 1982a, b, 1985) people are motivated to behave in ways that are congruent with their self-concepts: they buy products whose symbolic meanings are consistent with their self-concepts. Several empirical studies of car ownership provide support for this theory (Eriksen 1996; Grubb and Stern 1971; Heath and Scott 1998; Sirgy 1985).

Thus, in addition to their instrumental motivations, car users will be more motivated to acquire and use a PEV to the extent that its symbolic meanings are congruent with their self-concepts. To apply self-congruity theory to the study of PEV uptake we need therefore to be able to characterise the symbolic meanings of PEVs. Heffner et al. (2007) used qualitative interviews to investigate the symbolism of the first hybrid electric cars in California in the early 2000s. However it is difficult to use qualitative descriptions of meanings to estimate the extent of self-congruity in individuals. For this purpose a more quantitative measure of symbolic meaning is required.

Miller (2009) proposed that consumers are motivated to acquire and use goods that signal the user's five-factor personality traits and general intelligence, because these traits function as reproductive fitness indicators. Thus the principal symbolic meanings of products are the users' personality traits and intelligence. From this theoretical perspective other meanings are interpreted as proximate ways in which products signal these ultimate meanings. If we accept Miller's theory, then self-congruity can be measured by comparing an individual's self-reported five-factor personality and general intelligence with the five-factor personality and general intelligence that are attributed to the typical user of a product. Later we shall discuss research in the U.K. that has explored the influence of symbolic motivations, characterised in terms of self-congruity, alongside instrumental motivations in mainstream consumers' willingness to consider adopting PEVs.

The Bed of Procrustes

The majority of research on PEV adoption, in the U.K. as elsewhere, has been quantitative in nature. Despite its many benefits, quantitative research has one substantial limitation: it runs the risk of making the 'Procrustean' error of requiring participants to try to fit their own experiences onto conceptual frameworks that the researchers supply, no matter how good or bad the fit (Salmon 2003). Procrustes was an innkeeper in ancient Greek literature, who claimed to have a bed that was a perfect fit for any guest. He enforced this perfect fit by stretching his shorter guests, and chopping off the feet of his taller guests. The perspectives of participants in quantitative research are at risk of being force-fitted into researchers' own theoretical frameworks in a metaphorically similar manner. Questions are framed and posed by researchers, and participants' own frames of reference may not be explored, or valued. Quantitative findings that are subject to the Procrustean error can be misleading.

There is a simple way to mitigate this risk in research into mainstream consumer adoption of PEVs: carry out *qualitative* research to understand the ways that mainstream consumers construe PEVs and their own frames of reference for evaluating them. Subsequent quantitative research can then be framed in ways that are meaningful to mainstream consumers.

Understanding Mainstream Consumers' Own Perspectives: Findings from Qualitative Research

Having reviewed the various challenges involved in researching the potential for adoption of PEVs in the U.K., in the rest of the chapter we shall discuss in depth two studies that have attempted, in different ways, to address them.

The previous section argued that the starting point for understanding potential PEV adoption by mainstream consumers should be qualitative research aimed at exploring their own ways of construing PEVs, thus avoiding the bed of Procrustes problem. That is where we begin.

Graham-Rowe et al. (2012) carried out a qualitative study as a precursor to the quantitative survey discussed earlier in this chapter (Anable et al. 2011, 2014). Their study aimed to explore mainstream consumers' attitudes towards PEVs and beliefs about them. Recognising that mainstream consumers were psychologically distant from PEVs, the study sought to reduce participants' psychological distance by enabling each of them to experience using either a PHEV or a BEV for a period of seven days. The appropriate duration of the direct experience needed to reduce psychological distance from PEVs was not known, although Skippon and Garwood (2011) had found that their participants in a workplace-based trial could express clear and consistent attitudes about BEV performance and symbolic meanings after only a single drive of a BEV. In Graham-Rowe et al.'s study it was considered important that participants experienced recharging their vehicles, encountered the range restrictions of the vehicles, and experienced how the vehicles fitted in with their lifestyles. Seven days was perhaps too short to enable stable patterns of usage and recharging behaviour to emerge, but it was thought sufficient to enable participants to form mental representations of what PEV usage and recharging would be like, and how far these would be consistent with their lifestyles.

Qualitative research does not aim to be statistically representative, but rather to capture thematically the full range of ways members of the target population construe the topic under investigation. The study therefore adopted an illustrative sampling method (Turrentine and Kurani 2007), seeking a group of 40 participants differing in terms gender, relationship status, age, annual household income, number of cars in household, and location (rural/urban). All the participants were ICEV drivers who had purchased or leased a car less than two years old within the last five years, were currently looking for such a car, or were currently driving a company-owned car less than two years old. None of the participants had any direct experience of driving or being a passenger in a PEV.

Twenty participants were given direct experience of a BEV for seven days, and twenty a PHEV. There were two types of BEV: a Mitsubishi iMiev, and a Citroen C1 converted to BEV by the Electric Car Corporation. Both had a manufacturer's claimed range of around 90 miles on a full charge. By modern standards (2019) these ranges are low, but they were competitive at the time of the research. No PHEVs were yet available in the U.K. market so the research PHEVs were Toyota Prius HEVs that had been converted to PHEV operation by Amberjac Projects, with an expected all-electric range (AER) of around 30 miles (a range which is typical of 2019 PHEVs in the U.K. market).

Semi-structured interviews followed immediately after participants had completed their experience of using the vehicle. Interview data was analysed using an abbreviated form of grounded theory analysis (Strauss and Corbin 1998) to identify themes in participants' experiences of PEVs.

The researchers identified six emergent themes in participants' accounts of their experiences with PEVs that are discussed in turn below. In the spirit of academic reporting of qualitative research, descriptions of emergent themes include verbatim quotations from the data [drawn from those included by Graham-Rowe, et al. in their original (2012) paper] that are intended to demonstrate the evidence on which interpretations were based (Mays and Pope 1995).

Cost Minimization

Many of the participants had investigated the costs and savings associated with PEVs and compared these to their experiences with ICEVs. Participants recognised that PEVs offered fuel cost savings, and some of the PHEV participants even began to modify their driving styles to avoid engaging the ICE powertrain. However the high purchase costs of PEVs were seen as a major barrier. Participants expected something in return for the extra cost of PEVs (e.g. better design, performance, durability, or driving experience) but the vehicles they had driven did not meet this expectation:

> *There is no way I would pay £22,000 for an £8000 motor car… The numbers don't stack up.* (BEV participant)

Participants recognised that the fuel cost savings might offset higher initial purchase costs, but most were uncertain about these calculations, and even those who were confident about their savings estimates felt that it would take many years to offset the high purchase cost:

> *It'd probably save me about, I think £2000 worth of fuel in a year… I'd be running it for a few years before I saw any money back on that.* (PHEV participant)

Participants also expressed concerns that the lifespan of the battery might not be sufficient for them ever to offset the purchase cost; and that PEVs might suffer from higher depreciation rates, again negatively impacting on the cost/benefit ratio of PEVs compared to ICEVs.

Vehicle Confidence

Participants generally evaluated the performance of their PEVs as inferior to the ICEVs they were used to, leading to a sense that driving a PEV was less pleasant:

> *You always want your next car … to be better than your last. So it's almost like a back step to driving something which just doesn't feel quite as safe and powerful and comfortable as what I've got at the moment.* (BEV participant)

Confidence in BEVs was also undermined by experiencing how fast the battery depleted in normal driving. Some participants expressed this in terms similar to the industry's "range anxiety" construct. Others were less concerned about whether they could complete a particular journey on the charge remaining than they were about restricted utility compared with an ICEV. They saw a BEV as only suitable for journeys to a restricted set of close destinations:

There's just some journeys I don't think I'd do. (BEV participant)

For this reason BEVs were seen as being more suitable as second cars in households, to be used for short, local journeys. This was not the case for PHEVs which were perceived as being suitable for all journeys, and thus as potential main cars for households.

Vehicle Adaptation Demands

Participants frequently mentioned needing to adapt their behaviours to accommodate the particular features of PEVs. The first adaptation was the transition from occasional refuelling at a petrol station to regular recharging at home. Many experienced this as a benefit, but for some it was a source of unwanted extra hassle:

You can actually plug it into your own mains ... whenever it suits you... I think it was possibly easier because you've got your facilities there on tap. (BEV participant)

It was, like, "oh, my God, this is something extra I've got to do" (BEV participant)

The second was adapting to the combination of short range of BEVs, the sparsity of public recharging points, long recharging times necessitating long breaks in journeys, a higher degree of journey planning, and journeys forsaken because the destination was too far away or because the PEV was unavailable when recharging:

While the car's plugged in it's marooned, it's in an undriveable state (PHEV participant)

For some these adaptations amounted to a substantial impact of lifestyle:

It just made me a lot more aware of my lifestyle (BEV participant)

The third adaptation was changes in driving style to fit in with the characteristics of the PEV. BEV participants in particular had to adjust to lower acceleration at high speeds:

You think you're putting your foot down and it doesn't seem to be going anywhere (BEV participant)

Finally, the absence of engine noise that could warn pedestrians of the presence of the PEV was perceived as a safety risk that required drivers to take extra care.

Environmental Beliefs

The perceived environmental benefits of PEVs gave some participants and their passengers a positive feeling about using them, a "feel-good factor" as one PHEV participant described it. For others the feeling was more one of assuaged guilt for using a car at all.

I think you can feel a little less guilty maybe when you go out on your Sunday drives if you're in an electric car than you are in a petrol car. (PHEV participant)

On the other hand some participants expressed scepticism about whether PEVs genuinely contributed to reduced emissions:

Unless you can convince me that the power station was green I don't believe you've saved anything. (BEV participant).

Impression Management

Participants recognised that the PEVs they were driving projected signals about themselves to others. This was seen as beneficial by those with a pro-environmental identity to project. However others were much less self-congruent and described driving a PEV as acutely embarrassing:

The initial reaction was always the same – complete and utter ridicule of the whole concept. It was a huge PR problem. (BEV participant)

Participants discussed three types of people who they felt other members of the public would expect to be driving PEVs: first, people with limited needs for mobility ("*young mums, housewives, retired couples that just tootle around*" according to one BEV participant) who did not derive pleasure from driving; second, those for whom signalling their pro-environmental identity mattered more than functionality, and who were viewed negatively ("*hugging trees and going to public meetings about saving the planet*"); third (mainly applicable to PHEV drivers) those who adopted viable new technologies at an early stage ("*a forward-thinking person, modern, and perhaps on the pulse of technology*"), which was generally viewed as a positive social identity.

PEVs as a Work in Progress

Most participants' views towards PEVs were underpinned by a strong sense that they were a "work in progress", still being developed and not yet up to the standards of ICEVs in terms of performance, design, purchase cost, or pleasurable driving experience. As a result most participants were not interested in PEVs at the time of the study. They would prefer to wait and see what further development of the vehicles might bring.

I think there's more development work that has to be done on them. (BEV participant).[2]

[2]Bunce et al. (2014) found more positive responses in a somewhat later U.K. study with PEV Innovators (though that might also in part be explained by later vehicle models being available for their study).

A 2019 Perspective on Graham-Rowe et al.'s Qualitative Research

Some of the themes in participants' construals of PEVs in Graham-Rowe et al.'s study reflected the frames of reference of researchers at the time—for instance, that PEV adoption would be influenced positively by their lower cost of refuelling, but negatively by their higher purchase costs and restricted range. But much seemed new at the time. The finding that many mainstream consumers rejected the symbolic meanings they attributed to PEVs, and talked of being embarrassed to be noticed trying one out, was received with something close to shock by some of our colleagues. It drew attention to the importance of considering symbolic motivations when thinking about PEV adoption, and suggested that a "green" symbolic meaning for a class of vehicles might be a hindrance rather than (or as well as) a help in facilitating their widespread adoption. The finding that mainstream consumers responded negatively to the behavioural adaptations they would have to make if they were to integrate PEVs into their lives was also disturbing to those used to the enthusiastic responses of PEV Innovators. That is exactly why qualitative research that gives voice to participants' own perspectives is so important: it challenges researchers' frames of reference.

Qualitative researchers point out that any individual qualitative study is inevitably both socially and temporally situated. Since Graham-Rowe et al.'s study was conducted, PEVs have advanced substantially. The ranges possible with modern (2019) BEVs are substantially greater than those of the vehicles used in the study, and the problem of perceived under-performance is no longer an issue. Participants were right to see PEVs as a work-in-progress, but the time may be approaching when many of their reservations about PEVs in 2011 have been addressed.

Graham-Rowe et al.'s study was one of the first to explicitly research the potential for PEV adoption by mainstream consumers; it also directly addressed the problems of psychological distance and the bed of Procrustes. In the next section we shall discuss another U.K. study with mainstream consumers that addressed some of the other methodological issues we have discussed.

Measuring Mainstream Consumers' Likelihood to Adopt a BEV After Direct Experience of Using One

Graham-Rowe et al.'s qualitative study formed an initial stage of a wider programme that also included the segmentation analysis that was discussed at the start of this chapter (Anable et al. 2011, 2014). The questionnaire stage of this study (Energy Technologies Institute 2013) found that 98% of participants (all seven of the consumer segments other than Plug-in Pioneers) would require a substantial discount in purchase price to interest them in buying a BEV. Three segments, Plug-in Pioneers (the earliest adopters), Zealous Optimists (potential early adopters of both BEVs and PHEVs), and Willing Pragmatists (willing to consider PHEVs but not BEVs), together making up slightly over a quarter of the sample, were willing to pay a premium for a PHEV. The remaining segments required a much smaller discount over the price of an ICEV to interest them in a PHEV than they did to interest them in a BEV. The conclusion was clear: mainstream consumers were much more likely to adopt PHEVs than BEVs.

When the findings of the survey were disseminated, a common response was that many more mainstream consumers would be interested in BEVs if only they had the chance to try one out (i.e. if only they were psychologically closer to BEVs). The study we shall now describe was carried out specifically to test that argument.

Skippon et al. (2016) tested the effect of direct experience of a BEV on mainstream consumers' willingness to consider a BEV in a randomised controlled trial with 393 mainstream consumer participants. The study addressed many of the methodological issues outlined at the beginning of this chapter. It used a stratified sample of mainstream consumers, controlled for Hawthorne effects, provided participants with direct experience of using a BEV to reduce psychological distance, and measured both the symbolic meaning of BEVs and the effect of self-congruity on willingness to consider a BEV. By the time the study was conducted (2012) a new generation of BEVs had reached the U.K. market, with much better performance than the generation Graham-Rowe et al. had used in their research (conducted in 2011). The study therefore took the opportunity to measure mainstream consumers' evaluations of the performance of a 2012-generation BEV.

The research was conducted by the U.K.'s Transport Research Laboratory (TRL). Participants were recruited from the area around TRL's base in Crowthorne, England, to the west of London's Heathrow airport. The sample was stratified in terms of gender, age, annual mileage, and location (urban or rural) to be as representative as possible of the U.K. driving population. Potential participants were excluded if they did not have access to off-street parking and an off-road electricity supply at home, so that vehicles could be recharged safely at home.

Participants were randomly allocated to one of two groups: an experimental group who were given direct experience of a modern BEV, and a control group who were given direct experience of an equivalent ICEV. Random allocation of a sufficiently large sample ensured that other differences between the groups were controlled for. All aspects of participation in the trial apart from the vehicle were common to both groups, so controlling for the Hawthorne effect and enabling causal attributions to be made for any observed between-groups differences.

The BEV used in the study was a 2012 model medium family hatchback BEV,[3] which had a manufacturer's claimed range of >100 miles. In tests on a chassis dynamometer this model was found to have a range of 83 miles in repeated NEDC cycles, and 44 miles at a constant speed of 75 mph.

Independent variables in the study were vehicle (BEV or ICE) and measurement time-point (before or after the vehicle experience). We shall focus here on two types of dependent variables: (1) willingness to consider a BEV as a main or second car in a

[3]The authors withheld details of the specific models used since the research included evaluations of performance that might have conferred commercial advantage or disadvantage to manufacturers.

household, if the BEV had a range on a full charge of 50, 100, 150, 200, or 250 miles; and (2) symbolic meanings of BEVs (measured using the attribution-vignette method outlined below) and self-congruity.

Participants were given the use of the vehicle they received for a period of 36 h, with freedom to use the vehicles as they wished. This was substantially less than the seven-day experience in the Graham-Rowe et al. study, and was largely the result of time and resource constraints. However it was substantially longer than the experiences of participants in the studies by Skippon and Garwood (2011) and Axsen et al. (2013), and was considered "long enough to enable participants to experience and understand the capabilities and limitations of BEVs, and to reflect on how BEVs might or might not fit with their lifestyles and vehicle usage patterns" (Skippon et al. 2016, p. 30).

Symbolic meaning of BEVs was measured by asking participants to characterise a typical user of a BEV by answering ten questions drawn from the IPIP-NEO personality inventory; pairs of these questions measured the five-factor traits of openness to experience, conscientiousness, extraversion, agreeableness, and neuroticism. In addition they were asked to characterise the typical user of a BEV in terms of other factors relevant to their reproductive fitness: their gender, age, status, relationship investment, and physical attractiveness. Participants also self-reported their own personalities using the same IPIP-NEO items; self-congruity was calculated for each participant by comparing these with the personality they attributed to a BEV user.

Willingness to Consider Having a BEV

It is first worth noting that participants evaluated the performance of the BEVs more positively than that of the equivalent ICEV. This shows the great technical strides that the industry made in only a few years since the Graham-Rowe et al. study.

The percentage of participants who would choose a BEV as their main car is shown in Fig. 2. The question was asked of both groups, before and after the usage experience, and was repeated for various ranges on a fully charged battery. The percentage was similar for both groups before the usage experience at all values of range, increasing with range. The percentage decreased for both groups after the usage experience, but the decrease was always larger for the experimental (BEV) group compared to the control (ICEV) group (the between-participants difference was statistically significant only for the 150 mile range). Figure 3 shows similar data for willingness to consider a BEV as a second car in the household. The general pattern was similar, but a greater percentage of participants were willing to consider a BEV as a second car than as a main car. Among participants who had directly experienced a BEV, 64% would be willing to consider one if the range was 150 miles.

The researchers concluded that (a) more than 50% of mainstream consumers who had experienced using a BEV would consider having one as a main car if its range was 200 miles; (b) direct experience of using a 2012 model BEV *reduced* mainstream consumers' willingness to consider having one; (c) there was a Hawthorne effect

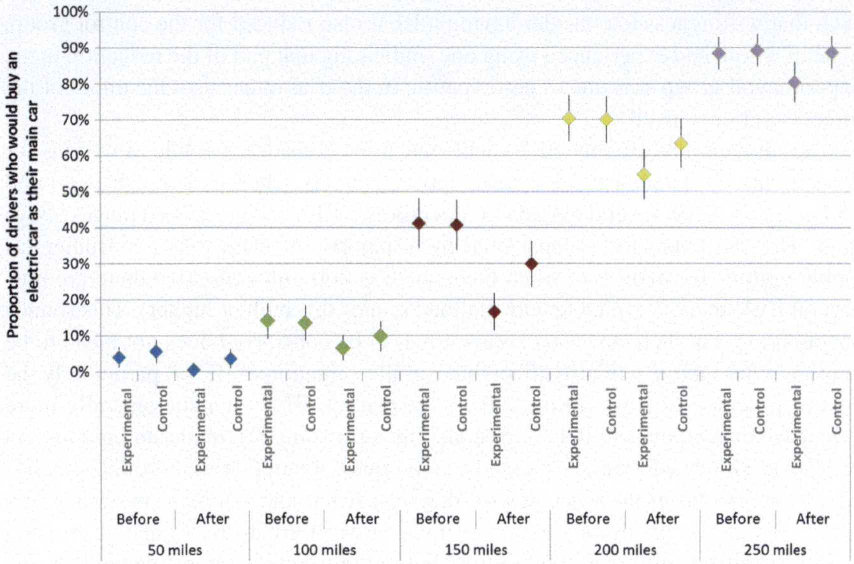

Fig. 2 Percentage of participants who would consider owning a BEV as a *main* household car for various possible ranges on a fully charged battery, before and after the usage experience, in each group

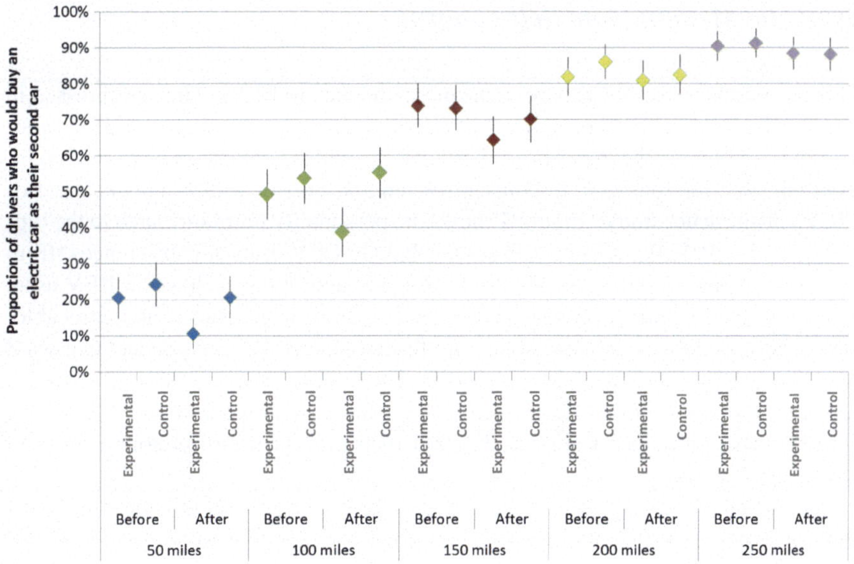

Fig. 3 Percentage of participants who would consider owning a BEV as a *second* household car for various possible ranges on a fully charged battery, before and after the usage experience, in each group

such that willingness to consider having a BEV also reduced for the control group, none of whom had experienced using one (indicating that part of the reduction in the experimental group was due to participation in the trial rather than the effect of the direct experience itself).

The implications are that BEVs will start to be considered viable as options for second cars by a majority of mainstream consumers when their range is around 150 miles (in 2019, several manufacturers produce BEVs with claimed ranges above this). They will start to be considered by a majority of mainstream consumers as viable options for main cars when their range is 200 miles (in 2019 there are now several BEV models available with claimed ranges this high or higher). This sounds encouraging, but there are two caveats. First, to be considered does not mean to be chosen; other factors will also affect the overall evaluation of BEVs, particularly the cost of purchase or lease. In the 2019 U.K. market, BEVs remain generally more expensive than equivalent ICEVs. Second, the term "range" remains ambiguous. As we saw in Graham-Rowe et al.'s study, some mainstream consumers conceptualise it in terms of the furthest destination they can reach and return from on a single charge. This conceptualisation typically involves extra-urban driving at high cruising speeds, when the range of a BEV is often shorter than manufacturers' claimed ranges. Participants may have been interpreting "range on a full charge" in this sense.

Symbolic Meaning and Self-congruity

This section discusses the symbolic meaning attached to BEVs. This is expressed in terms of the five-factor personality traits attributed to a typical user of a BEV (after having used one, by the experimental group). Results are expressed as z scores relative to the distributions of trait scores attributed to users of a wide range of classes of ICEVs in an earlier study.[4] Figure 4 shows the personality traits attributed to the typical user of a BEV. Participants did not attribute to BEV users any differences (from the norm means) for the traits of extraversion and neuroticism. However BEV users were attributed significantly elevated openness, conscientiousness, and agreeableness. These findings were consistent with those obtained by Skippon and Garwood (2011) using the same method with a different sample. Skippon et al. (2016) also reported findings for other characteristics associated with reproductive fitness. We note here that participants attributed higher than average socio-economic status to typical BEV users.

Given that the results indicated that BEVs have a clear symbolic meaning, does self-congruity with that meaning lead to greater motivation to consider a BEV?

---

[4]A z score expresses position of a value in relation to the mean and standard deviation of the distribution of scores in a "norm" group. A score of +1 in a particular trait indicates a position one standard deviation (of the norm data) higher than the mean of the norm data. In this case the norm group was a sample of around 3000 participants in an earlier study (Skippon 2014) in which participants attributed personality traits to a wide range of ICEV types (e.g. small hatchback, executive saloon, van, sports car).

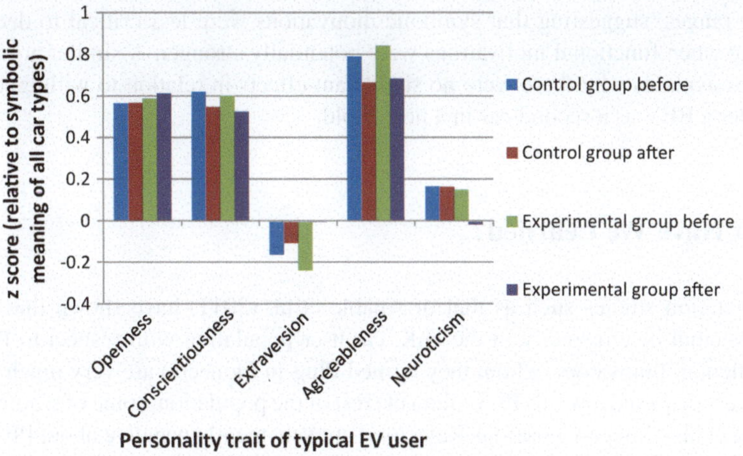

Fig. 4 Symbolic meaning of a BEV: personality traits attributed to a typical BEV user

Figure 5 compares the mean normalised self-congruity of participants in the experimental and control groups who responded "yes" and "no" to questions concerning willingness to consider a BEV as a main car if it had various ranges. Participants who were willing to consider a BEV with a range of 50, 100, or 150 miles had higher self-congruity than those who answered no, suggesting that self-congruity did indeed play a role in their willingness to consider. The effect disappeared for

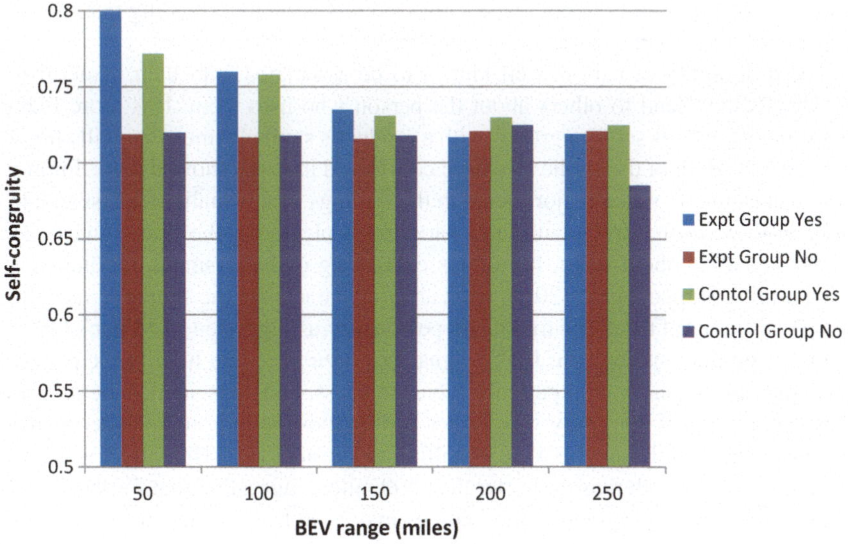

Fig. 5 Mean normalized self-congruity of participants in each group willing to consider (yes) and unwilling to consider (no) a BEV as a main car for different values of its range on a full charge

longer ranges, suggesting that symbolic motivations were less critical to decision making when functional motivations were potentially stronger. A similar argument perhaps explains why there were no significant effects in relation to willingness to consider a BEV as a second car in a household.

What Have We Learned?

Segmentation studies such as that of Anable et al. (2011) have shown that there is substantial heterogeneity in the U.K. car user population with respect to PEVs. In particular, Innovators (whom they termed Plug-in Pioneers) are very much more positively disposed towards PEVs than the rest of the population, some of whom (e.g. Anable et al.'s Image-Conscious Rejectors) are vehemently negative about PEVs. If we want to understand the potential for widespread PEV adoption accurately, we need to understand the attitudes of mainstream consumers.

As we have described earlier, this is where we hit a substantial problem. Understanding the potential for adoption of PEVs by mainstream consumers is more difficult than understanding the attitudes and behaviours of PEV Innovators because of the problem of psychological distance. Mainstream consumers have no direct experience of PEVs to draw on when evaluating them. Construal Level Theory predicts that they will therefore give more weight to higher-level construals, i.e. to the desirability of PEVs, than to the feasibility of having and using one. This suggests that PEV survey research with mainstream consumers will likely have a systematic bias, unless steps are taken to reduce psychological distance. A few studies, including the two that we have explored in depth, have addressed this issue by providing mainstream consumers with direct experience of using PEVs.

The desirability of cars is well known to be associated with their symbolism: the signals they send to others about the person who uses them. It is more likely that someone who is self-congruous with a product's symbolic meaning will choose that product. Both of the studies we have considered in depth showed this. "Impression management" was a major theme in the findings of the qualitative research by Graham-Rowe et al. (2011), with some participants pleased by the "green" messages a PEV signalled about them, but others expressing embarrassment at being seen using a PEV. Skippon et al. (2016) were able to characterise the symbolic meaning of BEVs in terms of the user's five-factor personality, as suggested by Miller's (2009) theory of product symbolism: BEVs signal that their users are high in the personality traits of openness to experience, agreeableness, and conscientiousness. They also found that self-congruity was higher among those willing to consider a BEV as a main car than those who were unwilling to consider one. That effect was only present for BEVs with ranges shorter than 200 miles, suggesting that for those with higher self-congruity, symbolic motivations may have been sufficient to overcome the practical limitations of having a shorter-range BEV.

The two studies we have explored in depth each addressed the methodological problems we highlighted at the start of the chapter. Graham-Rowe et al. addressed psychological distance and the problem of the bed of Procrustes through qualitative research that enabled participants to express their own ways of framing PEVs, finding themes that have not often been included in quantitative research based on researchers' frames of reference. Skippon et al. addressed psychological distance and the Hawthorne effect, finding that participating in a PEV study was in itself enough to change mainstream consumers' willingness to consider a PEV.

Where Next?

Overall, we hope we have shown that careful attention to research methodology is important in researching the potential for PEV adoption by mainstream consumers. This brings us to the question of where next for this kind of research? At the time of writing gaps still remain in understanding the role of PHEVs in the transition to a PEV parc. Specifically, studies are required with mainstream consumers which address the methodological challenges highlighted in this chapter and provide an appropriate sample of mainstream consumer participants with direct experience of both BEVs and PHEVs, so they are able to make adoption choices from a psychologically close perspective. Another specific topic that merits further research is the meaning of the term "range" for mainstream consumers using BEVs. Graham-Rowe et al. (2012) found evidence of range being interpreted in terms of restricted utility, expressed as destinations that were possible or not possible using a BEV. In addition, we hope soon to see more research on mainstream consumer adoption of PEVs in other markets.

A recent study (Beard et al. 2019) commissioned by the Energy Technologies Institute (2019) as part of its Consumers, Vehicles and Energy Integration (CVEI) Project has investigated mainstream consumers' attitudes to PEV adoption after reduction of psychological distance, using a design adapted from that of Skippon et al. (2016). The CVEI study used a within-participants design in which each participant experienced the use of equivalent BEV, PHEV, and ICE versions of the same model of car for four days each. Findings will be published in an academic journal in the near future. Broadly, the results indicated a greater willingness to consider PHEVs than BEVs as main cars in households; about a quarter of mainstream consumers said they would be likely to choose a BEV in the next five years whilst about half indicated they would choose a PHEV. Similar proportions of participants to those found by Skippon et al. (2016) were willing to consider having a BEV as a main or second car if the range was 200 miles and higher. The CVEI study also generated new data for willingness to consider having a PHEV as a main or second car, suggesting an improvement on present electric ranges of PHEVs (typically 25–30 miles) may be required for them to have mass-market appeal.

Even when more complete data on mainstream consumer adoption of PEVs is obtained, the challenge will not be over. In order to prepare for and sustain a U.K. parc dominated by PEVs, the impact of their charging needs on the electricity system must be understood. The impact of substantial extra demand from PEV charging, at a time when decarbonised electricity generation may lead to less consistent supply availability, may require 'managed' PEV charging that shifts charging times to periods when other demands are lower. Research is needed to explore the effectiveness and acceptability of 'hard' solutions which directly curtail energy transfer at peak times, and 'soft' solutions based on financial incentives to charge at off-peak times. Two recent trials in the U.K. that are in the process of reporting results at the time of writing have begun to address these questions: The Electric Nation (2019) project, using a sample of existing PEV users (who were Innovators in terms of the perspective of this chapter) has been investigating the technical feasibility of employing 'hard' solutions to manage local network demand, while the CVEI Project's Consumer Charging Trials (Energy Technologies Institute 2019), using a Mainstream Consumer sample who were provided with either a PHEV or a BEV, has been testing consumer responses to alternative 'soft' solutions in a randomised controlled trial.

As PEV usage and charging becomes more mainstream, the responses of mainstream consumers to other systems such as Vehicle-to-Grid and wireless charging must also be understood. In addition, the research described in this chapter has used participants who were able to park and recharge a PEV off-street at their own home. An estimated 40% of U.K. households do not have access to off-street parking at home and this may be a major barrier to their uptake of PEVs. Research is needed to identify managed charging solutions for such households that are sufficiently compelling to remove this barrier, including for cases where charging infrastructure may be shared across multiple households in a residential street. The technologies of the future are not set in stone, but one thing is clear; well-designed consumer research is a vital part of the journey.

References

Anable, J., Kinnear, N., Hutchins, R., Delmonte, E., & Skippon, S. (2011). *Plug-in vehicles infrastructure project. Consumers and vehicles: Consumer segmentation and demographic patterns* (WP 1.3.5 Project Report). Wokingham, U.K.: TRL.

Anable, J., Schuitema, G., & Stannard, J. (2014). *Consumer responses to electric vehicles literature review* (Report for energy Technologies Institute plug-in vehicles infrastructure project (June 2010). Transport Research Laboratory Published Project Report PPR728). Wokingham, U.K.: TRL.

Axsen, J., & Kurani, K. (2013). Hybrid, plug-in hybrid, or electric—What do car buyers want? *Energy Policy, 61,* 532–543.

Axsen, J., Orlebar, C., & Skippon, S. (2013). Social influence and consumer preference formation for pro-environmental technology: The case of a U.K. workplace electric-vehicle study. *Ecological Economics, 95,* 96–107.

Beard, G., Durrell, L., Kent, J., Skippon, S., Kinnear, N., Al-Katib, H., et al. (2019). *Deliverable D5.2—Consumer uptake trial report: Mainstream consumers' attitudes and willingness to adopt BEVs and PHEVs. Consumers, vehicles and energy integration project* (TRL Published Project Report PPR899). Crowthorne, U.K.: Transport Research Laboratory.

Belk, R. W. (1985). Materialism: Trait aspects of living in the material world. *Journal of Consumer Research, 12,* 265–280.

Bunce, L., Harris, M., & Burgess, M. (2014). Charge up then charge out? Drivers' perceptions and experiences of electric vehicles in the U.K. *Transportation Research Part A: Policy and Practice, 59,* 278–287.

Carley, S., Krause, R. M., Lane, B. W., & Graham, J. D. (2013). Intent to purchase a plug-in electric vehicle: A survey of early impressions in large U.S. cites. *Transportation Research Part D: Transport and Environment, 18,* 39–45.

Carroll, S., Walsh, C., Burgess, M., Harris, M., Mansbridge, S., King, N., & Bunce, L. (2013). *The ultra low carbon vehicle demonstrator programme—Final report: Assessing the viability of EVs in daily life*. Loghborough, U.K.: Cenex.

Carver, C. S., & Scheier, M. F. (1998). *On the self-regulation of behavior*. Cambridge, England: Cambridge University Press.

Choo, S., & Moktarian, P. (2002). *The relationship of vehicle type choice to personality, lifestyle, attitudinal and demographic variables* (Report number UCD-ITS-RR-02-06. Davis, CA). Davis: University of California.

Delang, C. O., & Cheng, W.-T. (2012). Consumers' attitudes towards electric cars: A case study of Hong Kong. *Transportation Research Part D: Transport and Environment, 17,* 492–494.

Department for Transport. (2019a). *VEH0101: Licensed vehicles by body type (quarterly): Great Britain and United Kingdom*. Retrieved from: https://www.gov.uk/government/statistical-data-sets/all-vehicles-veh01.

Department for Transport. (2019b). *VEH0130: Licensed ultra low emission vehicles by body type: United Kingdom*. Retrieved from: https://www.gov.uk/government/statistical-data-sets/all-vehicles-veh01.

Dimitropoulos, A., Rietveld, P., & van Ommeren, J. N. (2013). Consumer valuation of changes in driving range: A meta-analysis. *Transportation Research Part A: Policy and Practice, 55,* 27–45.

Dittmar, H. (1992). *The social psychology of material possessions: To have is to be*. Hemel Hempstead, England: Harvester Wheatsheaf.

Dittmar, H., & Brown, R. (2008). *Consumer culture, identity and well-being: The search for the 'Good Life' and the 'Body Perfect'*. Hove, England: Psychology Press.

Dolfsma, W. (2008). *Consuming symbolic goods: Identity and commitment, values and economics*. Abingdon, England: Routledge.

EAFO. (2018). Retrieved 27/6/18 from: http://www.eafo.eu/content/united-kingdom#country_pev_sales_vols_graph_anchor.

Electric Nation. (2019). *Electric nation: Your electric vehicle, your smart charge*. Retrieved from http://www.electricnation.org.U.K.

Energy Technologies Institute. (2013). *Transport: An affordable transition to sustainable and secure energy for light vehicles in the U.K.* (Loughborough, England: Energy Technologies Institute).

Energy Technologies Institute. (2019). *Consumers, vehicles, and energy integration (CVEI)*. Retrieved from https://www.eti.co.uk/programmes/transport-ldv/consumers-vehicles-and-energy-integration-cvei.

Eriksen, M. (1996). Using self-congruity and ideal congruity to predict purchase intention: A European perspective. *Journal of Euro-Marketing, 6*(1), 41–56.

Eyal, T., Liberman, N., & Trope, Y. (2009). Psychological distance and consumer behavior: A construal level theory perspective. In M. Wanke (Ed.), *Social psychology of consumer behavior*. New York, NY: Psychology Press.

Eyal, T., Liberman, N., Trope, Y., & Walther (2004). The pros and cons of temporally near and distant action. *Journal of Personality and Social Psychology, 86*(6), 781–785.

Focus Economics. (2019). *The World's top 10 largest economies*. Retrieved 17/7/19 from https://www.focus-economics.com/blog/the-largest-economies-in-the-world.

Fournier, S. (1991). A meaning-based framework for the study of consumer-object relations. *Advances in Consumer Research, 18,* 736–742.

Garling, T., & Steg, L. (2007). Threats from car traffic.

Geels, F. W. (2005). *Technological transitions and system innovations: A co-evolutionary and socio-technical analysis*. Cheltenham, U.K.: Edward Elgar.

Geels, F. W., Kemp, R., Dudley, G., & Lyons, G. (2012). *Automobility in transition? A socio-technical analysis of sustainable transport*. New York, NY: Routledge.
Goodwin, P. B. (1997). Mobility and car dependence. In T. Rothengatter & E. Carbonell Vaya (Eds.), *Traffic and transport psychology: Theory and application*. Oxford, England: Pergammon.
Graham-Rowe, E., Gardner, B., Abraham, C., Skippon, S., Dittmar, H., Hutchins, R., et al. (2012). Mainstream consumers driving plug-in battery-electric and plug-in hybrid electric cars: A qualitative analysis of responses and evaluations. *Transportation Research Part A: Policy and Practice, 46,* 140–153.
Graham-Rowe, E., Skippon, S., Gardner, B., & Abraham, C. (2011). Can we reduce car use and if so, how? A review of available evidence. *Transportation Research Part A: Policy and Practice, 45,* 401–418.
Griskevicius, V., Tybur, J. M., Sundie, J. M., Cialdini, R. B., Miller, G. F., & Kenrick, D. T. (2007). Blatant benevolence and conspicuous consumption: When romantic motives elicit costly signals. *Journal of Personality and Social Psychology, 93*(1), 85–102.
Grubb, E., & Stern, B. (1971). Self-concept and significant others. *Journal of Marketing Research, 8*(3), 382–385.
Hackbarth, A., & Madlener, R. (2013). Consumer preferences for alternative fuel vehicles: A discrete choice analysis. *Transportation Research Part D: Transport and Environment, 25,* 5–17.
Heath, A., & Scott, D. (1998). The self-concept and image congruence hypothesis: An empirical evaluation in the motor vehicle market. *European Journal of Marketing, 32*(11/12), 1110–1123.
Heffner, R., Kurani, K. S., & Turrentine, T. S. (2007). Symbolism in California's early market for hybrid electric vehicles. *Transportation Research D: Transport & Environment, 12,* 396–413.
Hoeffler, S. (2003). Measuring preferences for really new products. *Journal of Marketing Research, XL,* 406–420.
Jeekel, H. (2003). *The car-dependent society: A European perspective*. Farnham, England: Ashgate.
Jensen, A. F., Cherchi, E., & Mabit, S. L. (2013). On the stability of preferences and attitudes before and after experiencing an electric vehicle. *Transportation Research Part D: Transport and Environment, 25,* 24–32.
Keller, K. (1993). Conceptualizing, measuring, and managing customer-based brand equity. *Journal of Marketing, 57*(1), 1–22.
Landsberger, H. A. (1958). *Hawthorne revisited: Management and the worker, its critics, and development of human relations*. Ithaca, NY: Cornell University Press.
Lebeau, K., van Mierlo, J., Lebeau, P., Mairesse, O., & Macharis, C. (2012). The market potential for plug-in hybrid and battery electric vehicles in Flanders: A choice-based conjoint analysis. *Transportation Research Part D: Transport and Environment, 17,* 592–597.
Liberman, N., & Trope, Y. (1998). The role of feasibility and desirability considerations in near and distant future decisions: A test of temporal construal theory. *Journal of Personality and Social Psychology, 75,* 5–18.
Liberman, M., Trope, Y., & Stephan, E. (2007). Psychological Distance. In A. W. Kruglanski & E. T. Higgins (Eds.), *Social psychology: Handbook of basic principles*. New York, NY: Guilford Press.
Lieven, T., Muhlmeier, S., Henkel, S., & Waller, J. F. (2011). Who will buy electric cars? An empirical study in Germany. *Transportation Research Part D: Transport and Environment, 16,* 236–243.
Ligas, M. (2000). People, products and pursuits: Exploring the relationship between consumer goals and product meanings. *Psychology and Marketing, 17*(11), 983–1003.
Mays, N., & Pope, C. (1995). Qualitative research: Rigour and qualitative research. *British Medical Journal, 311,* 109.
McCracken, G. (1988). *Culture and consumption*. Bloomington, IN: Indiana University Press.
Mick, D. G. (1986). Consumer research and semiotics: Exploring the morphology of signs, symbols, and significance. *Journal of Consumer Research, 13,* 196–213.
Miller, D. (2010). *Stuff*. Cambridge, England: Polity Press.
Miller, G. (2009) *Spent. Sex, evolution and consumer behaviour*. New York, NY: Viking.

Nelissen, R. M. A., & Meijers, M. H. C. (2011). Social benefits of luxury brands as costly signals of wealth and status. *Evolution and Human Behavior, 32,* 343–355.

Next Greencar. (2019). *Electric car market statistics*. Retrieved from https://www.nextgreencar.com/electric-cars/statistics/.

Park, C. W., Jaworski, B., & MacInnis, D. (1986). Strategic brand concept-image management. *Journal of Marketing, 50,* 135–145.

Puttaswamaiah, K. (2002). *Econometric models: Techniques and applications*. New Delhi, India: Indus Publishing Company.

Rogers, E. (2003). *Diffusion of innovations* (5th ed.). New York: Free Press.

Salmon, P. (2003). A psychology for teachers. In F. Fransella (Ed.), *International handbook of personal construct psychology*. New York, NY: Wiley.

Schuitema, G., Anable, J., Skippon, S., & Kinnear, N. (2013). The role of instrumental, hedonic and symbolic attributes in the intention to adopt electric vehicles. *Transportation Research Part A: Policy and Practice, 48,* 39–49.

Shin, J., Hong, J., Jeong, G., & Lee, J. (2012). Impact of electric vehicles on existing car usage: A mixed multiple discrete-continuous extreme value model approach. *Transportation Research Part D: Transport and Environment, 17,* 138–144.

Sirgy, M. J. (1982a). Self-concept in consumer behaviour: A critical review. *Journal of Consumer Research, 9*(3), 287–301.

Sirgy, M. J. (1982b). Self-Image/Product-Image Congruity and Advertising Strategy. In V. Kothari (Ed.), *Developments in Marketing Science*, Vol. 5. (pp. 129–133). Marquette, MI: Academy of Marketing Science.

Sirgy, M. J. (1985). Using self-congruity and ideal congruity to predict purchase motivation. *Journal of Business Research, 13*(3), 195–206.

Skippon, S. M. (2014). *The psychology of vehicle performance* (Ph.D. Thesis). Milton Keynes, U.K.: The Open University.

Skippon, S., & Garwood, M. (2011). Responses to battery electric vehicles: U.K. consumer attitudes and attributions of symbolic meaning following direct experience to reduce psychological distance. *Transportation Research Part D: Transport and Environment, 16,* 525–531.

Skippon, S. M., Kinnear, N., Lloyd, L., & Stannard, J. (2016). How experience of use influences mass-market drivers' willingness to consider a battery electric vehicle: A randomised controlled trial. *Transportation Research Part A: Policy & Practice, 92,* 26–42.

Steg, L. (2005). Car use: Lust and must. Instrumental, symbolic and affective motives for using a motor car. *Transportation Research Part A: Policy and Practice, 39,* 147–162.

Strauss, A., & Corbin, J. (1998). *Basics of qualitative research: Techniques and procedures for developing grounded theory*. Thousand Oaks, CA: Sage.

Sundie, J. M., Kenrick, D. T., Griskevicius, V., Tybur, J. M., Vohs, K. D., & Beal, D. J. (2010). Peacocks, Porsches, and Thorstein Veblen: Conspicuous consumption as a sexual signalling system. *Journal of Personality and Social Psychology, 100*(4), 664–680.

Trope, Y., & Liberman, M. (2003). Temporal construal. *Psychological Review, 110,* 403–421.

Tucker, W. T. (1957). *Foundations for a theory of consumer behavior*. New York, NY: Rinehart & Winston.

Turrentine, T. S., & Kurani, K. S. (2007). Car buyers and fuel economy? *Energy Policy, 35,* 1213–1223.

Veblen, T. (1899). *The theory of the Leisure Class*. New York, NY: Penguin.

Ziegler, A. (2012). Individual characteristics and stated preferences for alternative energy sources and propulsion technologies in vehicles: A discrete choice analysis for Germany. *Transportation Research Part A: Policy and Practice, 46,* 1372–1385.

Electric Vehicle Adoption in Germany: Current Knowledge and Future Research

Patrick Plötz and Elisabeth Dütschke

Abstract This chapter provides an overview of the current knowledge of plug-in electric vehicles (PEVs) adoption as well as of their adopters and the use they make of the vehicles in Germany. This includes a separate discussion of drivers and barriers in the different segments of the car market (private, fleet and company cars) as well as of the role of charging infrastructure. The chapter summarizes previous studies by the authors on PEV diffusion and usage in Germany and compares the results obtained with those of other studies as well. The chapter concludes with a discussion of key areas for future research on PEV adopters in Germany.

Keywords Electric vehicles · Early adopter · Incentives

Introduction

The German Car Market and Automotive Industry

Germany is the biggest national car market in Europe and home to global automotive players like BMW, Daimler, Opel, and Volkswagen (incl. Audi and Porsche), with a worldwide annual passenger car production of more than twelve million cars.

Passenger car sales in Germany are about 3–3.5 million vehicles annually (cf. Fig. 1). The sales can be divided into about one million vehicles used as private, fleet and company cars each. The most important segments are medium and compact size passenger cars, but SUV market shares have grown substantially over the last years, similar to the development in other countries. The total stock of cars in Germany is about 45 million vehicles with an average lifetime of about 14 years. Sales of light-duty, medium-duty and heavy-duty commercial vehicles are much lower than passenger cars.

P. Plötz (✉) · E. Dütschke
Fraunhofer Institute for Systems and Innovation Research ISI, Breslauer Strasse 48, 76139 Karlsruhe, Germany
e-mail: patrick.ploetz@isi.fraunhofer.de

M. Contestabile et al. (eds.), *Who's Driving Electric Cars*, Lecture Notes in Mobility,
https://doi.org/10.1007/978-3-030-38382-4_10

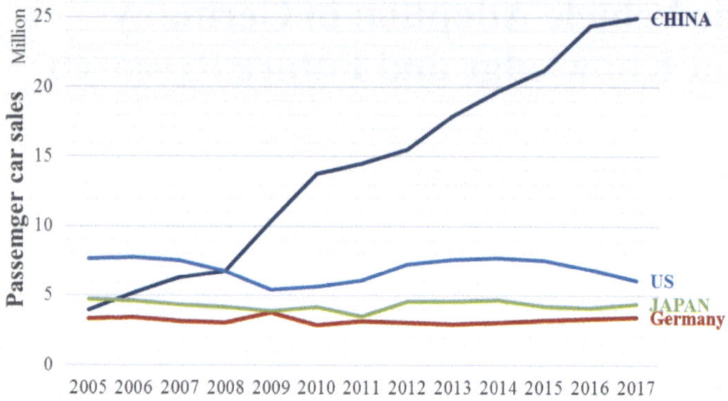

Fig. 1 Annual sales of passenger cars in Germany, China, Japan and the US. *Source* Own illustration based on OICA (2018)

The automotive industry is an important sector of the German economy with more than 800,000 employees, which accounts for one third of employees in this sector in the European Union (VDA 2018). The industry has received strong support from local and federal governments and there are strong personal links between the car industry and policymakers (Greenpeace 2016).

Chapter Overview

The chapter starts with a picture of the current level of market diffusion of plug-in electric vehicles (PEVs) in Germany and of the best available knowledge on the current adopters (section "Current state of PEV adoption"). This is followed by an overview of the present PEV usage in terms of annual mileage and electric driving (section "PEV usage"). Furthermore, a section is dedicated to discussing drivers and barriers for PEV usage, particularly with reference to charging infrastructure and support policies. The analysis discussed in this chapter strongly draws on the authors' own research on PEVs in Germany but also refers to publications by others. We close with a brief critical discussion of priorities for future research on PEV adopters in Germany.

The chapter focuses on battery electric vehicles (BEVs) and plug-in hybrid electric vehicles (PHEVs), both of which can be generically referred to as plug-in electric vehicles (PEVs). Other electric vehicles such as fuel cell electric vehicles are not discussed as they have only been adopted in very small numbers to date and face different challenges such as the implementation of a hydrogen infrastructure. We focus on passenger cars but discuss light and heavy-duty commercial vehicles in the concluding section on future research.

The present contribution is original in various ways. Firstly, it provides a review of existing studies on PEV diffusion and usage in Germany which has not been available so far. Secondly, it covers PEV usage of all three user groups: private, company, and fleet vehicles. Lastly, it discusses barriers and drivers with respect to PEV market diffusion with special focus on Germany.

The Current State of PEV Adoption

PEV Market Overview

Germany, France, the UK and Italy are the biggest national car markets in Europe, with Germany leading in total passenger car sales. However, for PEVs the situation differs. In terms of PEV sales share, the Nordic countries are leaders in Europe—and also worldwide. In 2018 the average PEV sales share in Europe was about 2% and Germany was close to that average (cf. Fig. 2). Please note that Norway has been left out to not distort the scale, i.e. if Norway were included the development in other markets would hardly be visible.

Compared to other major car markets, the diffusion of PEVs in Germany started slowly. Market shares stayed well below 0.5% until 2014 and were strongly shaped by federal support for field trials. A notable growth in sales shares has been perceivable only since 2015. In 2016 sales underwent a phase of stagnation, but then financial subsidies of about €4000 per BEV became available and more PEV models by German manufacturers were introduced. Since then, sales shares accelerated and have risen above 2%.

In comparison to other large automotive markets such as China, the US and Japan, the PEV sales share in Germany is not exceptional (cf. Fig. 2). The countries differ, however, in the distribution between BEVs and PHEVs. For several years, German manufacturers only offered a limited set of models and these mainly included BEVs. Accordingly, BEVs dominated the PEV sales in Germany until 2014. Nevertheless,

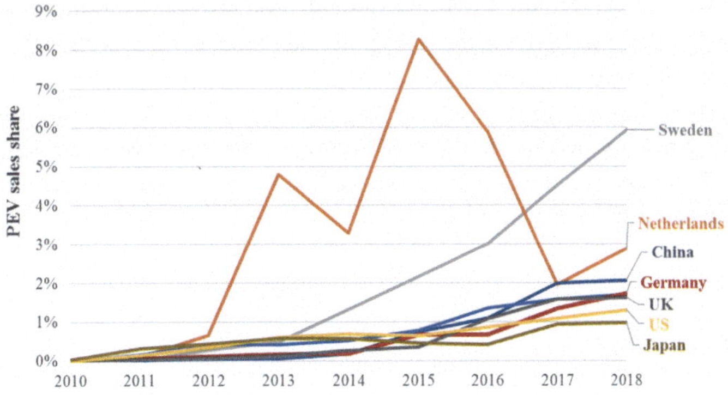

Fig. 2 PEV sales shares in selected PEV markets excluding Norway. *Source* Own illustration based on IEA (2018)

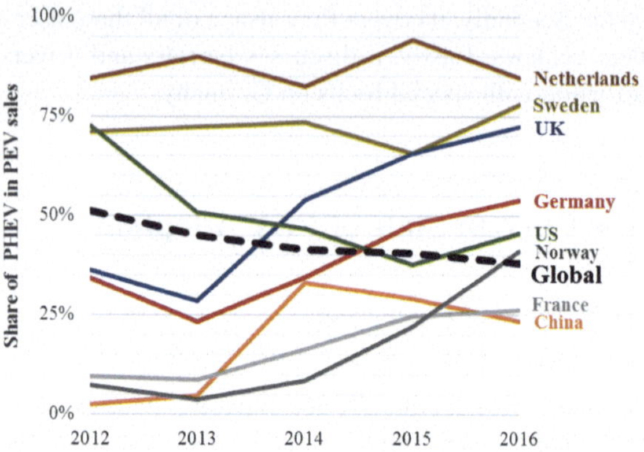

Fig. 3 Share of PHEV sales in several major PEV markets. *Source* Own illustration based on IEA (2018)

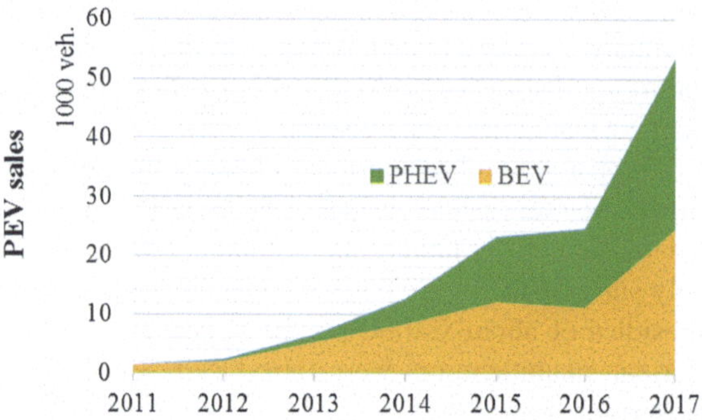

Fig. 4 Annual PEV sales in Germany broken down by BEVs and PHEVs. *Source* Own illustration based on IEA (2018) and EAFO (2018)

with the introduction of PHEV models by major automakers in 2015, we now observe about an equal share of PHEV and BEV sales in German PEV sales (cf. Figs. 3 and 4).

The most popular PEV models in Germany are made by German manufacturers. For BEVs these are the Daimler Smart, VW e-Golf and BMW i3. However, global bestsellers such as the Nissan Leaf and Renault Zoe also play an important role in German BEV sales. For PHEVs, the VW Golf and Passat GTE as well various PHEV models from Mercedes are important (cf. Fig. 5).

This leads to the question of who is buying PEVs. In 2016, about one third of PEVs was owned by private individuals whereas two thirds were owned by legal entities, incl. fleet usage and company cars (DDI 2017). The two user groups will be discussed in the following sections in turn.

Private PEV Users

The early adopters of PEVs have received a considerable amount of attention in the literature (cf. Frenzel et al. 2015, 2016, Plötz et al. 2014; Peters and Dütschke

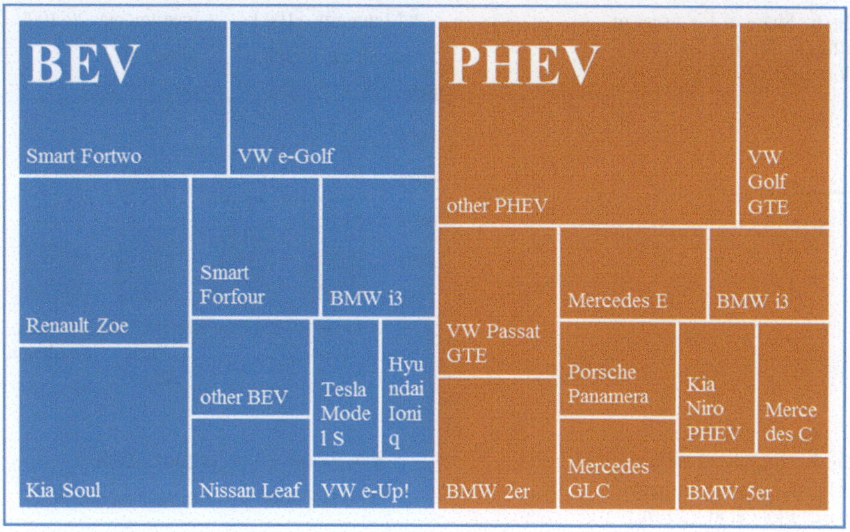

Fig. 5 BEV and PHEV sales in Germany in 2017 by model. Box size corresponds to market share. *Source* Own illustration based on EAFO (2018)

2014; Jarass et al. 2014; Egbue and Long 2012; Hidrue et al. 2011; Ozaki and Sevastyanova 2011; Rezvani et al. 2015 for a review). For Germany, the private early adopters of PEVs can be broadly characterized as middle-aged men, with technical and environmental interests and higher socio-economic status, living in rural or suburban areas (Peters and Dütschke 2014; Plötz et al. 2014).

A very early study for Germany from 2011 surveyed German car users (see Peters et al. 2011; Agosti 2010; Peters and Dütschke 2014). The participants of the survey of car owners were recruited by various means (internet platforms, online forums, blogs, and newspapers) in order to obtain a broad sample including persons who actually drive a PEV, those with a high interest in purchasing a PEV as well as other consumers with less knowledge of and/or interest in PEVs. The sample of this survey consists of 81.4% men, the mean age is 40.9 years, the mode of monthly disposable household income is €2001–3000 (categorical scale), the average household size is 2.5 persons and on average there are 1.4 cars per household. 51% of the respondents hold a university degree (masters or higher) and 42% work in technical professions.

A second source of data on more recent PEV users is Frenzel et al. (2015). We will refer to this source as “today’s user” below as it is the largest and most representative study for Germany so far. Frenzel et al. (2015) approached all German PEV users via mail and a paper-based questionnaire. Their sample includes users of private and commercial PEVs as well as PEVs smaller than passenger cars, e.g. the Renault Twizy quadricycle. A detailed description of the sample can be found in Frenzel et al. (2015).

Figure 6 shows the distributions of PEV adopters by size of town or city they live in for today’s PEV users, for a sample of PEV users from 2011 and for potential early

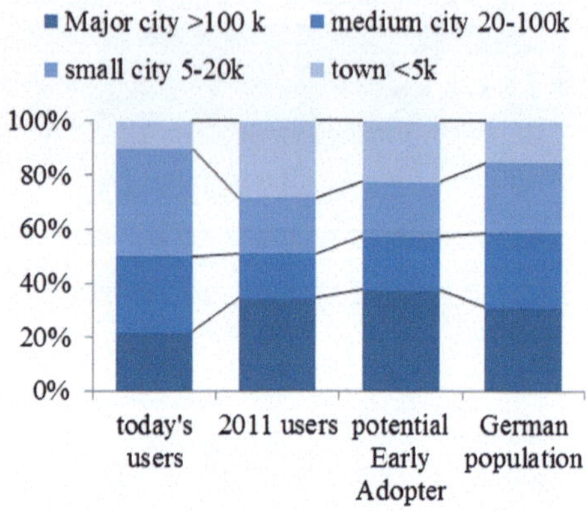

Fig. 6 Share of EV users in different city sizes. *Source* Plötz et al. (2017c)

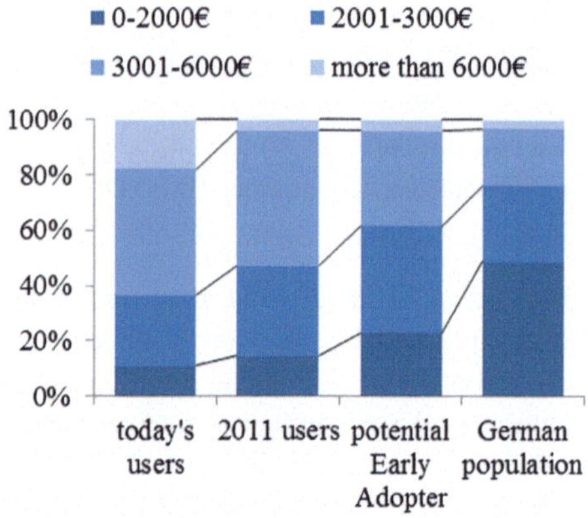

Fig. 7 Household income of early adopters in Germany. *Source* Plötz et al. (2017c)

adopters in 2011 (cf. Plötz et al. 2014), and compares them with the same distribution for the whole German population (including non-car owners) as a reference. Figure 7 shows a similar analysis for the distributions of PEV adopters by household income (from Frenzel et al. 2015)

Concerning the 2011 users and potential early adopters, the share from major cities and towns is noteworthy. Note, however, that the sample size of the '2011 users' and 'potential early adopter' was very limited so that individual shares have to be interpreted with caution. The share of 2011 users from major cities (more than 100,000 inhabitants) varies between the different sources and user groups and could be higher or lower than the total share in the German population. Yet, it is clear that today's users are more likely to live in medium sized cities of 20–100,000 inhabitants than would be expected from the overall German population. According to the data, we expect the next generation of adopters to be closer to the distribution of German population than the current generation of adopters. Note, however, that a further differentiation in cities with 100,000–500,000 and more than 500,000 inhabitants could slightly change the picture.

With respect to net household income, the different estimates agree that first users of PEV have an above average household net income with large shares from high- and very high-income groups (3001–6000 € and >6000 €). This finding is consistent with Rogers' general theory on the diffusion of innovation indicating that innovators and early adopters have a high income and a high social status. The available data for Germany also suggests that the average income of the PEV adopters has increased in recent years.

A closer look at the differences in socio-economic criteria between (1) actual users and (2) likely early adopters, (3) likely later adopters and (4) likely non-adopters can be taken with a survey in one of the German showcase regions for PEVs. The German federal government offered strong (monetary) support for 90 projects in four so-called showcase regions from 2012 to 2016 to support the development and implementation of electric mobility. Projects included supporting the installation of charging infrastructure, procurement of PEVs by companies and municipalities, and establishing sharing of electric vehicles. Our study to which we refer here was conducted in the showcase region around Stuttgart, home to Daimler and Porsche. Details on this study which includes 1510 respondents from a sample that is representative of the population in this area can be found in Dütschke et al. (2017). As shown in Table 1, the adoption groups also differ in some socio-demographic criteria (tested with ANOVA and Chi2 tests, results categorized as significant if $p < .05$). Group 4 lives in significantly smaller households than groups 2 and 3 and car-ownership is significantly more prevalent in group 1 compared to group 3 and 4, i.e. the PEV seems to be an additional vehicle. Additionally, men and respondents with a university degree are overrepresented in groups 1–3 and underrepresented in group 4. Due to the very small number of group members, all results involving

Table 1 Adoption groups for EVs (Dütschke et al. 2017)

Category	(1) Users	(2) Intending to buy	(3) Interested in PEVs	(4) Not-interested in PEVs
		Early adopters	Later adopters	Non-adopters
Sample size N	9	64	858	579
Share of sample (%)	0.6	4.2	55.9	37.7
Average age in years	36.3	43.0	43.1	44.3
Average household size	3.0	2.7	2.5	2.3
Average cars per household	2.2	1.6	1.4	1.3
Share female (%)	33.3	32.8	49.1	57.3
Share with university degree (%)	11.1	39.1	32.8	23.7

group 1 have to be treated with caution. Nevertheless, these findings also underline the trend that PEVs are likely to be adopted by groups of higher socio-economic status and they point to male dominance in PEV adoption—also in the future—in Germany.

Fleet PEV Users

In addition to private users, this section takes a look at PEVs as part of company fleets, i.e. at vehicles that are owned by corporations and used by a number of employees for work-related travel. Typically, these cars are parked on the premises of the company and are usually not used for private travel. To describe fleet PEV users in Germany in more detail we draw on findings from two projects which gathered data on early adopter organizations in Germany. In the first project, survey data on BEV early adopter organizations were collected between 2011 and 2014 as part of a project supporting the implementation of field trials funded by the German Ministry for Transport (BMVI 2015). This database consists of responses from 380 fleet users of PEVs. 78 of these drivers were also involved in the decision to include a PEV in the fleet while the majority of 302 were not involved in this decision. In the survey, one organization can be represented by several respondents.[1] In the second project, we look at data from a large field trial not included in the above study that covers 109 organizations (Ensslen et al. 2018) and was part of the Stuttgart showcase region mentioned above.

From the first study we find that the typical user of a PEV in an organization is younger (39 years on average) than the early adopters from households (cf. the discussion of private drivers above) while decision-makers in organizations are of the same age as drivers from households (50 years old on average). German drivers of fleet PEVs are also mainly men: 87% of the decision-makers and 79% of the drivers are men. The second study only provides data on the fleet managers responsible but not on the drivers, however they are also predominantly men (85%).

To get an idea of which employees are more likely to drive a PEV in this early stage of market development, we compare the share of industries represented in the sample of study 1 to the distribution of workers across sectors in Germany. This identifies over-representations from the energy sector (18% compared to 8%), the free-lance service sector (23% compared to 6%) as well as the public sector (23% compared to 6%). Regarding organizational size, the study shows that especially organizations with more than 500 employees are more likely to have already integrated PEVs in their fleets compared to smaller organizations (around 50% compared to 33% in Germany). However, the second study paints a different picture, as in this case 38% of the participating organizations belong to the manufacturing sector, 14% to the wholesale and retail sector, 13% to public administration, 8% to information and

[1]For reasons of data privacy, it was not possible to associate respondents with organizations in this survey. Thus, the number of organizations that are represented is not known.

communication and 6% to the construction sector, which is in line with the shares of firms in these sectors. Furthermore, in this study, 80% of participants came from organizations with less than 250 employees.

Taking these findings together leads to the conclusion that the early adopter market for organizations in Germany is strongly dominated by contextual factors such as the availability of suitable support schemes. Beyond this, the socio-demographic structure seems to be similar to the one of private drivers with one exception: company use of PEV involves a larger share of young men than private PEV use. This leads to the conclusion that fleet usage of PEVs could partially contribute to bringing more people into contact with PEVs, however, these groups are very similar to other early adopters.

Company Car PEV Users

The third group of car buyers in Germany are company car users i.e. employees who have one car to use for private and business purposes that is owned by their company. This group accounts for about one third of annual passenger car sales in Germany, making it an important group, however up till now very little has been known about this group. In official statistics, these cars are categorized as company-owned, the same as fleet vehicles, thus, there are not even precise numbers of their shares available. Taxation for company cars in Germany usually follows the rule that the user has to pay income tax on an additional monthly income of 1% of the official purchase price of the vehicle plus 0.03% per kilometer distance between home and workplace. This regulation is unfavorable for PEVs which have higher initial prices. However, a new regulation has entered into force in 2019 and which cuts the taxes for alternatively fuelled vehicles by 50%. Since many company car users typically travel many kilometers per year, one could expect that BEVs with their limited range are not interesting for them (except perhaps, for some Tesla Model S drivers). Accordingly, PHEVs could be an option for company car drivers as has been the case in the Netherlands with high PHEV company car subsidies until 2015.

In summary, company car drivers are an important car buyer group in Germany, but further research is required as to their role in PEV purchase and usage.

PEV Usage

The last section introduced the characteristics of several important PEV user groups in Germany but did not discuss their PEV usage. In this section, we will review existing PEV usage data from Germany.

Annual Vehicle Kilometers Travelled

The average German passenger car shows about 14,000 km annual vehicle kilometers travelled (VKT). In comparison, the findings for BEVs and PHEVs regarding the average annual VKT in Germany are mixed (Table 2). Plötz et al. (2017a) report higher annual VKT for all BEVs and PHEVs yet with limited sample sizes, whereas Frenzel et al. (2015) report below-average annual VKT for BEVs. The following table shows average values for various BEVs from both studies. Unfortunately, Frenzel et al. (2015) do not differentiate between car models.

Table 2 Average annual VKT of different PEVs in Germany

Model	Sample size	Mean VKT in km	Source
BEV			
BMW i3	20	18,923	Plötz et al. (2017a)
Tesla Model S	37	34,581	Plötz et al. (2017a)
Nissan Leaf	31	16,511	Plötz et al. (2017a)
Renault Zoe	63	16,698	Plötz et al. (2017a)
VW eGolf	14	15,582	Plötz et al. (2017a)
BEV private car	1020	11,700	Frenzel et al. (2015)
BEV fleet car	820	11,410	Frenzel et al. (2015)
BEV mean		12,334	
PHEV			
Opel Ampera	25	15,380	Plötz et al. (2017a)
Mitsubishi Outlander	46	21,940	Plötz et al. (2017a)
Toyota Prius	89	20,860	Plötz et al. (2017a)
Volvo V60	15	23,130	Plötz et al. (2017a)
PHEV private car	253	13,950	Frenzel et al. (2015)
PHEV fleet car	170	16,930	Frenzel et al. (2015)
PHEV mean		16,730	

It could be hypothesized that PEV users drive more than the average since we are in an early market phase and the present-day early adopters have a higher than average income which correlates with annual mileage. While this explanation is in line with the findings from Plötz et al. (2017a), it is not supported by Frenzel et al. (2015). Partially this could be explained by the different sources and samples. Plötz et al. (2017a) rely on users reporting their fuel economy which might bias the sample towards high VKT users for whom fuel economy is more important. On the other hand, the sample of Frenzel et al. (2015) contains many small and compact cars which typically show lower annual mileage.

Electric Driving of PHEVs

The share of kilometers driven electrically—the utility factor (UF)—is characteristic of the usage of PHEVs. A high UF implies driving mainly powered by electricity whereas a low UF implies driving mainly with conventional fuel. The UF is influenced by several factors, including the all-electric range (AER) of the PHEV, the frequency of long-distance trips, recharging frequency but also the use of auxiliaries (for heating or cooling) and the individual driving behavior. Tables 3 and 4 show summary statistics of UFs from PHEVs used in Germany.

Table 4 demonstrates the wide range of actual UF in real world driving. For example, the range of observed UF for 89 Toyota Prius PHEVs in the sample of Plötz et al. (2017a) varies from 2 to 80% at an all-electric range of 25 km according to the New European Driving Cycle (NEDC). Furthermore, the average UF increases with the average AER: the Prius has a 25 km AER, the Outlander and Volvo have about 50 km and the Opel Ampera 82 km. Similar ranges and averages are shown in the following table for a different sample from Germany reported in Plötz et al. (2017b).

The average UF (with 66% confidence interval, i.e. plus/minus one standard error—symbols in Fig. 8) are shown against the NEDC AER in Fig. 8 together with a non-parametric regression (solid line) and the UF according to NEDC (dashed line). There is a clear increase in average UF with growing AER as expected from a larger AER. However, most mean UF are clearly below the NEDC values given by $UF_{NEDC} = 25/(25 + AER)$. Interestingly, the deviation between average UF and the NEDC

Table 3 Summary statistics of PHEV UF in Germany acc. to Plötz et al. (2017a)

Vehicle	N	Min (%)	Mean (%)	SD (%)	Max (%)
Opel Ampera	25	27	72	21	100
Mitsubishi Outlander PHEV	46	0	47	20	94
Toyota Prius PHEV	89	2	30	18	80
Volvo V60 PHEV	15	29	49	14	73

Table 4 Summary statistics of PHEV UF in Germany acc. to Plötz et al. (2017b)

PHEV Model	Sample size	NEDC AER (km)	UF mean (%)	UF SE (%)	UF NEDC (%)
Toyota Prius	71	23	35	2	48
Mercedes C350e	2	31	47	6	55
Porsche Cayenne	2	36	28	17	59
Volvo XC90	4	40	43	10	62
Audi A3 e-tron	27	50	41	4	67
Golf GTE	40	50	40	3	67
Volvo V60	19	50	48	4	67
VW Passat	10	50	30	7	67
Mitsubishi Outlander	84	52	50	2	68
Opel Ampera	46	83	73	4	77
BMW i3 REX	8	170	83	6	87

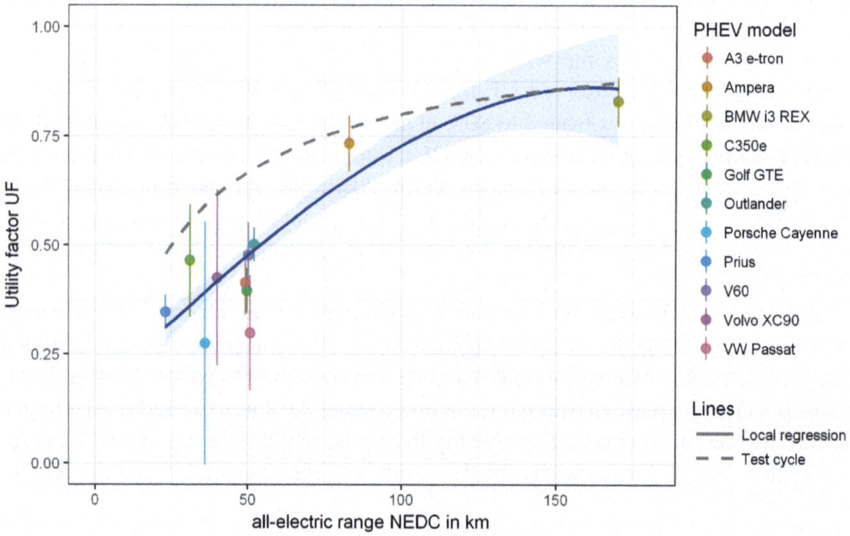

Fig. 8 Average UF (symbols), local regression (solid line) and NEDC UF values as a function of AER in km according to NEDC. *Source* Plötz et al. (2017b)

estimate decreases with higher AER. This should be verified with other data on short and long-ranged PHEVs.

Apart from the UF as such, the total annual electric kilometers are interesting as they indicate the GHG mitigation potential of PHEVs (cf. Plötz et al. 2017b). Figure 9 shows the annual electric VKT of different PHEVs and BEVs with data from Germany and the US (Plötz et al. 2017b). Note that AER values differ from Fig. 10 because, to be able to compare US and German data, we have used AER values based on the US Environmental Protection Agency (EPA) test cycle for the

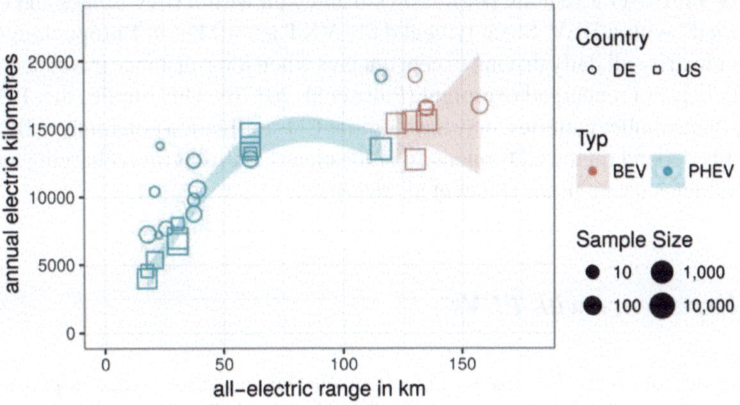

Fig. 9 Annual electric vehicle kilometers of different PHEVs and BEVs in Germany (DE—circles) and the US (squares) as well as local regression incl. 95% confidence bands (shaded area) for PHEV(green) and BEV (red). *Source* Plötz et al. (2017b)

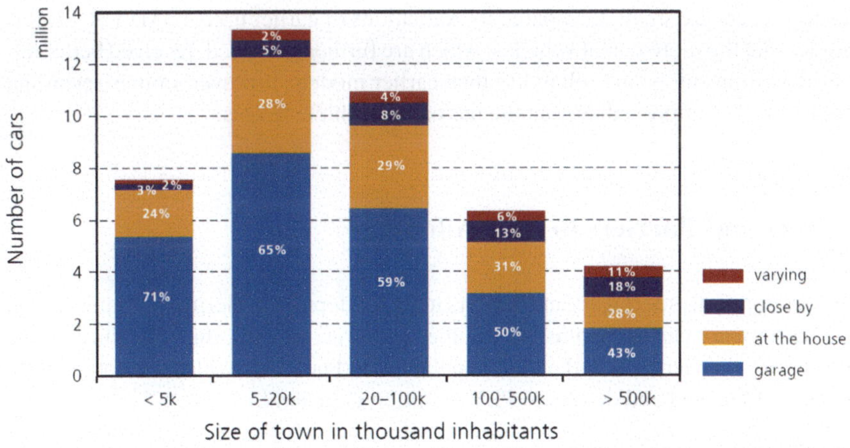

Fig. 10 Distribution of passenger cars in Germany by city size and type of parking used overnight. *Source* Plötz et al. (2013)

US and applied a correction factor of 0.75 to the NEDC AER values for the German data.

Figure 9 shows that, as expected, there is a clear increase in electric VKT with growing AER, similar to the trend observed for UF. However, the increase in electric annual VKT levels off at about 60 km real-world AER. With this range, the best estimate for electric annual VKT is about 13,000–15,000 km. Therefore, BEVs with longer ranges do not achieve higher electric annual VKT than long-ranged PHEVs. The reason is that PHEVs use electric power for some km of every daily VKT whereas BEVs use electric power for all km of some days. Typically, about 80% of the daily VKT over a year are below 100 km and well within BEV ranges and can be "electrified" with a BEV. Since typical daily VKT are 40 km in Europe, long-AER PHEVs electrify all daily driving except on days when long-distance travel occurs, or the vehicle is not recharged overnight (Plötz et al. 2017b). This implies that PHEVs, despite their smaller batteries, may have similar CO_2 mitigation potentials to BEVs—the details depend on the CO_2 intensity of the electricity used for recharging and the actual vehicles usage times (Plötz et al. 2017b).

User Experience with PEVs

Drawing on data from 500 users of electric vehicles including usage in private and business context (BMVI 2015), it is consistently found that German early adopters rate PEVs overall positively: this includes usability, safety and reliability as well as pleasure of driving, maximum speed and perceived environmental impact of the vehicles. However, specific aspects like charging duration, capacity for luggage transport etc. and overall range only receive medium ratings. Differences between private and business users are small. Compared to evaluations of earlier users (BMVI 2012) this implies that the more recent vehicles, which are further advanced, received better ratings regarding safety and reliability than earlier models, however, ratings regarding range have not improved despite the increasing vehicle ranges.

Drivers and Barriers to PEV Adoption

This section analyses drivers and barriers to PEV adoption in Germany. Many drivers and barriers are clearly applicable to other countries as well, thus the aim is to be specific for the German market and population and not to provide a comprehensive account of barriers and drivers to PEV adoption in general.

Barriers and Drivers by User Group

Private Users

What makes German households finally buy an electric vehicle? What attributes of the buyers and the vehicle are important for the decision to purchase? To provide answers to these questions we draw on findings from two studies that we conducted and that were both previously mentioned. The first one draws on survey data from citizens (n = 120 private users) involved in field trials of electric mobility (BMVI 2015), the second one refers to the survey data from a representative sample of inhabitants (n = 1510) of one of the German showcase regions, i.e. an area in which several initiatives to promote electric vehicles have been implemented.

Both studies indicate a high relevance of perceived fit of electric driving with personal needs in everyday driving but also for special occasions. This fit refers to actual mobility patterns, but also to personal fit, i.e. that the PEV is in line with the respondent's personality. In addition, as results from the first study point out, for actual users the comparison with a conventional vehicle plays a role—those PEV drivers that are less likely to associate independence and flexibility with a conventional car are more enthusiastic about continuing to drive a PEV in the future.

In the second study, as mentioned above in Table 1, participants were divided into four groups. Those two groups that are currently less likely to adopt PEVs, attribute a higher relevance to social norms, i.e. later adopters and non-adopters have more positive opinions on PEVs if they think that their peers hold more positive attitudes towards them. This implies the conclusion that these people orient themselves towards others in their opinion on PEVs. Thus, if PEVs become more widespread these people might also change their intentions and buy a PEV.

Taking these findings together, they point to the fact that, beyond the challenge for PEVs to fulfill current mobility needs, the image of PEVs will be decisive in their diffusion.

Fleet Users

Based on a survey of 229 car pool managers consisting of adopters and non-adopters, Globisch et al. (2017) explore why some organizations adopt PEVs for their fleets while others do not. They find that a key variable is the car pool manager's personal interest in PEVs which is highly related to interest in technology in general as a personal characteristic. Beyond personal characteristics, other drivers for PEV adoption are organizational innovativeness and the expectation of environmental benefits and positive effects on employee motivation, while the fear of mobility constraints and doubts about the reliability of PEVs act as barriers.

A further study by Globisch et al. (2018) analyzed which factors are related to PEV acceptance in corporate fleets relying on survey data of actual users of electric vehicles in organizations (N = 575). They find similar but not identical factors: perceived organizational usefulness and reactions of company employees to PEVs in general are found to be highly important. Regarding the perception of vehicle characteristics, environmental benefits and perceived ease of use are identified as relevant antecedents to PEV acceptance. Overall both studies point out that PEV adoption by organizations is dependent on personal perceptions and characteristics of the decision maker, but also on the organizational climate regarding innovations in general and PEVs in particular, as well as on the attributes of PEVs, with environmental benefits playing a major role.

Company Cars

Company car drivers are an important market for new car sales in Germany with about one million vehicles annually. Yet, PEVs seem to have played only a minor role in this segment (there is hardly any statistical data available since company cars are not differentiated from fleet cars in official vehicle sales statistics). Several reasons could be responsible for this. Firstly, company car drivers have high annual mileages with frequent long-distance business travel. Thus, the limited all-electric range of both BEVs and PHEVs is a challenge for this user group. Secondly, company cars are subject to income tax in Germany (benefit-in-kind) and 1% of the vehicle's list price is counted as additional monthly income. Thus, the higher purchase prices of PEVs make them less attractive compared to other vehicles. Furthermore, a substantial number of company car users have the possibility of refueling for free with no equivalent for free charging for PEVs.

Two PEV incentives specifically for company car users are available in Germany. The higher purchase price of BEVs (which increases the tax payable for a company car) has been slightly reduced (to account for high battery prices) for several years now without known effects. However, this only covered the disadvantage in taxation without really being an advantage compared to conventional vehicles. Recently though the German federal government decided to lower the company car tax by 50%, i.e. only 0.5% instead of 1% of the vehicle's list price are taxable on the monthly income (Spiegel.de 2018). Overall, the subject of company car drivers and motivations to choose alternatively fuelled vehicles in this segment clearly needs to be researched more in depth.

Charging Infrastructure

Charging infrastructure for PEVs is often discussed as a major hurdle for widespread PEV adoption. It is important to note, however, that this did not come up as a major issue in the acceptance studies cited above. Nevertheless, we discuss the availability of charging infrastructure in Germany separately which points out potential reasons why this is not a strong issue in Germany.

For private PEV users, regular recharging overnight is probably the most important infrastructure requirement. Slow-charging infrastructure at home is most easily installed in garages. Available data shows that about 65% of privately-owned German cars are parked in garages overnight (cf. Fig. 10). However, the data does not specify what share of the garages are privately owned small garages and what share belong to larger garages in multi-dwelling units.

Even in large cities with more than 100,000 inhabitants almost half of the vehicles are parked in garages. Thus, many passenger car users park in privately owned garages overnight, where slow-charging infrastructure could probably be installed. However, particularly in large cities, part of these garages are probably underground garages with individual parking spots but shared by many. Here, the installation is more problematic since these garages are often joint property and all owners have to agree to changes. In more rural areas, the vast majority of garages are likely to be owned by single households.

Public slow-charging infrastructure has substantially grown over the past years. At present there are about 11,000 public slow chargers (up to 22 kW) in Germany (DDI and Noerr 2018). The number of slow-charging points is higher since a charger can contain several charging points. Public charging points are subsidized through several programs from the federal government. These programs have been very successful, i.e. many communities applied, and the available budget was soon used up.

Fast-charging stations (>22 kW) have also been built all over Germany in parallel with the slow-charging infrastructure. These are mainly built along highways to make a dense grid for long-distance travel available for PEVs. At present about 400 fast-charging stations are accessible in Germany with a concentration in densely populated areas. With a highway network of about 13,000 km this implies an average distance of roughly 50 km between fast-charging stations. These are mostly DC charging stations typically with a charging power of 50 kW. Furthermore, an industry consortium of several car makers (IONITY) has announced that they will build an additional fast-charging network with 150–350 kW power.

In summary, many car owners have easy access to home charging in Germany which facilitates PEV market diffusion in such early market phases as it is currently in in Germany. A few studies have shown that public charging infrastructure generally has a positive impact on PEV sales, however, the actual size of the effect is difficult to estimate, due among other reasons to the methodological difficulties in determining the direction of the effect (does charging infrastructure lead to more PEVs or vice versa?). Compared to other PEV markets such as France or Sweden, the number of charging stations per million PEV is of similar magnitude (but lower than in Norway or the Netherlands).

Policies

Over the past years, different policy measures have been adopted to foster the spread of electric vehicles in Germany.

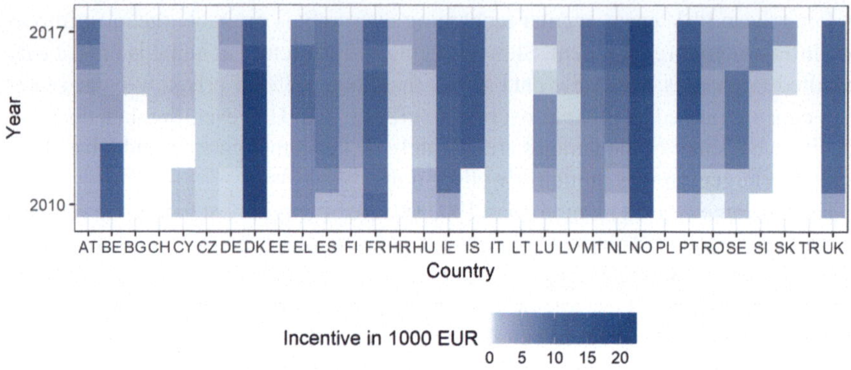

Fig. 11 Financial incentives in European countries over time. *Source* Münzel (2018)

For many years public subsidies focused on R&D programs for electric vehicles and particularly traction batteries. The basic rationale is the importance of the German automotive industry in Germany and PEVs being regarded as the long-term future for the car industry, so that high added value manufacturing can be kept in Germany in the long term. Additionally, some minor financial incentives for purchase and usage were introduced in 2010. These included no registration tax for BEVs, which amounts to a financial benefit of a few hundred euros per year.

More recently in 2016, higher purchase incentives were introduced in Germany. Car buyers now receive a subsidy of 4000 euros for the purchase of a BEV, of which 2000 euros come from the federal government and 2000 from the car industry. Car buyers can apply for a 2000 euro re-imbursement from the government and have to negotiate the 2000 euros from industry with the car salesperson. Thus, in practice, only about 2000 euros are directly perceived as an incentive and hence the effect on BEV uptake has been limited in comparison to other countries. Figure 11 shows the value of the financial incentives in Germany over time in comparison to other European countries (Münzel 2018).

Furthermore, several support schemes focused on the procurement and installation of charging infrastructure. Additionally, the federal government supports German municipalities in the transition towards electric transportation. This includes a special law ("Elektromobilitätsgesetz") that allows giving PEVs special rights e.g. free or exclusive parking, driving in bus lanes or driving within restricted areas. So far, municipalities have mainly applied the parking priorities. Additionally, municipalities can ask for financial support to develop strategies for electric driving and to identify promising measures. They can also apply for financial support in case of PEV procurement. In the autumn of 2017, the "Sofortprogramm saubere Luft" (immediate action program for clean air) was introduced to improve air quality in a number of German cities and comply with a European directive, namely 2008/50/EC on ambient air quality and cleaner air for Europe. This program is supported with one billion euros and focuses on the electrification and digitalization of transport. However, it is still too early to evaluate potential outcomes of this measure.

Recommendations for Future Research

Market Diffusion

The PEV market is still in an early phase in Germany and future growth of sales is widely expected. Early on, in 2009, the German federal government announced a target of one million PEVs in use in Germany by 2020 (Bundesregierung 2009). This target will not be achieved as there were only about 100,000 PEVs in use at the beginning of 2018 (KBA 2019), increasing by 34,000 in the first half of 2018. However, many experts believe that with continued market growth and the introduction of new German PEV models, the target will be reached, only a few years later than originally planned. Furthermore, the breaking of "Dieselgate", i.e. the scandal concerning the installation of defeat devices in diesel exhaust after-treatment systems for these to operate only under laboratory testing conditions, has led to dropping market shares of diesel vehicles in Germany. Since diesel cars used to have very high market shares among company and fleet cars, Dieselgate could actually lead to a faster PEV sales diffusion, in particular for PHEVs since these do not have fundamental range limitations and are more suited to company car range needs.

In summary, many actors are disappointed that the target of one million PEVs will not be reached by 2020, but many are also aware of the importance of the long-term evolution. However, knowledge is still limited concerning the best long-term phase out of subsidies but also concerning the potential contribution of PEVs to reach GHG mitigation targets in the transport sector in Germany. Furthermore, the role of diesel engines in GHG mitigation in Germany should be made clearer in future research as the historical view of their lower GHG emissions has been questioned recently.

Next Adopters

A few findings in this chapter pointed out that early adopter groups come from specific segments of society, typically combining a higher socio-economic status with an interest in technology, assigning multiple benefits to PEV driving. While further improvement of the vehicles and the available charging infrastructure will certainly support market diffusion, research has also emphasized that diffusion of PEVs takes place in a socio-technical system: individual perceptions on how PEVs fit with the personal life-style and social influence of what others think about PEVs play an important role as well—also in organizational decision making. On the one hand this opens additional pathways for PEV diffusion by supporting positive perception through campaigns or role models. On the other hand, this also implies the risk of slowing down diffusion if societal discourse takes a critical stance, possibly fed by those benefitting from the current regime.

Open questions for future research include structural issues for private users, e.g. availability of charging possibilities at home and in the workplace, a more in depth look at organizational decision-making, and especially closing the many gaps in research on company cars. Furthermore, the usage behavior in terms of electric miles driven or recharging and its determinants are not fully understood yet.

Freight Vehicles

Most PEV studies focus on passenger cars. However, freight vehicle categories have received more attention in Germany recently. Light, medium and heavy-duty freight vehicles could see noteworthy growth over the next years.

In the light-duty segment, parcel services have been testing small BEV transporters for several years now. Interestingly, the biggest German parcel service, the German mail (Deutsche Post AG) contacted major German car manufacturers for a small BEV delivery van. However, they were unable to provide such a vehicle and Deutsche Post partnered with a university start-up named "street scooter" instead to build their own BEV delivery vans. By 2018 the company had an annual production capacity of 20,000 vehicles (golem.de 2018), offered three vehicle models with various configurations and had become the biggest German electric light-duty vehicle producer.

A major driver in the electric freight vehicle market is the potential ban of diesel vehicles in German cities. Currently 79 German cities have low emission zones and 13 have access regulations (cf. www.urbanaccessregulations.eu). Hamburg has banned diesel vehicles from several streets and the city of Stuttgart has announced a ban of diesels with EURO 1–5 from the city center. The access limitations force logistics companies to transition to electric vehicles. The same trend is starting for medium and heavy-duty vehicles up to 26 t of gross vehicle weight. However, there are only very few models commercially available so far.

For heavy-duty vehicles in long-distance transport, the Federal Ministry for the Environment, Nature Conservation and Nuclear Safety (BMU) has funded the development of trucks powered by overhead lines. The system is similar to the electric road system in Sweden. Currently, three test tracks on public roads are under construction and planned to go into operation in 2019 or 2020 (BMU 2018).

Several technologies currently compete as long-term solutions for heavy-duty road transport: battery electric trucks with or without a catenary system, fuel cell hydrogen electric vehicles, and synthetic fuels created with renewable electricity. It is still highly uncertain which technology is most advantageous and what the specific implications for the energy and transport system are. Much more research is required on the future of carbon neutral heavy road transport.

Other Technologies and Mobility Services

PEVs are only one of the paths of innovation in the automotive sector that are currently gaining momentum. Other developments include alternative fuels such as compressed natural gas, liquefied petroleum gas and synthetic fuels as well as hydrogen fuel cell vehicles. The latter require building up new hydrogen infrastructure for their diffusion to be possible. Currently, 44 hydrogen stations are operating in Germany and 49 are under construction (cf. http://h2tankstellen.cleanenergypartnership.de/). This process is driven by a public-private partnership, the Clean Energy Partnership. While some progress has been made, the pace of development is slow—originally the goal was to reach the number of 50 stations by the end of 2015. In parallel, research on the diffusion and acceptance of hydrogen mobility has pointed out that experts perceive PEVs as the main competitor and therefore their market diffusion as potentially the main barrier to a further development of hydrogen mobility in the medium term (Dütschke et al. 2017).

Additionally, more disruptive developments in the mobility sector are evolving such as progress in the field of autonomous driving and shared mobility. In both cases the rapid advancement of information and communication technologies is a major factor which could lead to substantial changes in the mobility sector—with or without electrified drivetrains. Staying competitive within these developments will be a major challenge for the German industry in the years to come. The impact of shared mobility solutions or autonomous vehicles on electrification of road transport and its potential to mitigate GHG emissions in transport remain yet to be determined.

References

Agosti, R. (2010). *Nutzerakzeptanz von Elektroautos. Untersuchung eines frühen Stadiums der Innovations diffusion bei verschiedenen Nutzergruppen* (Master's Thesis). Zurich: Swiss Federal Institute of Technology.

BMU (Bundesministerium für Umwelt, Naturschutz und nukleare Sicherheit). (2018). *Elektro-Lastwagen an der langen Leine*. https://www.bmu.de/themen/luft-laerm-verkehr/verkehr/elektromobilitaet/elektro-lastwagen/. Accessed June 2, 2018.

BMVI. (2012). *Roadmap zur Kundenakzeptanz – Zentrale Ergebnisse der sozialwissenschaftlichen Begleitforschung in den Modellregionen*. With assistance of Elisabeth Dütschke, Uta Schneider, Andreas Sauer, Martin Wietschel, Jana Hoffmann, Sabine Domke. Karlsruhe: Fraunhofer ISI 2012.

BMVI. (2015). Elektromobilität in Haushalten und Flotten: was beeinflusst die Kauf- und Nutzungsbereitschaft. In BMVI (Ed.), *With assistance of Elisabeth Dütschke, Martin Wietschel, Joachim Globisch, Uta Schneider, Christian Schlosser, Dominique Sevin, Tilman Wilhelm.*. Berlin. http://www.now-gmbh.de/de/publikationen.html. Accessed June 2, 2018.

Bundesregierung. (2009). *Nationaler Entwicklungsplan Elektromobilität der Bundesregierung*. https://www.erneuerbar-mobil.de/sites/default/files/2016-08/nep_09_bmu_bf.pdf. Accessed June 2, 2018.

DDI (Deutsches Dialog Institut GmbH, Ed.). (2017). *Abschlussbericht der Begleit- und Wirkungsforschung 2017*. Begleit- und Wirkungsforschung Schaufenster Elektromobilität (BuW). Frankfurt a.M.: DDI.

DDI., & Noerr. (2018). *Elektromobilitätsgesetz (EmoG) Gesetz zur Bevorrechtigung der Verwendung elektrisch betriebener Fahrzeuge Berichterstattung 2018*. Im Auftrag von: Bundesministerium für Verkehr und digitale Infrastruktur. Bundesministerium für Umwelt, Naturschutz und nukleare Sicherheit. Frankfurt am Main.

Dütschke, E., Upham, P., & Schneider, U. (2017). Report on results of the stakeholder survey. Deliverable 5.1 in the HYACINTH project. *With assistance of Christian Oltra, Roser Sala*. http://hyacinthproject.eu/wp-content/uploads/2017/12/HYACINTH-D5_1-Report-on-results-of-the-stakeholders-survey_v02_DEF.pdf. Accessed June 2, 2018

EAFO (European Alternative Fuels Observatory). (2018). *EV sales in Germany*. http://www.eafo.eu/content/germany.

Egbue, O., & Long, S. (2012). Barriers to widespread adoption of electric vehicles: An analysis of consumer attitudes and perceptions. *Energy Policy*, 48, 717–729 (*Special Section: Frontiers of Sustainability*).

Ensslen, A., Gnann, T., Jochem, P., Plötz, P., Dütschke, E., & Fichtner, W. (2018). Can product service systems support electric vehicle adoption? *Transportation Research Part A: Policy and Practice*. https://doi.org/10.1016/j.tra.2018.04.028.

Frenzel, I., et al. (2015). *Erstnutzer von Elektrofahrzeugen in Deutschland*. Nutzerprofile, Anschaffung, Fahrzeugnutzung, DLR-Forschungsbericht, Ergebnisbericht der Nutzerbefragung von Elektrofahrzeugen in Deutschland. Köln: DLR.

Frenzel, I., Müller, S., & Dzhimova, M. (2016). Electric mobility in Germany: Under-standing pioneers and market niches in commercial traffic. *Transportation Research Board* 95th Annual Meeting (No. 16-3306).

Globisch, J., Dütschke, E., & Schleich, J. (2018). Acceptance of electric passenger cars in commercial fleets. *Transportation Research Part A: Policy and Practice, 116*, 122–129. https://doi.org/10.1016/j.tra.2018.06.004.

Globisch, J., Dütschke, E., & Wietschel, M. (2017). Adoption of electric vehicles in commercial fleets. Why do car pool managers campaign for BEV procurement? *Transportation Research Part D: Transport and Environment*. https://doi.org/10.1016/j.trd.2017.10.010.

Golem.de. (2018). *Elektrolastwagen: Streetscooter eröffnet weitere Fabrik*. https://www.golem.de/news/elektrolastwagen-streetscooter-eroeffnet-weitere-fabrik-1805-134679.html. Accessed June 2, 2018.

Greenpeace. (2016). *Schwarzbuch Autolobby - Wie die Autolobby die Zukunft einer Branche verbaut*. Berlin. https://www.greenpeace.de/sites/www.greenpeace.de/files/publications/s01841_web_greenpeace_schwarzbuch_autolobby_04_16.pdf. Accessed June 2, 2018.

Hidrue, M. K., et al. (2011). Willingness to pay for electric vehicles and their attributes. *Resource and Energy Economics, 33*, 686–705.

IEA (International Energy Agency). (2018). *Global electric vehicle outlook 2018*. Retrieved from https://www.iea.org/gevo2018/.

Jarass, J., et al. (2014). Die early adopter der Elektromobilität in Deutschland – wer sie sind und wie sie fahren. *Internationales Verkehrswesen, 2*, 70–72.

KBA (Kraftfahrtbundesamt). (2019). *Bestand an Pkw am 1. Januar 2018 nach ausgewählten Kraftstoffarten*. https://www.kba.de/DE/Statistik/Fahrzeuge/Bestand/Umwelt/2018_b_umwelt_dusl.html?nn=663524. Accessed June 2, 2018.

Münzel, C. (2018). *Analyzing the impact of financial incentives on the market penetration of PEVs in Europe using linear regression models* (Master thesis). Karlsruhe: Karlsruhe Institute of Technology and Fraunhofer Institute for Systems and Innovation Research ISI.

OICA (International Organization of Motor Vehicle Manufacturers). (2018). *World motor vehicle salesby country and type 2005–2017*. Online: http://www.oica.net/category/sales-statistics/

Ozaki, R., & Sevastyanova, K. (2011). Going hybrid: An analysis of consumer purchase motivations. *Energy Policy*, 39(5), 2217–2227.

Peters, A., & Dütschke, E. (2014). How do consumers perceive electric vehicles? A comparison of German consumer groups. *Journal of Environmental Policy & Planning, 16*(3), 359–377, (Special Issue: Sustainable Mobility—Challenges for a Complex Transition).

Peters, A., Agosti, R., Popp, M., & Ryf, B. (2011). Electric mobility—A survey of different consumer groups in Germany with regard to adoption. In *Proceedings from ECEEE 2011 Summer Study*, Belambra Presqu'Ile de Giens, France.

Plötz, P., et al. (2014). Who will buy electric vehicles? Identifying early adopters in Germany. *Transportation Research Part A: Policy and Practice, 67,* 96–109.

Plötz, P., Funke, S. Á., & Jochem, P. (2017a). Empirical fuel consumption and CO_2 emissions of plug-in hybrid electric vehicles. *Journal of Industrial Ecology*. https://doi.org/10.1111/jiec.12623. (Online First).

Plötz, P., Funke, S. A., Jochem, P., & Wietschel, M. (2017b). CO_2 mitigation potential of plug-in hybrid electric vehicles larger than expected. *Scientific Reports, 7*(1), 16493.

Plötz, P., Gnann, T., Kagerbauer, M., & Heilig, M. (2017c). Can models predict electric vehicle users? In EVS30 (Ed.), *International Electric Vehicle Symposium & Exhibition*.

Plötz, P., Gnann, T., Wietschel, M., & Kühn, A. (2013). *Market evolution scenarios for electric vehicles-detailed version*. Commissioned by acatech, German National Academy of Science and Engineering and Working Group, 7. https://www.isi.fraunhofer.de/content/dam/isi/dokumente/cce/2013/NPE_TCO_Bericht_en.pdf. Accessed June 3, 2018.

Rezvani, Z., Jansson, J., & Bodin, J. (2015). Advances in consumer electric vehicle adoption research: A review and research agenda. *Transportation Research Part D: Transport and Environment, 34,* 122–136.

Spiegel.de. (2018). *Kabinett beschließt Steuervorteil für Elektro-Dienstwagen*. http://www.spiegel.de/auto/aktuell/elektro-dienstwagen-bundeskabinett-beschliesst-steuervorteil-gesetzentwurf-a-1221161.html. Accessed June 2, 2018.

VDA (Verband der Automobilindustrie): *Zahlen und Daten*. Online: https://www.vda.de/de/services/zahlen-und-daten/zahlen-und-daten-uebersicht.html 8/8/2018.

China

China's Booming Plug-in Electric Vehicle Market—How Will It Continue?

Shiqi Ou, Zhenhong Lin, Xin He, Steven Przesmitzki, Zhixin Wu, Jihu Zheng, Renzhi Lv and Liang Qi

Abstract The plug-in electric vehicle (PEV) is recognized as a pivotal component in transportation energy security and technology innovation in China. With government incentives and substantial capital investments, the Chinese PEV market has become one of the largest in the world. However, is this market boom ephemeral or will it be continuing? This chapter provides a general review of the Chinese PEV market mainly from the consumer perspective, by summarizing the characteristics of the vehicle policies and consumer concerns in the PEV market in recent years. Moreover, it introduces detailed examples on how the policies, technologies, and infrastructure shape the market. In addition to employing large amounts of reliable data from the China Automotive Technology and Research Center for the review, the authors also adopted a quantitative analysis tool—the New Energy and Oil Consumption Credits

S. Ou · Z. Lin (✉)
Oak Ridge National Laboratory, National Transportation Research Center, Knoxville, TN 37932, USA
e-mail: linz@ornl.gov

S. Ou
e-mail: ous1@ornl.gov

X. He · S. Przesmitzki
Aramco Services Company: Aramco Research Center—Detroit, Novi, MI 48377, USA
e-mail: xin.he@aramcoservices.com

S. Przesmitzki
e-mail: steven.przesmitzki@aramcoservices.com

Z. Wu · J. Zheng · L. Qi
China Automotive Technology and Research Center, Tianjin 300300, China
e-mail: wuzhixin@catarc.ac.cn

J. Zheng
e-mail: zhengjihu@catarc.ac.cn

L. Qi
e-mail: qiliang@catarc.ac.cn

R. Lv
Equipment Industry Development Center, Ministry of Industry and Information Technology, Beijing 100846, China
e-mail: lvrenzhi@163.com

M. Contestabile et al. (eds.), *Who's Driving Electric Cars*, Lecture Notes in Mobility,
https://doi.org/10.1007/978-3-030-38382-4_11

model—to project the potential industrial impacts of the government policies in multiple scenarios. Based on the findings, it seems that the prosperity of the PEV market, incentivized by the government, will continue in the short run. The direct subsidies to PEV from the government will gradually phase out, and the supply-side push policies such as the dual-credit policy and other charging infrastructure incentives will work as the invisible hand to force the automakers to invest more in the PEV segment. However, in the long run, innovation in PEV technology and business models, together with consumer acceptance will be the major factors that can sustain further market growth.

Keywords China · Plug-in electric vehicle · Vehicle policy · Vehicle industry · Consumer behavior

Introduction—Recent Fast Growth

China's PEV Market Is Growing Rapidly

Although the promotion of plug-in electric vehicles (PEVs) faces technology challenges and an insufficient charging infrastructure, the global automotive industry is arguably sprinting toward the electrification era. Annual sales and stocks of PEVs have been hitting new records yearly in almost every major vehicle market in the world since 2011, as shown in Fig. 1. By 2017, the global annual sales of light-duty PEVs reached over one million units. The three largest PEV markets by annual sales are: China, the United States, and Western Europe. In an effort to mitigate energy security concerns and to promote a more sustainable and environmentally friendly

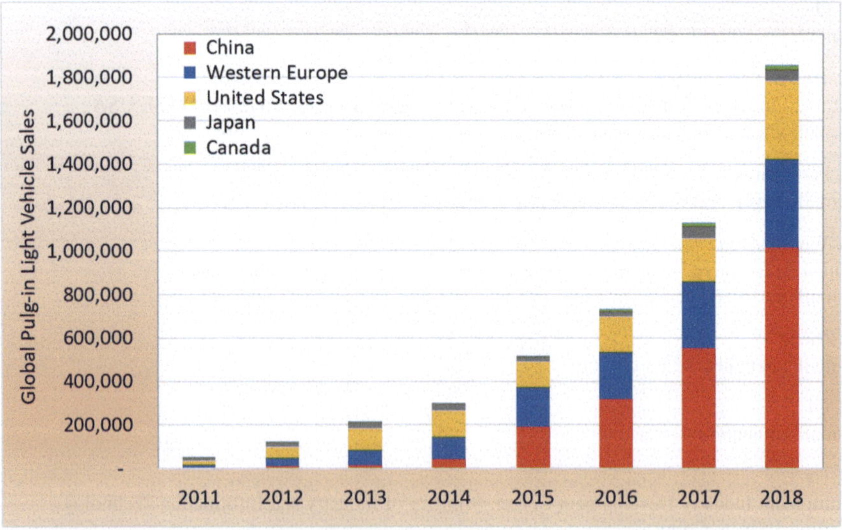

Fig. 1 Global light-duty PEV annual sales, 2011–2018 (Office of Energy Efficiency & Renewable Energy 2016, 2017; Schmidt 2018; CATARC 2019; International Energy Agency 2017)

transport system, governments have worked to bolster the PEV market. Even more aggressively, countries such as France and the UK have pledged to ban the sale of gasoline and diesel fueled vehicles by 2040 (Castle 2017), in order to make room for PEVs.

The Chinese government has long been convinced of the significance of PEVs in the passenger car market. As a developing country, China wants to seize the emerging market opportunity and to leapfrog the manufacturing industry of other countries: the government has included PEV manufacturing in its industrial upgrading plan "*Made in China 2025*", a government blueprint to promote the transition of the Chinese industry to hi-tech manufacturing (Howell et al. 2014). More importantly, imported crude oil made up more than 60% of total oil consumption in China by 2015 (Jiao et al. 2017). Facing issues of energy security and greenhouse gas emissions from fossil fuels, the government is eager to diversify the energy sources used in road transport, which accounts for 80% of transport energy consumption (Chai et al. 2017). Moreover, trying to be a responsible international stakeholder, China vowed to cap its carbon emissions by 2030 and curb worsening air pollution (The Strategic Advisory Committee, SAE-China 2016). Promoting the PEV market is an effective action by China's government to meet its commitments.

PEVs have attracted the attention of the Chinese government since 2001. The "*National High-Tech R&D Program* of the government's *Tenth Five-Year Plan (2001–2005)*" included a research budget for PEVs. In the period of the *Eleventh Five-Year Plan (2006–2010)*, to encourage more widespread adoption of PEVs, the government started a large-scale PEV demonstration project, named *"Ten Cities—Thousand Vehicles Demonstration Program"* during 2009 and 2012. Since then, the institutional environment for the PEV industry has been increasingly favorable, evidenced by frequent high-level official statements, major policy implementation, massive governmental and industrial investments, and generous purchase and adoption incentives (Ou et al. 2017). As a result, the PEV market in China began an explosive growth.

According to the Automotive Data Center of the China Automotive Technology and Research Center (CATARC), PEV sales in 2018, including both passenger and commercial vehicles, reached a record high of 1,017,040 (CATARC 2019). The PEV sales share in the vehicle market has increased from 0.26% in 2014 to 4.38% in 2018, shown in Fig. 2. After the slow growth of the first few years, PEV sales grew massively in 2013 when the Chinese government dedicated more financial support to the PEV market and enacted other regulations heavily in favor of PEVs, such as the vehicle license plate registration privilege in Beijing, Shanghai, and other megacities.

The current fast-growing Chinese PEV market exhibits several features so far (Ou et al. 2019):

- Small sized cars (including minicompact, subcompact, and compact) made up the greatest share of PEV sales in recent years. In the light-duty PEV market segment, small PEVs garnered more than 80% of sales in 2016; on the other hand, sales of large sized PEVs (e.g., SUV) are fairly small. Analyzing the ten-best-selling light-duty PEV models in the 2015 and 2016 markets, all of the popular PEV models

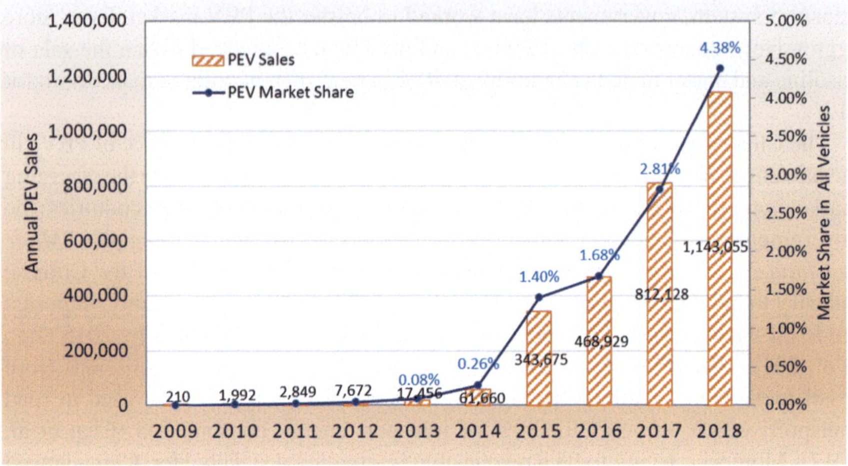

Fig. 2 Annual PEV sales in China, 2009–2018 (CATARC 2019)

are found to be compact or small sedans. As the PEV market expands, larger sized PEVs, equipped with longer electric range, could be more popular.

- Though overall PEVs are more expensive than their counterpart internal combustion engine vehicles (ICEVs, gasoline or diesel fueled), their price differences in China are much smaller than they are in the second-largest light PEV market, the United States. The sales-weighted average manufacturer suggested retail price (MSRP) for all light PEVs sold in China in 2016 is about 216,250 CNY ($31,295 USD[1]) before subsidies, which is about $9000 USD higher than the sales-weighted average light ICEV MSRP ($22,000 USD) in China in the same year. The MSRP range of PEVs in the U.S. varies from $23,800 to $140,700 USD (basic model MSRP before tax credits) with a sales-weighted average MSRP of $49,037 USD. This is approximately $16,000 USD higher than the average passenger vehicle MSRP ($32,000 USD) in 2016, as reported by Davis et al. (2016). All the numbers are calculated before considering the PEV monetary and non-monetary incentives. The actual price paid by the Chinese consumer for a PEV is very likely to be lower than its benchmarked ICEV.
- Currently, many demands for PEVs come from the fleet companies or government instead of individual buyers, as shown in Fig. 3. Integrated with the technologies in shared mobility, many PEVs are bought by companies for short-term rental or ridesharing for consumers in urban areas in China. One business model is for companies to establish PEV rental locations in busy areas such as shopping malls, office buildings, university campuses, and major transit stations in cities. Consumers can lease a PEV for short-term driving with a membership card or

[1]The currency exchange rate of $1.000 USD = 6.910 CNY which is the average exchange rate in 2016. Source from the U.S. IRS "Yearly average currency exchange rates", https://www.irs.gov/individuals/international-taxpayers/yearly-average-currency-exchange-rates (visited in May 2018).

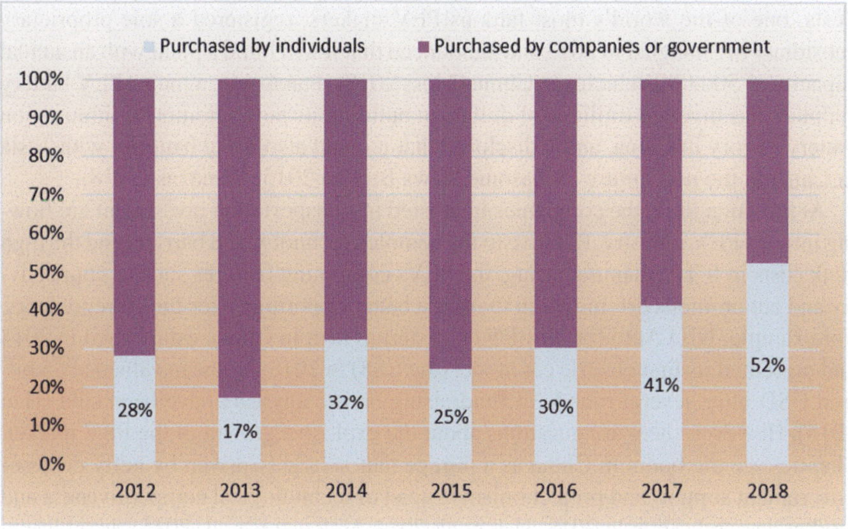

Fig. 3 PEV share purchased by individuals in China, 2012–2018 (CATARC 2019)

an application registration via smartphone apps (Marquis et al. 2013). Another popular business model is to use PEV fleets for peer-to-peer ridesharing. For example, CAOCAO is a start-up fleet company set up by Geely, one of the largest privately-owned automakers in China. Its operation mode is very similar to other famous ridesharing companies like Uber: customers use a smart phone app to book a ridesharing car. One idiosyncratic feature of CAOCAO is that all of its ridesharing vehicles are PEVs made by Geely. In this way, the automaker can introduce its own PEV products to the passengers.

Will It Continue?

China's goals for the PEV industry and market are ambitious. Activities in policymaking and industry, and recent PEV demand indicate a strong and sustainable momentum. Based on the "*Planning for the Development of the Energy-Saving and New Energy Automobile Industry (2012–2020)*" issued by the State Council, the government expects the PEV stock to reach five million units and the production capacity of the PEV industry to ramp up to two million a year by 2020 (The State Council of P.R China 2012). PEV manufacturing is regarded as a pillar industry in the "*Made In China 2025*" (Howell et al. 2014).

Moreover, the prosperous PEV market is attracting both domestic and foreign automakers. Many major vehicle original equipment manufacturers (OEMs) and suppliers in the world have been preparing for the growing PEV market in China.

Tesla, one of the world's most famous PEV makers, registered a sole proprietary subsidiary in Shanghai in 2018 and announced that it will build a plant with an annual capacity of 500,000 vehicles in China (Moss 2018). Panasonic, a major PEV battery supplier, has invested millions of dollars in building its new automotive lithium–ion battery factory in China, and it disclosed that it could also make batteries with Tesla in China in the near future (Panasonic News Bureau 2015; Yamazaki 2018).

At the same time, the companies interested in transportation investment are flowing into the PEV industry. Relative to the complex technological barriers and the high sunk costs in ICEV manufacturing, the PEVs combining features such as connectivity and autonomous driving seem to offer a better opportunity for the entrepreneurs. For example, NIO Auto (or NextEV) is a startup firm in China, established in 2014 and presented its first electric car model (the ES8) in 2018, has been valued at $5 billion USD after several rounds of fundraising before any cars have been sold (Hao 2018). However, there are questions about the explosive growth of the PEV market. Skeptics see the boom in China as a mirage that is largely driven by hefty doses of government support and protectionism instead of technological competitiveness and market demand (Temple 2018; Hao et al. 2014; Helveston et al. 2015). In addition, the government's generous subsidies draw criticism over the financial burden and unfair competition that result from them (Howell et al. 2014). Insufficient supervision of PEV regulations creates opportunities for subsidy frauds by automakers, and the public doubts the sustainability of the unusual growth of PEV sales (Schmitt 2016). Moreover, from an environmental sustainability perspective, some researchers argue that PEVs cannot reduce life-cycle emissions of air pollutants given China's current power generation mix (Huo et al. 2010, 2015).

Different stakeholders, for different reasons, are asking the same questions—will the fast growth of PEV sales in China continue? What are the consumer concerns in the PEV market, and how will policy and technology shape the market? This chapter attempts to provide a better understanding of the underlying driving factors of the current Chinese PEV market, particularly with emphasis on consumers. As one of the leading research institutes in the Chinese vehicle industry, CATARC obtains the most up-to-date and reliable data from government, automotive OEMs, and consumers. Table 1 summarizes the data items provided by CATARC and used in this study. The data on consumer behavior, passenger car market and infrastructure is used to simulate the impact of policy on consumer choices in the vehicle market with the vehicle policy analysis tool—the New Energy and Oil Consumption Credits (NEOCC) model—that was developed by Ou et al. (2018b) to analyze vehicle market dynamics by quantifying consumer's vehicle ownership costs in China.

NEOCC was developed by the Oak Ridge National Laboratory (ORNL) with support from Aramco Services Company to simulate the automotive industry's compliance scenarios under multiple vehicle policies. The model version adopted by this chapter is the 2nd version released in 2017, which is calibrated to 2016 sales, prices and fuel economies of the China passenger vehicle market. The general picture of the 2016 market in China is shown in Table 2 (Ou et al. 2018b). NEOCC maximizes industry profits by optimizing internal subsidies for plug-in electric vehicles

Table 1 PEV data for the Chinese market provided by CATARC

Data	Disaggregation	Source
Annual sales of PEVs	By year from 2009 to 2018 By province By automaker and brand	China Insurance Group
Sales prices of PEVs	By year from 2013 to 2018 By automaker and brand	Dealers self-reporting through CATARC
Government policies	By year from 2013 to 2018 By province	Ministry of Industry and Information Technology (MIIT) and CATARC
Vehicle classification	By vehicle size	CATARC
Vehicle performance	By weight, battery size, electric consumption, etc.	MIIT

Table 2 China's 2016 passenger vehicle market (Ou et al. 2018b)

Segment	Technology	Sales	Weight[a] (kg)	Electric range[a] (km)	FC-CS[a] (L/100 km)	EC-CD[a] (kWh/100 km)	MSRP[a] ($2016)
Sedan	CV-High	9809,923	1302.7	–	6.57	–	20,317
	CV-Med	1952,849	1302.7	–	5.58	–	22,566
	CV-Low	457,884	1302.7	–	4.59	–	25,786
	PHEV-50	24,736	1725.3	55.77	4.77	13.61	39,176
	PHEV-70	18,952	1720.0	70.00	5.08	20.00	30,354
	PHEV-85[b]	0	1900.0	85.00	6.23	20.50	35,100
	BEV-150	166	1296.5	101.61	–	14.56	29,192
	BEV-250	150,881	1104.5	159.93	–	14.40	25,916
	BEV-350	62,799	1677.0	280.81	–	16.92	35,063
	BEV-400	13,584	2420.0	400.00	–	20.50	53,517
SUV	CV-High	8208,582	1550.2	–	7.73	–	25,005
	CV-Med	651,270	1550.2	–	6.26	–	30,279
	CV-Low	4098	1550.2	–	4.78	–	38,806
	PHEV	26,499	2388.0	84.61	8.75	20.49	38,995
	BEV	2620	1575.3	211.40	–	18.82	31,659

[a]The values are all sales-weighted. FC-CS refers to fuel consumption in charge sustaining mode, EC-CD refers to electricity consumption in PEV charge-depleting mode, MSRP refers to manufacturing suggested retail price

[b]The 2016 market in China has no records of PHEV-85 sales; the featured elements for PHEV-85 are presumed based on unlisted vehicle models from OEMs

(PEVs) and efficient conventional vehicles, subject to the policy compliance constraints. NEOCC takes into account: PEV sales share and sales-weighted average fuel consumption requirements, incremental cost of fuel efficiency, range limitation of battery electric vehicles, charging infrastructure, consumer choice and central and local government incentives for PEVs. NEOCC can be used to analyze potential compliance strategies conducted by the auto companies and profit impacts by the government incentives and policy constraints, including pricing strategies, technology challenges, government subsidy, and infrastructure needs.

Chapter Structure

While recognizing that policy, technology, industry, infrastructure and consumers and inextricably linked, in this chapter we particularly emphasize the role of policies and how they impact consumer decisions in the market. The chapter is structured as follows. The next section summarizes and discusses the government's vehicle policies and incentives, which have provided the main driver of China's PEV sales so far. The section also presents the results of the NEOCC model simulations of likely automaker compliance strategies out to 2020 and discusses whether the changing policy is likely to result in sustained PEV market momentum. The following section focuses on consumers and particularly addresses the costs to consumers that are associated with PEV range anxiety and charging infrastructure issues in China, the current limitations of battery technology as well as business operation prospects and how these might impact consumer's acceptance. The chapter concludes with a section discussing the strengths and limitations of the analysis conducted, lessons learned, and remaining gaps in knowledge on consumer choices analysis in the PEV market in China.

Policy Has Been the Biggest Driver

The Chinese passenger car buyers have been very receptive to the government policies and incentives in the vehicle market (Ou et al. 2019). For example, in August 2015 the government enacted an incentive that reduces the ICEV purchase tax by half for vehicles with an engine displacement of less than 1.6 L. It had an immediate impact on both vehicle sales and consumer choice of vehicle classes: vehicle sales increased much faster later in the fourth quarter of 2015 (2015 Q4), as evidenced by its growth rate compared with previous years. It also shows that nearly two-thirds of the sedans sold during 2015 Q4 had an engine displacement of <1.6 L (CATARC 2019). This remarkable market responsiveness might reveal a strong latent demand from first-time car buyers with limited disposable income. An appropriate policy or incentive could bring a significant influence on the consumer purchase behavior in the Chinese vehicle market. In particular, many researchers believe that the PEV sales in recent years were largely, if not entirely, driven by the government policies that influence the buyer decisions in the vehicle market (Howell et al. 2014; Zheng et al. 2012; Li and Sun 2011; Ma et al. 2017).

Government Policy Phases

The policy support to PEVs provided by the Chinese government can be divided into four distinct phases chronologically, as shown below.

- 2009–2012: During this period, the government enacted the "*Ten Cities—Thousand Vehicles Demonstration Program.*" This program called for thousands of PEVs to be purchased, with financial incentives, and driven as demonstration vehicles, in about 10 selected cities each year. However, individual consumers were not eligible for the subsidies in most cities. Before 2013, more than 80% of the PEVs sold in China were public fleet vehicles, mostly buses, purchased under this program (Howell et al. 2014).
- 2013–2015: In 2013, the Ministry of Finance approved new PEV purchase subsidies for the next three years (2013–2015) (CATARC 2019). Central government incentives were open to all vehicle purchasers, including public fleets and individual buyers, and were much more comprehensive than previous ones, including tax credits, deployment of charging infrastructure, discounts for vehicle charging, among others.
- 2016–2020: In the current phase, the government is gradually moving away from direct purchase subsidies and on to supply-push strategies that regulate the vehicles that the automotive OEMs are allowed to sell on the Chinese market. With the "*Measures on Parallel Administration of Passenger Car Enterprise Average Fuel Consumption and New Energy Vehicle Credits*" (dual-credit policy, which will be discussed in a later section) issued in September 2017, the government regulates the corporate average fuel consumption of new vehicles produced and forces every automaker to meet steadily increasing PEV quotas in the vehicle products. At the same time, the government has been gradually reducing the PEV purchase subsidies, and the technical criteria for determining whether PEV models are eligible for subsidies have become more comprehensive and practical (Ministry of Science and Technology of P.R.C 2014).
- After 2021: The central government has pledged to phase out the PEV purchase subsidies. Meanwhile, the government also expects the dual-credit policy and the charging infrastructure could play more important roles in the development of PEV market. From the supply side, the dual-credit policy can effectively render the industry to meeting the average fuel economy standards while producing more PEVs for the market; from the demand side, with reasonable charging infrastructure incentives from the central and local governments, the rapid development of charging infrastructure will offer PEV drivers better using experience and less inconvenience.

Monetary Incentives—Purchase Subsidies

Table 3 shows the central government subsidies allowed for PEVs (passenger car segment) from 2013 to 2015. During this period, the only technical criterion for the purchase subsidies in the passenger car segment is the vehicle's electric range.

Before 2013, policymakers were more willing to encourage government organizations and public transit systems to purchase PEVs as demonstration vehicles. After 2013, however, government purchase subsidies were open to the public and individual consumers, motivated by the high subsidies, started to buy PEVs. Unlike the monetary incentives in the United States, where the consumers earn tax credits for their purchased PEVs, the PEV purchase subsidies in China are claimed directly by the PEV automaker, not by the consumers. The prices paid by consumers are after subsidies, which mitigates the consumers' financial burden and might elevate the attractiveness of the PEVs.

In April 2015, the Ministry of Industry and Information Technology (MIIT) issued "*Financial Support Policies on the Promotion and Application of New Energy Vehicles (2016-2020)*" (MIIT 2015) amending previous incentives to include new technical specifications that regulate PEV subsidies. According to this regulation, the central government expects to gradually reduce subsidies for PEVs, while increasing subsidies for FCVs. As Table 4 shows, FCV subsidies increased from 180,000 CNY ($27,739 USD) in 2015 to 200,000 CNY ($30,821 USD) in 2016.

However, after 2015, motivated by the need to eliminate the subsidy frauds (Jiang 2016) and regional protectionism (Liu and Yu 2018) and to promote technology advancement, the government decided to speed up phasing out subsidies for PEVs and added more technical requirements. In particular, between 2016 and 2018, MIIT

Table 3 Central government subsidies to PEVs and fuel cell vehicles (FCVs) (passenger vehicle segment), 2013–2015 (Ministry of Science and Technology of P.R.C 2014; MIIT 2013)

Type	Electric range	2013	2014	2015
Plug-in hybrid electric vehicles (includes EREVs[a])	>50 km	35,000 CNY ($5394 USD)	33,250 CNY ($5124 USD)	31,500 CNY ($4854 USD)
BEV	80–150 km	35,000 CNY ($5394 USD)	33,250 CNY ($5124 USD)	31,500 CNY ($4854 USD)
	150–250 km	50,000 CNY ($7705 USD)	47,500 CNY ($7320 USD)	45,000 CNY ($6935 USD)
	>250 km	60,000 CNY ($9246 USD)	57,000 CNY ($8784 USD)	54,000 CNY ($8322 USD)
FCV		200,000 CNY ($30,821 USD)	190,000 CNY ($29,280 USD)	180,000 CNY ($27,739 USD)

[a]EREV stands for range extended electric vehicle

Table 4 Central government subsidies to PEVs and FCVs (passenger vehicle segment), 2016–2020 (2015 version)

Type	Electric range	2016	2017–2018	2019–2020
PHEV (includes EREV)	>50 km	30,000 CNY ($4545 USD)	=2016 level × 80%	=2016 level × 40%
BEV	100–150 km	25,000 CNY ($3788 USD)		
	150–250 km	45,000 CNY ($6818 USD)		
	>250 km	55,000 CNY ($8333 USD)		
FCV		200,000 CNY ($30,303 USD)	=2016 level	=2016 level

presented a series of new PEV subsidy plans that included more technology requirements on PEVs such as the energy density of the battery system and electricity consumption in BEVs, and the fuel consumptions in PHEVs. Table 5 gives an example of the technology criteria on BEVs. The government expects the increasingly stringent subsidy criteria to drive improvements in PEV technology that will ultimately make them competitive with the ICEVs.

Firstly, the minimum driving range requirement to qualify a BEV for subsidies was increased from 80 to 100 km in 2016, and further increased to 150 km in mid-2018. Starting in 2016, the maximum speed of a BEV was set at a minimum of 100 km/h (62 mph) instead of 80 km/h (50 mph). As shown in Fig. 4, nearly all

Table 5 Central government subsidies for BEVs (passenger vehicle segment) as amended in 2017 to include more specific technical requirements

Maximum speed, V (km/h)	Electric range, R (km)	Weight, m (kg)	Electric consumption, Y (kWh/100 km)	Battery energy density, Q (Wh/kg)	Subsidy
$V \geq 100$	$R \geq 100$	$m \leq 1000$	$Y \leq 0.014 \times m + 0.5$	$Q \geq 90$	20,000 CNY ($2894 USD)
		$m \leq 1000$	$Y \leq 0.014 \times m + 0.5$	$Q \geq 120$	22,000 CNY ($3184 USD)
		$1000 < m \leq 1600$	$Y \leq 0.012 \times m + 2.5$	$Q \geq 90$	36,000 CNY ($5210 USD)
		$1000 < m \leq 1600$	$Y \leq 0.012 \times m + 2.5$	$Q \geq 120$	39,600 CNY ($5731 USD)
		$m > 1600$	$Y \leq 0.005 \times m + 13.7$	$Q \geq 90$	44,000 CNY ($6368 USD)
		$m > 1600$	$Y \leq 0.005 \times m + 13.7$	$Q \geq 120$	48,400 CNY ($7004 USD)

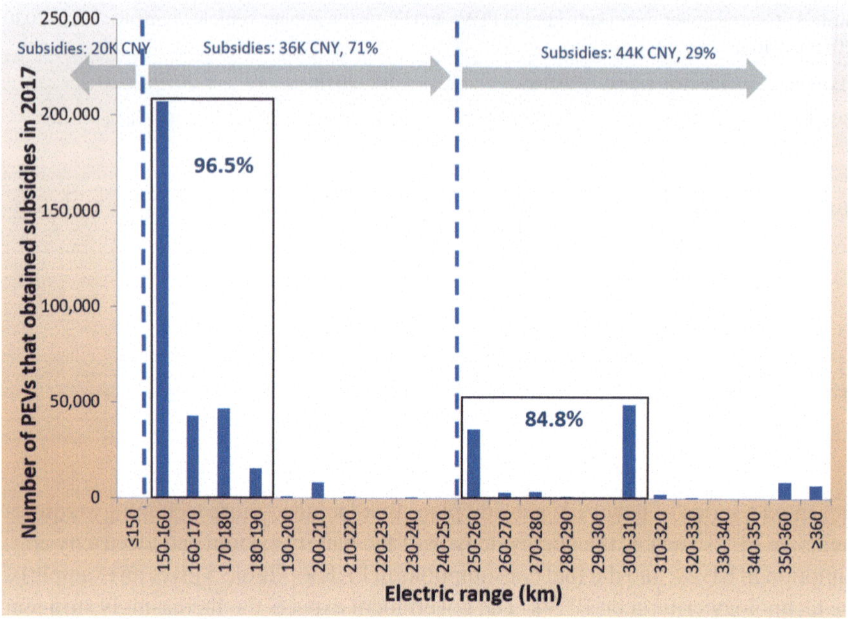

Fig. 4 Central government subsidies for BEVs (passenger vehicle segment) by electric range in 2017 (CATARC 2019)

PEVs produced in 2017 qualified for government subsidies and had an electric range greater than 150 km (CATARC 2019).

Also, in 2016, a technology criterion for energy density in the PEV battery system was added that also provided detailed standards on battery testing procedures and safety assessments. In 2017, BEVs with battery systems having an energy density over 120 Wh/kg could obtain extra government subsidies. However, the energy density of most BEVs produced in 2017 was less than the threshold for the extra "bonus" as shown in Fig. 5. More than 43% of BEVs produced in 2017 had a battery system energy density of 90–100 Wh/kg, which is far lower than the government's target of 260 Wh/kg by 2020 (CATARC 2019).

Furthermore, the new subsidy regulation after 2016 includes energy consumption as a technology criterion. This change allows PEVs with better energy-saving technology to receive higher subsidies and reduces the possibility of poor-quality PEVs entering the market. As shown in Fig. 6, the criterion for energy consumption (kWh/100 km) is an equation that depends on vehicle weight (kg) and that is progressively tightened every year. In 2018, the regulation has been made more detailed so as to distinguish the BEVs with different energy consumptions into different bands, so that the BEVs with low energy consumption can obtain higher amounts of subsidies. In this way, the government expects that the automakers, motivated by the high subsidies, will produce PEVs with more energy-efficient technologies.

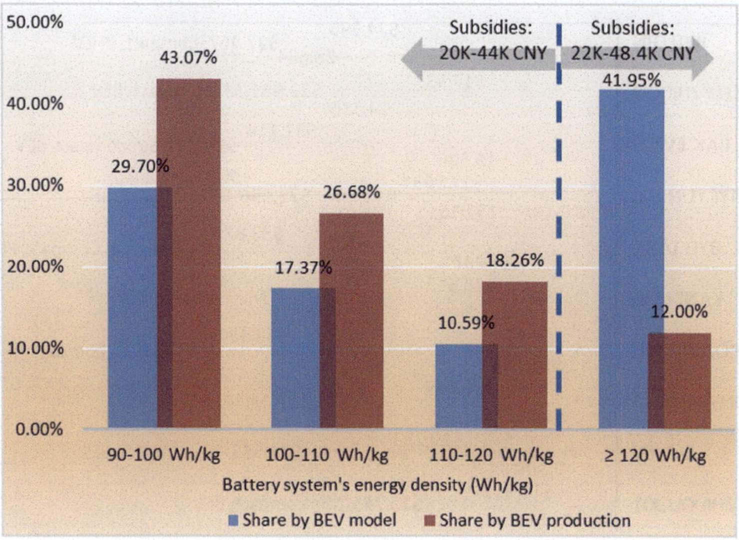

Fig. 5 Central government subsidies for BEVs (in passenger vehicle segment) by battery energy density in 2017 (CATARC 2019)

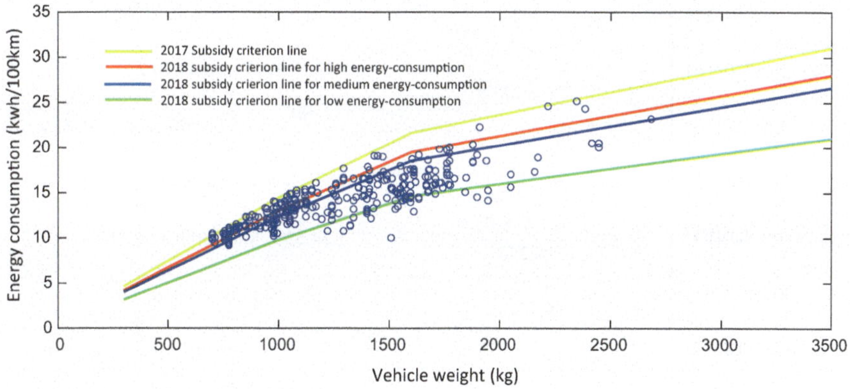

Fig. 6 Central government subsidy criteria for BEVs (passenger vehicle segment) by electricity consumption in 2017–2018 (CATARC 2019)

Because of the government's monetary incentives, the actual purchase prices for PEVs are reduced. Figure 7 shows the sales volumes and prices before and after central government subsidies for the ten best-selling PEV models in China in 2015 and 2016. The average ratio of central government subsidies to MSRP for these BEV models is about 26%, and the ratio for PHEV models is about 14%. In general, the Chinese government prioritizes the development of BEVs instead of the PHEVs, by providing fewer subsidies for the latter compared with the former. For

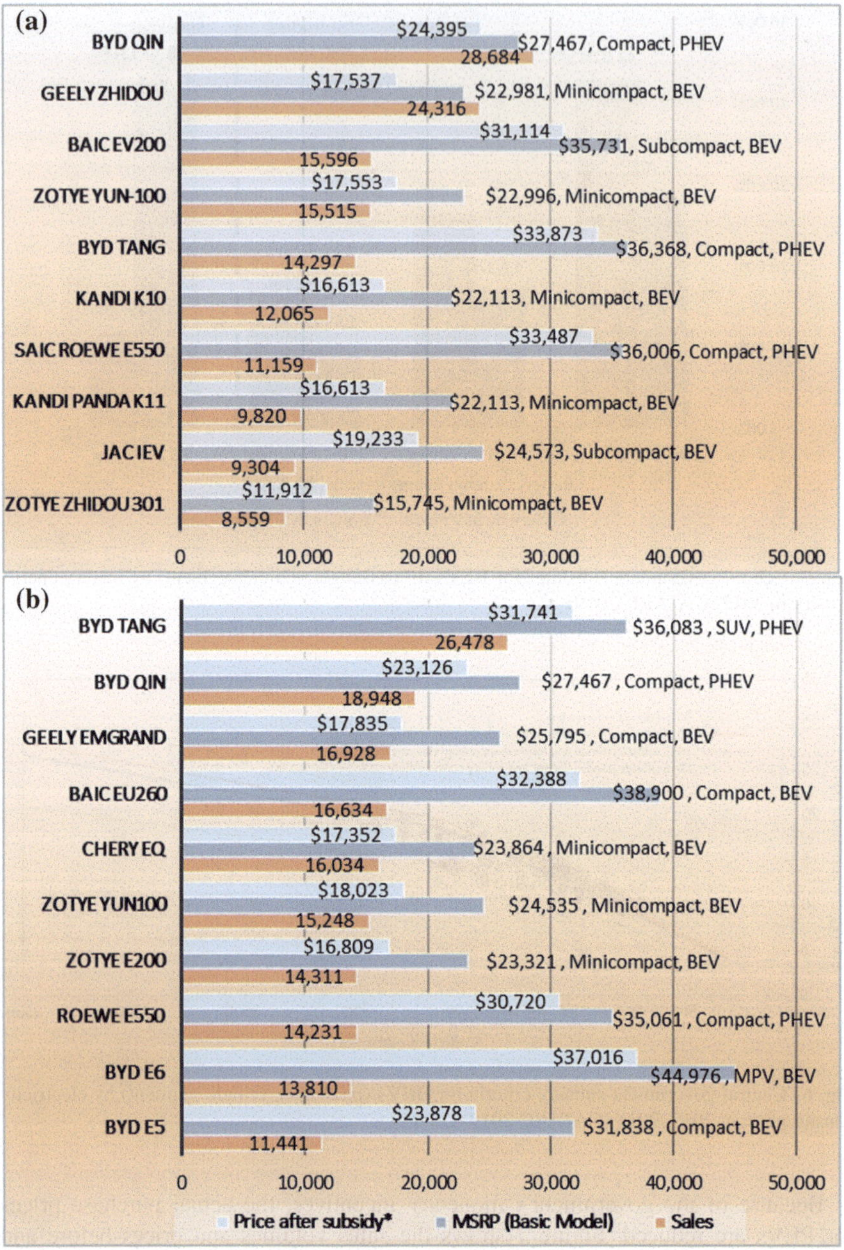

* Price after subsidy = MSRP – central government subsidy.

Fig. 7 Sales, prices, and subsidies for the ten best-selling PEV models in China in **a** year 2015, and **b** year 2016 (CATARC 2019; D1EV Web 2016; Chen 2016). *Price after subsidy = MSRP − central government subsidy

Table 6 PEV models qualified for government subsidies (passenger vehicle segment) in 2017 and 2018 (CATARC 2019)

Technology	Electric range, R (km)	PEV models qualified for subsidies	
		Year 2017	Year 2018
BEV	$100 \leq R < 150$	0	(No subsidies anymore)
	$150 \leq R < 200$	174	39
	$200 \leq R < 250$	27	15
	$250 \leq R < 300$	79	59
	$300 \leq R < 400$	62	55
	$R \geq 400$	19	19
PHEV (includes EREV)	–	43	32
Sum	–	404	219

instance, Beijing's municipality subsidizes only BEV models and excludes PHEV models from its city subsidy catalogue. This municipal regulation is criticized as regional protectionism that suppresses the sales of PHEVs from automakers outside of Beijing and favors BAIC Motors, one of the largest auto companies in Beijing, which produces only BEVs in its PEV market strategy (Sohu 2016).

The central government is becoming more experienced and cautious when approving the purchase subsidies on qualified PEVs. As there are more and stricter requirements, the number of PEV models qualified for subsidies largely decreased, as shown in Table 6.

The 2018 PEV purchase subsidy regulation stipulated that the maximum purchase subsidies will be reduced to 70% of the 2016 level (MIIT 2018). It is also possible that the central government could entirely terminate the purchase subsidies for the PEVs with short electric range earlier than originally planned. To promote the BEVs with a long electric range, the central government completely phased out the subsidies to the BEVs with electric range less than 150 km in 2018. In general, the government subsidizes more BEVs with a longer electric range, but the latter means a bigger battery and higher costs.

Moreover, in order to avoid conflicts of interest and regional protectionism, the central government decided that subsidies provided by local governments should be capped to 50% of those provided by the central government. Many municipal governments also stipulated that the total amount of PEV subsidies provided by the central and local governments must not exceed 60% of a vehicle's MSRP (China Automotive Technology and Research Center 2018). Starting in 2019, the local governments are not allowed to financially subsidize PEV (except for battery electric buses and fuel cell vehicles) purchases any more.

Nonmonetary Incentives

Vehicle License Plate Registration Privilege

Although the government has established generous purchase subsidies, the most powerful incentive to purchase a PEV is probably a special privilege for vehicle registration in the Chinese megacities. Cui et al. (2018a) quantified the local incentive policies in 30 Chinese cities in 2015 and found that the vehicle registration priority for PEV vehicles is the most valuable. These first-tier cities (such as Beijing, Shanghai, Guangzhou, and Shenzhen) have rigid regulations for vehicle license plate registration to control the fast growth of passenger cars ownership, but PEVs are exempt from ownership quotas. According to the report by Chi (2018), the result of 2018's first bimonthly license plate lottery in Beijing showed the odds of getting an ICEV plate at an astonishingly low of 1 in 1907, the lowest ever since the license plate lottery was introduced. Moreover, the ICEVs for which license plates were issued still face circulation restrictions in megacities, which intend to alleviate traffic congestion and mitigate air pollution. By contrast, PEVs are exempt from anti-congestion and anti-pollution rules.

Thus, granting vehicle license plate registration privileges is a critical motivator for potential buyers to consider PEVs in megacities. According to the research by the SHEVDC (2017), about 68% of PEV buyers in Beijing and 73% of the PEV buyers in Shanghai claim that the policy of prioritizing PEV plate licenses is the main reason they choose PEVs. Meanwhile, according to this investigation, the low operating cost is the main reason cited by PEV buyers from small cities.

Although the vehicle plate licensing privilege is a strong non-monetary incentive, this is extremely valuable, and its monetary value can be quantified. For example, in Shanghai, the average bid price from January 2017 to October 2017 for an ICEV plate ranged between 87,600 CNY and 93,500 CNY ($12,677–$13,531 USD), as shown in Fig. 8, and the average weighted price is 90,246 CNY ($13,060 USD) in 2017 (CATARC 2019). This can be considered as a shadow value of license plate regulation for PEVs in Shanghai because, compared to gasoline vehicle buyers, the PEV buyers can bypass the bidding process and avoid the additional cost involved. In other cities such as Shenzhen, which also has the PEV license plate regulation, the bidding price for license plates increases constantly. The average lowest bidding price increased to 51,658 CNY ($7476 USD) in 2017 (Shenzhen Municipal Government 2018). In Beijing, the shadow value of the license plate had already reached nearly 130,000 CNY ($20,425 USD) in 2015 (Cui et al. 2018a). After comparing with the government subsidies discussed in an earlier section, one can safely say that the shadow value of this privilege is on the same order of magnitude as the central and local government subsidies combined. Therefore, it is very enticing for new passenger car buyers that PEVs are exempt from the license plate lottery in Beijing, or bidding on license plates in Shanghai, Shenzhen, Hangzhou, and some other megacities (Ou et al. 2019). At the same time, with the rapid growth of the PEV purchases in the first-tier cities, the waiting time to obtain a free PEV license plate significantly extends.

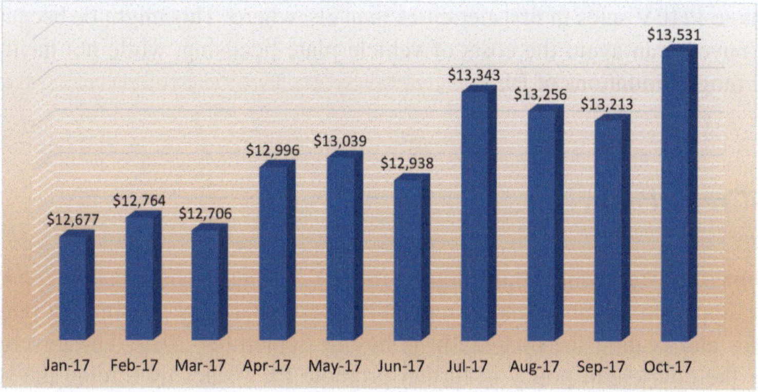

Fig. 8 Bidding prices in Shanghai during Jan. 2017 through Oct. 2017 (Shenzhen Municipal Government 2018)

For example, in Beijing, because of the limited quota for free PEV license plates in each year, the consumers now have to wait for a few years before they can obtain one. Therefore, the effectiveness of the PEV license plate registration privilege, one of the most influential non-monetary incentives in the first-tier cities, are gradually fading.

It is believed that the existence of such favorable policies in Chinese megacities is directly responsible for the high number of PEVs sold there as compared to other areas of China (Cui et al. 2018b). In order to motivate the local PEV market while controlling passenger car ownership, all the first-tier cities (Beijing, Shanghai, Guangzhou, Shenzhen) have vehicle plate license privileges for PEVs. As shown in Fig. 9, nearly one fourth of BEVs and over 60% of PHEVs were sold to these four megacities in 2017. It seems that the vehicle plate license privilege is more effective

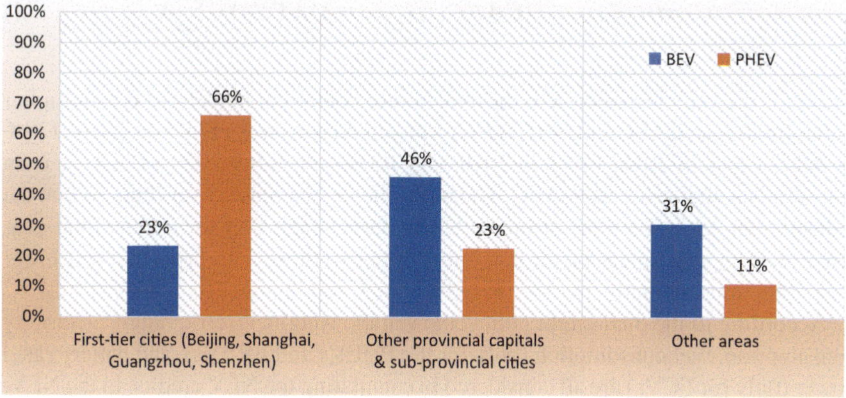

Fig. 9 PEV sales by city type in China, 2017 (CATARC 2019)

to promote PHEV sales in first-tier cities than elsewhere. This might be because the PHEV buyers can avoid the costs of vehicle plate licensing, while not having the driving range limitations of BEVs.

Dual-Credit Policy

The Chinese government is transitioning its PEV support from direct purchase subsidy to supply cross-subsidization regulation, particularly by phasing out purchase subsidies and issuing the dual-credit policy in September 2017. The government expects the dual-credit policy to further promote PEV sales by forcing the supply side (Ou et al. 2019).

The dual-credit policy has two complementary parts: Corporate Average Fuel Consumption (CAFC)-credit rules and New Energy Vehicle (NEV)-credit rules. The CAFC-credit rules require that the actual CAFC of an auto firm may be no larger than the target CAFC. The CAFC credits are calculated based on the difference between the actual CAFC and target CAFC. The NEV-credit rules require that an auto firm's ratio of ICEVs to produced or imported NEVs (including BEVs, PHEVs, EREVs, and FCVs) should be no smaller than the required minimum ratio. NEV credits are calculated based on the difference between achieved NEV scores and the NEV minimum ratios. To avoid penalties, the automakers or importers need to meet both the CAFC and NEV targets, starting in 2019. According to the policy, to reduce the difficulties, the auto suppliers in 2018 only needed to meet the CAFC targets.

In the CAFC credit rules, an auto company's actual CAFC and target CAFC are calculated based on the Chinese national standard, GB27999-2014. The CAFC credits earned in calendar year j ($Credit_{CAFC,j}$) for an auto company are calculated based on Eq. (1).

$$Credit_{CAFC,j} = \left(CAFC_T \cdot e_j \cdot s - CAFC_A\right) \cdot \sum_{i=1}^{K} V_i \tag{1}$$

where i is the vehicle model; K is the total number of the car models; e_j is a multiplier to the final-year CAFC target ($CAFC_T$) in calendar year j (134% in 2016, 128% in 2017, 120% in 2018, 110% in 2019, and 100% in or after 2020); $CAFC_A$ is an auto firm's actual achieved CAFC. The multiplier e_j decreases over time, so that the industry can achieve a more stringent CAFC target each year toward the final-year target; s is an adjustment factor that only applies to those auto firms that sell fewer than 2000 units annually, in order to reduce the CAFC burden for smaller firms.

According to the dual-credit policy, the vehicle weight, electric range, electricity consumption, fuel consumption (only for the PHEVs and EREVs), and battery rated power (only for FCVs) are all considered in calculating the NEV credits. In the NEV-credit rules, the total NEV credits earned in calendar year j ($Credit_{NEV,j}$) for an auto firm are calculated based on Eq. (2).

$$Credit_{NEV,j} = NEV_A - NEV_T = \sum_{i=1}^{M} S_{NEV,i} \cdot EV_i - R_j \cdot \sum_{i=1}^{N} ICEVI_i \quad (2)$$

where, NEV_A is the NEV score achieved by an auto firm; NEV_T is the NEV target score; M is the total number of the NEV models; N is the total number of ICEV models; $S_{NEV,i}$ is the NEV score earned by the car model i. EV_i is the annual produced or imported volume of the NEV model i; $ICEV_i$ is the annual produced or imported volume of the ICEV model i; R_j is the NEV required minimum ratio in the calendar year j (10% in 2019, and 12% in 2020).[2]

The dual-credit policy also stipulates the trading and transferring rules for CAFC and NEV credits. According to the policy, the NEV credits are tradable among auto firms; the CAFC credits are not tradable but can be transferred across shareholding or affiliated manufacturers and be carried forward to future years. More regulation details are available by downloading the dual-credit policy from the MIIT's website (MIIT 2017).

Market Dynamics Simulated by NEOCC

NEOCC was developed to capture the dual-credit policy in the Chinese passenger vehicle market. In NEOCC, a non-linear solving method combined with a multinomial logit model is adopted to explore the compliance strategies of the vehicle industry under dual-credit policy constraints. Moreover, the model studies the impact of manufacturing costs on the market of vehicle powertrain technologies, and investigates the impacts of driver travel patterns on the sales of powertrain technologies and related energy consumption (Ou et al. 2018b). Figure 10 shows the logic flow of the NEOCC model.

As mentioned by Ou et al. (2018b), some key assumptions of the model include:

- Vehicle performance (technology metrics, cost, markup, etc.), unexplained consumer preferences, and the overall vehicle market size in years 2017–2020 will remain the same as they were in year 2016. This is both for simplicity and for the consideration of the short time period.
- The model considers only sedans and SUVs because these two segments have taken over 85% of the Chinese passenger vehicle market (CATARC 2019).
- The characteristics of driver and energy infrastructure such as the driving pattern, recharging, refueling situations do not change significantly during 2017–2020.
- Consumer heterogeneity is currently ignored for simplicity.
- NEOCC uses the discrete choice model to estimate the percentage of consumer purchases by powertrain or fuel type.

[2] In the June 2017 draft version of dual-credit policy, the required NEV minimum ratio is 8% in 2018. However, the official version released in September 2017 postpones the implementation of NEV-credit rules one year later, which means only CAFC-credit rules took effects in 2018.

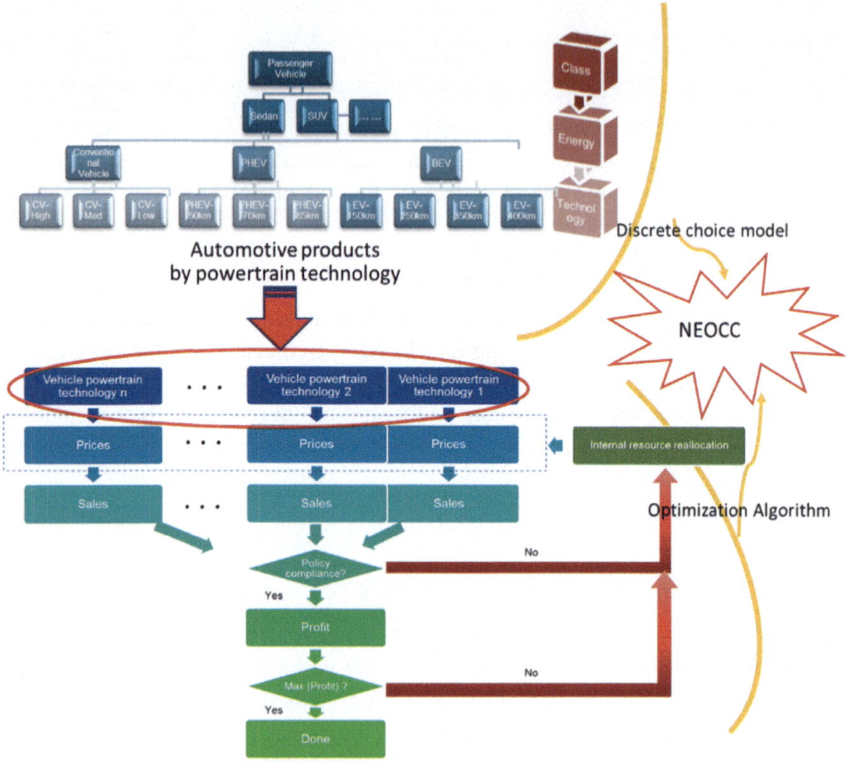

Fig. 10 Logic flow of the NEOCC model (Ou et al. 2018b)

- The model ignores quantifying the difference between production and sales, although technically the dual-credit policy is based on production (or imports).
- Gamma distribution is assumed as the probability distribution of the driver's daily travel, which is used for calculating energy costs and range anxiety costs of PEVs (Lin 2014).
- The market competition between firms inside the vehicle industry is assumed to be highly efficient.
- The conversion ratio between a CAFC credit and a NEV credit equals one, which is in accordance with the dual-credit policy.
- The vehicle industry is assumed to meet with certainty all the policy requirements, which are implemented as optimization constraints in NEOCC. The administrative penalties of non-compliance are not considered.

The NEOCC model is helpful for quantifying the impacts of the dual-credit policy on the Chinese passenger vehicle market. Figure 11 gives the simulated PEV sales under the dual-credit policy in 2016–2020. The annual sales of PEVs (passenger vehicle segment) are estimated to be about 2.33 million in 2020, resulting in a total greater than 4.75 million PEV units sold during 2016–2020, slightly under the

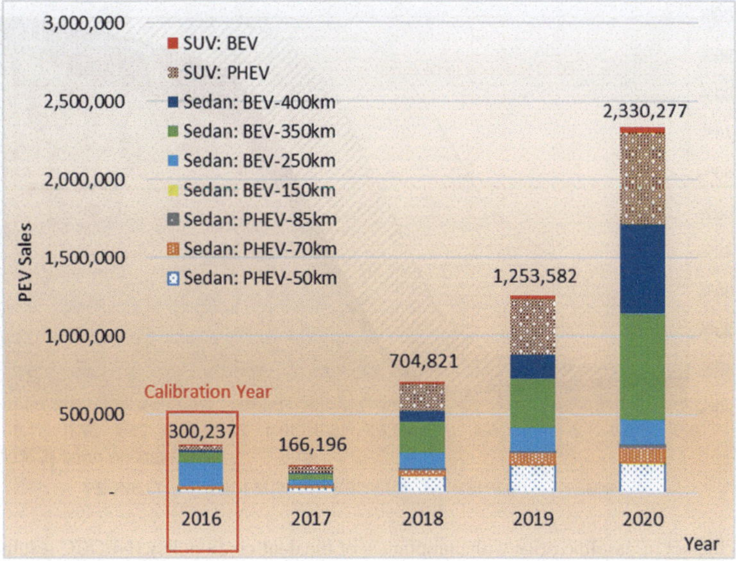

Fig. 11 PEV sales by vehicle type in 2016–2020 in the dual-credit scenario simulated by NEOCC (Ou et al. 2018b)

5 million-target announced by the government in "*The Planning for the Development of the Energy-Saving and New Energy Automobile Industry (2012–2020)*". Among these 4.75 million PEVs, about 60% are BEVs, and more than 70% are sedans. The BEV sedan with an electric range of 350 km will be the best seller among the 15 PEV types in the simulations. Considering that in reality the vehicle industry does not only pursue profit maximization for the current year, but is also concerned with future competition and profitability, it is expected that actual PEV annual sales in 2017 will be higher than the NEOCC estimations in the dual-credit policy scenario.

Further, the policy impacts on profit (PIP) of the automotive industry can be quantified by the NEOCC model. Compared to no rules in the market, as estimated by the NEOCC model, the automotive industry with the dual credit policy will "lose" around $2,122 per vehicle by 2020. Note that PIP is intended to reflect the effect of the policy, and by no means indicates the industry's actual profitability. 2016 is the calibration year, when the dual-credit policy is not yet implemented, thus the PIP in 2016 is set at zero. Nevertheless, the PIP magnitude gradually increases after 2016, especially starting in 2019 when both rules are implemented. This is because the gradual phasing-out of the PEV subsidies and the increasingly stringent CAFC and NEV targets force the automotive industry to internally subsidize the more efficient ICEV and PEV products in order to achieve the sales needed to comply with regulation. This causes a drop in the profit of the automotive firms.

However, the impacts of PIP could be afforded by the automakers through optimizing the PEV manufacturing costs or adopting new fuel-saving technology into the

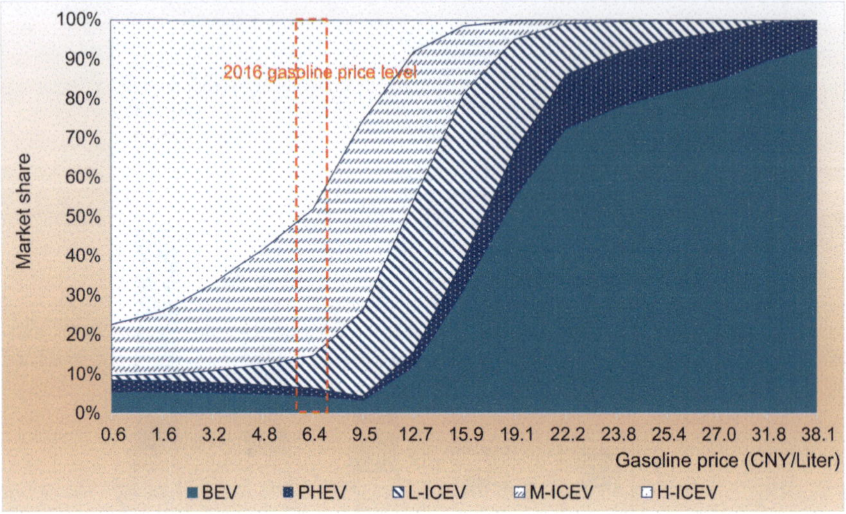

Fig. 12 Impact of gasoline price with compliance of the dual-credit policy (NEOCC results)

ICEVs. More importantly, because the dual-credit policy stipulates the rules for trading the NEV credits among companies, some OEMs who have surplus NEV credits could make up their losses or even profit by selling the credits to others. Moreover, along with the government incentives on the PEV industry, most automakers have already been transitioning their strategies toward PEVs and preparing for more stringent energy consumption and emission policies. Therefore, it is expected that the automotive industry will still be able to digest the temporary profit losses.

The fluctuation of the gasoline prices could also impact the buyer decision making on vehicle purchases, which is simulated by NEOCC. In the model, the gasoline price of 6.35 CNY/L, which corresponds to 3.48 USD/Gallon, and is the average gasoline price in China in 2015, is used as the baseline. Figure 12 presents the changes of vehicle shares with the gasoline prices in year 2019 under the dual-credit policy. The ICEVs are grouped into three fuel types: low-fuel consumption ICEVs (L-ICEVs), medium-fuel consumption ICEVs (M-ICEVs), and high-fuel consumption ICEVs (H-ICEVs). The PEV market share simulated in the NEOCC model decreases to about 4.4% when the gasoline price is around 9.53 CNY/L, or 5.22 USD/Gallon. This is because when the gasoline price has a minor increase, consumers tend to buy fuel-efficient ICEVs instead of the PEVs, as shown in Fig. 12. When the gasoline price keeps growing and obviously exceeds the current price level, the consumers start to consider PEVs, and PEV sales become very sensitive to gasoline prices. Although gasoline price is regarded as a critical element for consumers considering PEVs, the simulation results indicate that consumers are not urged to purchase PEVs under the current gasoline price range unless prices continuously rise.

Although the refined oil product price mechanism was enacted by the Chinese state agency—National Development and Reform Commission (NDRC) (NDRC 2018), the instructed refined oil product price by the government refers to the "international crude oil price" which is a weighted value measured based on the spot price of oil trade markets in Brent, Cinta (Indonesia) and Dubai. Although regulated by the government, the instructed price is adjusted frequently, so that it can follow the international oil market. However, to motivate the vehicle market transition to the energy saving technology, starting from 2009, the government has been increasing the gasoline consumption tax (GCT) which was added from 1.00 CNY/L in 2009 to 1.52/L CNY in 2015 (NDRC 2018). The GCT is now around 20% of the final gasoline price paid by the consumers.

The dual-credit policy plays a significant role in the vehicle industry (Ou et al. 2017). Under this policy, simulation shows that BEV sedans with long electric ranges are more acceptable by the consumers than other PEV models: the BEV-350 sedan will be the most popular among PEV models, followed by the SUV PHEV. The model also shows that the gasoline price could impact the consumer purchases in the vehicle market, but it seems that the consumers prefer to buy the fuel efficient ICEVs than the PEVs when the gasoline price increases within the range of fluctuation observed in 2016.

Though the NEOCC model is very helpful in analyzing the consumer decisions in the passenger vehicle market under compliance with the vehicle policies, some limitations exist in this version. This NEOCC version considers only vehicles purchased by individuals, while many PEVs are purchased as fleet cars by government or companies in China. Moreover, to obtain the largest profits in a particular year, in the model all available NEV credits are traded and all available CAFC credits are used and not assumed to be carried forward to future years.

Local Government Incentives

Monetary Incentives from Local Governments

In addition to the PEV incentives from the central government, local governments influence the local PEV market as well as the PEV industry. Since the beginning of the "*Ten Cities—Thousand Vehicles Demonstration Program*" in 2009, many provincial or municipal level governments have set their own PEV sales targets. By the end of 2015, the PEV demonstration cities had put in place more than 190 different incentives schemes in order to support adoption of PEVs. Figure 13 illustrates the cumulative PEV sales by province in China by 2016 and by 2018 (CATARC 2019).

Table 7 provides an overview of the local incentives available in the cities and provinces where most PEVs were sold in 2018. In general, the PEV market was much better developed in the coastal areas and megacities compared to inland areas and small cities. For example, Beijing, Shanghai, and Shenzhen (first-tier cities in China), and Jiangsu, Zhejiang (coastal provinces) have the most PEV sales, and their sales target ratios are the highest in China. This might be because higher fiscal revenues

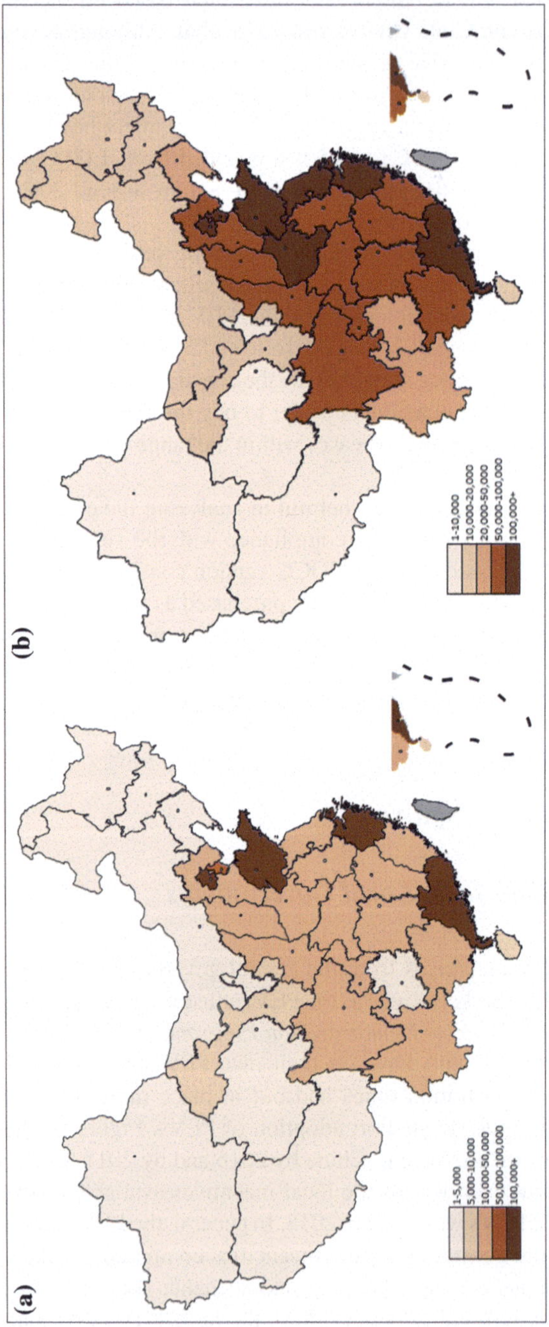

Fig. 13 Cumulative PEV sales by province in China in **a** 2011–2016; **b** 2011-2018 (CATARC 2019)

Table 7 Local government subsidies to PEVs (individual purchase), 2018 (Sohu Auto 2018)

City/Province	Subsidy target	Maximum subsidy
Provinces: Guangdong, Henan, Hainan, Qinghai, Shanxi, Yunnan	BEV/FCV/PHEV	CGS × 50%
Changsha (city)	BEV/FCV	CGS × 20%
	PHEV	CGS × 15%
Shanghai (city)	BEV/FCV	CGS × 50%
	PHEV	CGS × 30%
Wuhan (city)	BEV/FCV/PHEV	Wheel base > 2.2 m: CGS × 50%
		Wheel base ≤ 2.2 m: CGS × 20%
Xi'an	BEV/FCV/PHEV	CGS × 30%
Cities: Chengdu, Guangzhou, Hangzhou, Harbin, Wenzhou, Xiamen, Zhoushan	BEV/FCV/PHEV	CGS × 50%

CGS Central Government Subsidies

allow them to provide better monetary incentives and a more favorable institutional environment. After 2016, when the central government started to phase out subsidies, the direct monetary incentives from local governments were also reduced gradually. Besides, to avoid potential regional protectionism and fiscal waste, many cities stipulated that the monetary incentives from the local government should be less than 50% of the monetary incentives from the central government, and that the total subsidies cannot exceed 60% of the PEV's MSRP (Sohu Auto 2018). In second half of 2019, the direct monetary incentives from the local governments completely exited the PEV market (except for battery electric buses and FCVs).

Other Local Government Incentives

Local governments are more flexible than the central government and can provide PEV incentives that are tailored to the needs of local consumers and automakers. Therefore, in addition to direct purchase subsidies, municipal governments offer other types of incentives to PEV owners: road toll exemptions, discounted electricity prices, free public parking, access to high occupancy lanes, grants for installing home charging facilities, etc. The most common incentives provided by municipalities are further discussed below.

- Exemption from traffic restrictions for PEVs: In cities such as Beijing or Wuhan, road space rationing or traffic restrictions are enforced for passenger vehicles on workdays in attempts to mitigate traffic congestion and emissions. Because PEVs are exempt from the restrictions in these cities, it largely reduces the inconvenience to the PEV owners (Ou et al. 2019).

- Free toll roads and parking for PEVs: in Wuhan, the toll bridges and highways are toll-free for PEVs. The municipality of Shenzhen allows first hour free parking for PEVs in municipal parking lots.
- Subsidy and tax exemption for installing privately-owned charging outlets: in the city of Shenzhen, the government directly subsidizes PEV buyers 10,000 CNY to 20,000 CNY ($1541 to $3082 USD) for the car insurance, and the installation fee on the home charging outlets in 2016 (CATARC 2019).
- Subsidies to install and operate public charging stations. The municipality of Shenzhen increased the subsidies for public fast-charging outlet installation from 300 CNY/kW ($43 USD/kW) in 2016 to 600 CNY/kW ($87 USD/kW) in 2017, and public slow-charging outlet installation from 150 CNY/kW ($22 USD/kW) in 2016 to 300 CNY/KW ($43 USD/kW) in 2017 (Shenzhen Municipal Finance Commission 2017).
- Support for local PEV manufacturers and other auto parts suppliers. In order to support the development of the local PEV industry, local governments often offer automakers tax exemptions for PEV R&D and PEV sales expenses (Cui et al. 2018a).

Although the local governments have issued diverse incentives to encourage the PEV sales, the government subsidies and intervention would not be a sustainable means for developing a stable and competitive market in the long run. The central government has been thinking to limit the extra subsidies, which could potentially distort the market and bring regional protectionisms, to the PEV purchases (CATARC 2019). In the near future, it is expected that the non-monetary incentives will stay as a major means for the PEV market growth. For example, the local government will focus the non-monetary incentives on the charging infrastructure which can largely reduce the PEV charging inconvenience and range anxiety (Ou et al. 2019).

Consumer Concerns in China's PEV Market

BEV Range Anxiety

Range anxiety can be regarded as the biggest concern for consumers in the Chinese PEV market. The 2016 SHEVDC study has revealed that the first concern among the Chinese consumers that took part in the interview is the electric range of BEVs (39%) (SHEVDC 2017). On the one hand, a BEV with a longer electric range could largely relieve the consumer's driving anxiety. On the other hand, a longer electric range generally means a bigger battery and higher production costs for the BEVs.

Automakers have to consider the trade-offs between the vehicle production costs and electric range and think about the question—what is the best electric range for a BEV? In the U.S., researchers suggested an electric range with 100–200 miles should be probably more effective on the BEV promotion based on the real world consumer driving patterns (Lin 2014; Tamor et al. 2013). Though it is still unclear what is the best electric range in BEVs which are appropriate to most driving patterns in China, the consumer range anxiety costs and fuel costs can be quantified (Lin 2014) and compared for the BEVs with different electric ranges. The method to calculate the consumer range anxiety costs is based on the study by Lin (2014).

First of all, the vehicle lifetime kilometers (L) is estimated using Eq. (3).

$$L = \sum AVKT/(1 + DR)^i, \quad i \in Y_{PEV} \tag{3}$$

where AVKT, the annual VKT, is 19,681 km, or 53.92 km daily vehicle kilometers travelled (DVKT) mean (M_n) (CATARC 2014); DR is the discount rate which is assumed to be 7%; Y_{PEV} is a perceived lifetime used in calculation of energy costs (Lin and Greene 2011). Some scholars believe that consumers highly discount future energy savings and recommend 2–4 years for calculation, while some others believe in consumer full valuation of fuel economy and recommend 14–16 years, close to the average lifetime of a vehicle (Xie and Lin 2017). We adopt 10 years as a commonly used middle point value.

The variation of DVKT is characterized as a Gamma distribution, which has been validated using real-world GPS travel data (Lin 2014). The daily commute round trip is approximately 16.22 km, which is assumed as the DVKT mode (M_d) (CATARC 2014). The shape parameter (k) and scale parameter of the Gamma distribution (θ) are two important metrics to shape the Gamma distribution p(x) and are estimated based on Eqs. (4) and (5).

$$k \cdot \theta = M_n \tag{4}$$

and

$$(k - 1) \cdot \theta = M_d \tag{5}$$

If the electricity charging costs at home and in public are the same for the BEV owners, the daily electric VKT (EVKT) can be estimated using Eq. (6) (Lin 2014).

$$EVKT = \int_0^{L_E} xp(x)dx + \int_{L_E}^{+\infty} p(x)dx \tag{6}$$

where x is the random DVKT (km) that follows a Gamma distribution probability density function $p(x)$ and L_E is the electric range of a PEV (km). Therefore, the lifetime electricity consumption costs can be calculated with Eq. (7).

$$C_e = EVKT \cdot \frac{L}{M_n} \cdot \delta \cdot P_e \tag{7}$$

where C_e is the lifetime electricity costs for a BEV ($ USD), δ is the electricity consumption rate in a BEV (kW/km), and P_e is the electricity price ($ USD/kWh) in China in 2016, which is assumed to be $0.081 USD/kWh.

The quantitative value of the BEV range anxiety is assumed to be equivalent to the cost that the BEV driver has to pay to complete the trip when the BEV is out of its electric range. When electric range has run out, it is assumed that the driver will have to resort to alternative modes/means, such as renting a gasoline car to finish the trip that day. Therefore, the BEV range anxiety could be estimated as in Eq. (8).

$$C_a = [1 - p(x \leq L_E)] \cdot \frac{L}{M_n} \cdot P_{rental} \tag{8}$$

where, C_a is the lifetime range anxiety costs for a BEV ($ USD), and P_{rental} is the average price of a rental car in China in 2016, which is assumed to be $29 USD/day. Meanwhile, the measurement of range anxiety via the concept of backup vehicles is an assumption that needs to be further validated (Lin 2014). For example, an alternative way to look at the cost of range anxiety may be the willingness to pay for extra range, which would be highly dependent on the income level of the drivers and their attitudes to PEVs, could lead to different results altogether. However, the backup gasoline vehicle can be a viable substitute to the BEV when its electric range is not enough to finish the trip, and the value of the backup vehicle can be quantified (e.g., from the rental car price). Therefore, the concept of backup vehicles is adopted for quantifying the range anxiety in this study.

Considering the market data information collected in Table 2, the lifetime electricity costs and lifetime range anxiety costs borne by the average Chinese vehicle driver can be calculated and shown in Fig. 14. Since the driver travel pattern and driving costs vary by region, the calculation assumes an average Chinese driver scenario, which is used for the average cost level in China. The BEV lifetime is assumed to

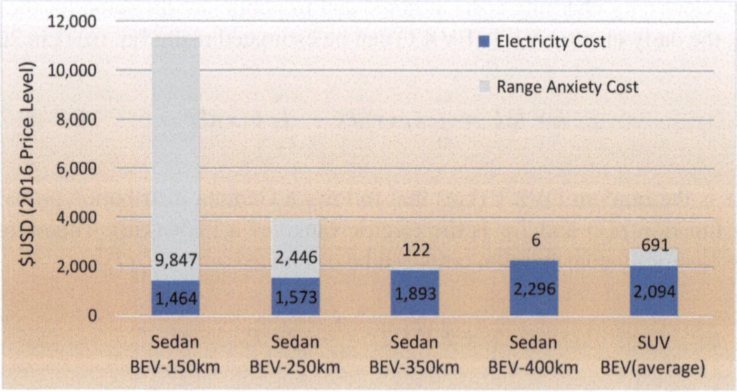

Fig. 14 Lifetime electricity cost ($) and lifetime range anxiety cost ($) for an average Chinese PEV driver in 2016

be ten years, and the discount rate of lifetime mileage is assumed to be 7%. When driving a battery electric sedan with 150 km electric range, the lifetime range anxiety cost can be nearly $10,000 USD, which can be as high as one third of the average MSRP ($29,192 USD). With the increase of the BEV electric range, the lifetime range anxiety cost declines quickly and becomes a small amount for the battery electric sedan with 350 km electric range, since the vehicle with such electric range could basically satisfy the average travel pattern of the Chinese drivers.

The range anxiety and electricity costs could vary with the driver travel patterns by geographic area and city type. For example, the annual VKT is an important indicator for quantifying the driver ownership costs. It shows that the annual VKT in Beijing is constantly decreasing: the annual VKT of privately-owned cars changes from 19,596 km in 2011 to 11,983 km (Beijing Transportation Research Center 2017) in 2016. The data on the annual VKT and daily commute distance adopted in this study is based on research conducted years ago, and therefore the accuracy of cost quantification may be impacted by changes in travel patterns that have occurred since. However, the methodology used in this study to quantify the effect of range anxiety on consumer adoption of PEVs remains relevant.

Charging Infrastructure

The concern of Chinese PEV adopters on the charging infrastructure is second only to the concern about the PEV electric driving range (SHEVDC 2017). To stimulate the PEV market, China has deployed a large-scale initiative to create public charging facilities and has also established monetary incentives for installing home charging outlets. According to the government charging infrastructure policy "*Guidelines for the Development of Electric Vehicle Charging Infrastructure (2015–2020)*" issued in 2015, more than 120,000 charging or battery swap stations will be built, and more than five million charging outlets will be installed in China by 2020. The "*Technology Roadmap for Energy Saving and New Energy Vehicles*" (Technology Roadmap), released by the technology roadmap advisory committee in 2016, proposed the targets of the charging infrastructure in China, as shown in Table 8 (The Strategic Advisory Committee, SAE-China 2016).

Table 8 Major targets for the charging infrastructure in China (The Strategic Advisory Committee, SAE-China 2016)

Indicators	Year 2020	Year 2025
Charging/battery swap stations	≥12,000	≥36,000
Charging outlets	≥5 million	≥20 million
Fast charging	Electric range ≥ 100 km in 15 min	Electric range ≥ 100 km in 10 min
Slow charging	Power ≥ 6.6 kW	Power ≥ 10 kW

Table 9 Charging service fees in cities and provinces from 2014 to 2017 (CATARC 2019)

Cities/Provinces	Charging service fee
Beijing	≤0.80 CNY/kWh
Shanghai	1.30–1.60 CNY/kWh
Guangdong	0.70–1.00 CNY/kWh
Anhui	0.53–1.7 CNY/kWh
Jiangsu	1.11–1.80 CNY/kWh
Shandong	0.50–1.35 CNY/kWh

Local governments and PEV manufacturers also offer incentives for the purchase and installation of charging outlets. Summarized by Cui et al. (2018), PEV owners can charge their vehicles with a subsidized price of electricity in public charging infrastructure in cities such as Shanghai and Shenzhen. In order to encourage private capital to participate in the construction and operation of the public charging infrastructure, the government allows public charging operators to directly collect a user charging service fees, in addition to the electricity costs, on both slow and fast charging. Table 9 presents the charging service fees in some cities and provinces from 2014 to 2017 (CATARC 2019). For example, the charging service fee is stipulated to be less than 0.45 CNY/kWh in Shenzhen (Cui et al. 2018a). At the same time, PEV manufacturers such as Beijing Auto, BYD, JAC, and even Tesla have similar policies to offer PEV buyers' free installation or free home charging outlets. These incentives could largely increase the willingness of potential consumers to purchase PEVs.

By October 2019, China had installed nearly 480,000 non-home charging outlets, which grow nearly 90% more than May 2018. As shown in Fig. 15, the number of non-home charging facilities (including both public and privately-owned charging outlets) in the United States is only one-third of the number in China, where the government has been investing heavily in the charging infrastructure (Ou et al. 2017). Specifically, in China there are more non-home DC fast-charging outlets available than slow-charging only outlets, with some charging outlets available for both DC fast-charging and slow-charging. Note that consumers in China rarely have a garage for charger installation. By May 2018, there were 304,412 home charging outlets in China, which is much fewer than the PEV population. This indicates that the public chargers in China could be, to a great extent, used as primary chargers. Where the PEV owner does not have access to a private garage or private parking space, the PEV's primary charger could be public infrastructure at on-street parking spaces. It is therefore reasonable to assume that public charging could be more convenient in China than in the U.S. Availability of public charging infrastructure seems to provide another explanation, besides cost-saving reasons, as to why Chinese consumers chose short electric range PEVs: widely accessible public fast-charging facilities (200, 281 by October 2019) might improve the willingness to purchase inexpensive PEV with lower electric ranges even without home charging facilities. For example, in Beijing, the charging outlets are provided in public parking lots that PEV drivers have priority

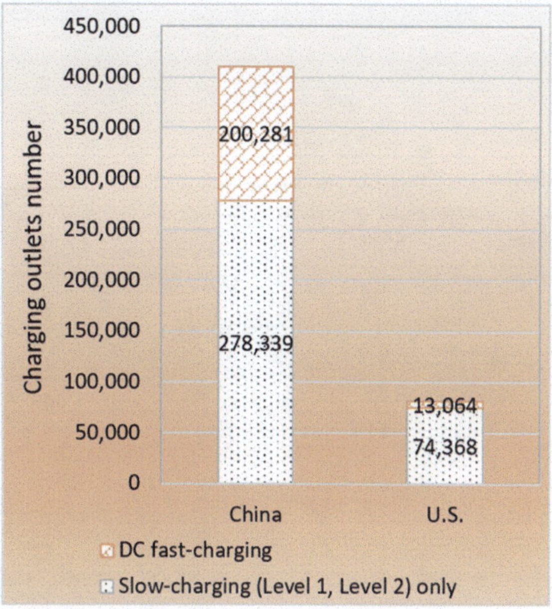

Fig. 15 Non-home charging outlets in China by October 2019 and in the United States by February 2020 (CATARC 2019; U.S. Department of Energy 2020)

to use, therefore guaranteeing the PEV owner a temporary public parking place (1–2 h) where ICEV owners would have to compete among themselves for limited parking places in crowded urban areas (Beijing Municipality Government. 2018).

Ownership Costs of Home-Charging

Residential parking indicates the off-street spaces owned by local residents. For example, the off-street parking lots include underground car parking below the building where individual parking spaces are owned by those who own flats in the building. Considering charging costs and convenience level, home charging is widely considered to be the most important charging option for PEV adoption, but it requires adequate and dedicated residential parking spaces. Unlike most new car buyers in the United States, Chinese car owners predominantly live in high-rise apartments where parking spaces are usually limited, which could largely increase the ownership costs on the PEV users in China.

Based on the estimations by Ou et al. (2018a), the residential parking availability, which is the ratio of the resident parking spaces to the number of households in the communities, in 2015 was severely insufficient, as Fig. 16 shows. For example, in Shanghai, one of the largest cities in terms of population and PEV stock, residential parking availability is only about 26%, which indicates that only 26 out of one hundred households in Shanghai have resident parking spaces that could accommodate

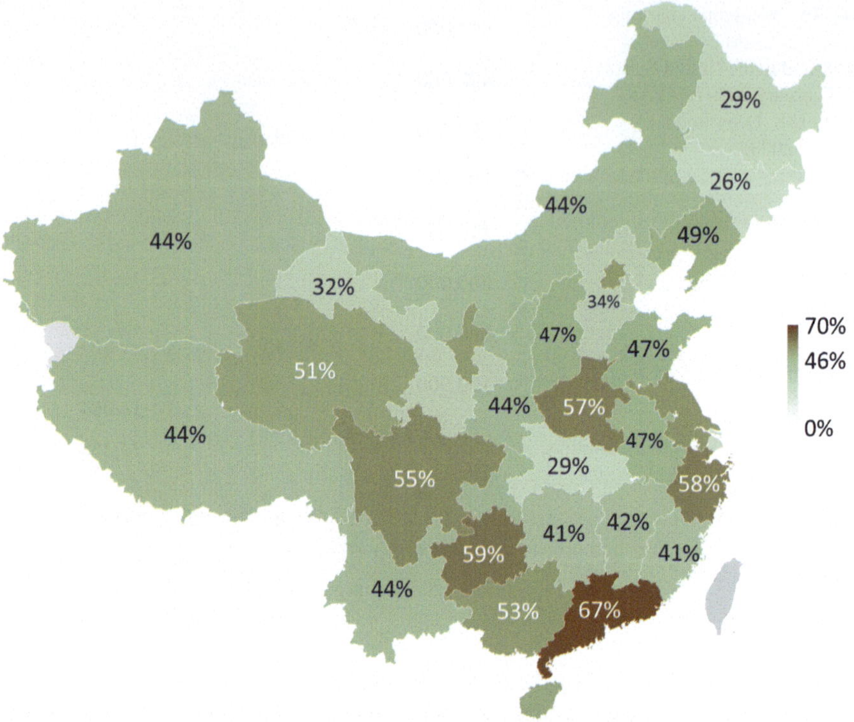

Fig. 16 Estimated residential parking availabilities in the Tier 1 cities of mainland China in 2015 (Ou et al. 2018a)

the installation of PEV charging outlets. The average national home parking availability was respectively 46.2% in Tier 1 and 43.6% in Tier 3 cities; however, it was 38.6% in Tier 2 cities in 2015.

The lack of residential parking availability could greatly increase the costs of PEV ownership and might severely restrict the convenience of owning PEVs among those who have no access to off-street parking. Ou et al. (2018a) quantified the ownership cost impacts of a residential parking space to a PEV owner. It is an additional monetary cost of the electricity used for charging, and also includes an estimate of the inconvenience/time cost of having to use public chargers as primary chargers. Therefore, the variation of the quantified values depends on the vehicle powertrain, drive patterns, charging habits, location by city type and region. The detailed information for calculating the shadow values can be referred to in Ou et al. (2018a).

The shadow value of the residential parking space for PEV owners ranges from $2376 to $10,687 USD, depending on electric range and powertrain type, as shown in Fig. 17. In particular, the highest mean value for a residential parking space is $10,687 USD for a BEV-150 km (a BEV with electric range of 150 km) sedan. This is because BEV-150 km sedan owners have to look for vehicle charging more

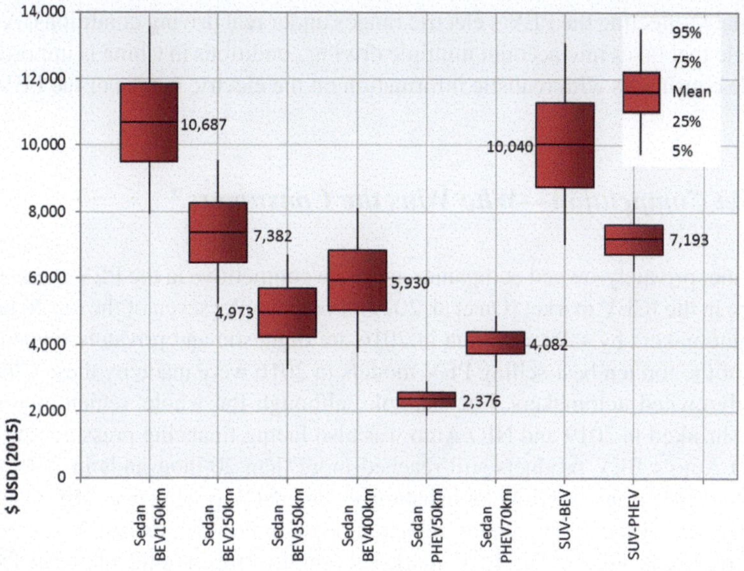

Fig. 17 The shadow values of home parking by vehicle type for PEV owners who have no home parking places in 2015 (Ou et al. 2018a)

frequently if they have no residential parking spaces for home charging. Meanwhile, the sales-weighted MSRP of a BEV-150 km sedan is $17,248 USD in China in 2015 (CATARC 2019), which is approximately $7000 USD more than the shadow value of the residential parking space. Considering the central government subsidy to BEV-150 km sedan at $6935 USD in 2015, the shadow value from the residential parking space to the BEV-150 km sedan is almost the same as their actual purchase price in the estimation.

Battery Performance

Battery performance issues, especially range deterioration, could dampen PEV values. A PEV's electric range in reality is often shorter than the MIIT's rated range, due to battery degradation over time or the PEV's usage conditions (e.g. driving with air conditioning on). These factors could exacerbate the range anxiety of PEV drivers. According to the D1EV website, there are approximately 330 reported complaints about PEVs in China in 2017; one of the most unsatisfactory issues is the "lost" electric range (Chen 2018). On the one hand, the intrinsic reason for "lost" electric range is the limitation of the battery technology: lithium battery (e.g. the NMC/NCA or LFP battery) performs worse at low temperatures. On the other hand, the New European Driving Cycle (NEDC), adopted by the MIIT as the standard driving cycle for testing energy consumption and electric range in China, has obvious flaws in accurately reflecting the PEV's electric ranges under real driving conditions. A driving cycle that takes into account multiple driving conditions in China is important to provide consumers with realistic information on the electric ranges of the PEVs.

Market Competition—Who Wins the Consumers?

Domestic, privately-owned companies are more competitive in the PEV market than they are in the ICEV market (Ou et al. 2019). For example, seven of the top 20 largest PEV automakers by sales in China in 2016 are domestic and privately-owned, and seven of the top ten best-selling PEV models in 2016 were made by these Chinese, privately-owned automakers. For example, although the whole vehicle marekt in China shrinked in 2019 and NIO Auto was also facing financing pressure, the sales of NIO Auto's BEV products still reached more than 20 thousands in 2019 (81% more than 2018), and its NIO ES8 (a battery electric SUV pricing over $55,000 USD) has been one of the most pouplar vehicle models in the high-end SUV segment in China (OFweek 2020). The PEV market seems to offer a solid platform for the small, privately-owned automakers to compete, unlike the ICEV market, which has higher entry barriers and is dominated by established large OEMs. The Chinese ICEV market is dominated by a few state-owned automotive OEMs that partner with foreign OEMs such as General Motors, Toyota, and Volkswagen. The incumbents who have integrated upstream suppliers, distribution channels and downstream dealers have strong advantages over new players. Other barriers to entering the ICEV market include strict industrial quality standards and mature safety requirements that entail professional labor training, aftermarket services, along with the need for complete R&D systems and sustainable investments. However, PEV manufacturing still lacks systematic and widely accepted industrial standards. To avoid over-investment in ICEVs and to promote the PEV industry, the government ruled that no more business licenses would be granted to new ICEV OEMs, while it welcomes capital investment in the PEV industry (Xinhua News Agency 2016).

The more established Chinese ICEV car companies seem to be smaller players in current PEV market (Ou et al. 2019). The top conventional car makers SAIC, Dongfeng, FAW, and Chang'an in China, all of which are state-owned, rank only at 4th, 11th, 29th, and 7th in the 2017 PEV sales (CATARC 2019). The incumbent automakers might be too cautious, even though they have PEV strategic plans on paper, or are reluctant to dilute their lucrative business in ICEVs. For example, FAW Auto, one of the largest state-owned automakers in China, has shown poor performance in their PEV sales even with its advantages in R&D, marketing, and mass-manufacturing plants. Nevertheless, not all incumbent automotive OEMs are slow to act in the PEV industry. For instance, BAIC (Beijing Auto), although not a leading state-owned automaker in ICEV manufacturing, held first position in the PEV market with 100,293 units sold in 2017 (CATARC 2019). Considering that the PEV market growth is still in its early stages in China, the incumbent state-owned auto firms still have a competitive edge in the long-term by relying on their advantages in government financing, abundant cash flows, and technological support from their multinational strategic alliances.

Foreign automakers who manufacture high-end PEVs are also extremely competitive on the Chinese market. On the one hand, leap-frogging to PEVs and cultivating a competitive domestic industry is one publicly intended goal for Chinese policymakers, as evidenced in policy documents such as the "*Made in China 2025*" published by the State Council, which states that energy-saving and new energy vehicles are the pillar industries to transform China into a leading manufacturing power (Ou et al. 2017). On the other hand, the state planners also understand how important openness and fairness are for a healthy and sustainable industry. China has laid out a plan to gradually open the vehicle industry by 2022, and has pledged to completely remove the foreign ownership cap for companies making PEVs by the end of 2018 (Moss and Colias 2018). Foreign car companies have been allowed to make vehicles in China without partnering with a Chinese automaker. For example, Tesla built its first oversea Gigafactory in Shanghai within a year in 2019, and it aims to deliver cheaper BEV products for the Chinese consumers and to compete with the local BEV manufacturers (such as NIO and Xpeng Motors) and the global giants (such as BMW, Daimler AG) (Bloomberg 2020). Furthermore, the government reduced the vehicle import taxes from 25 to 15% in 2018 (The Tariff Commission 2018). With mature production management in PEV manufacturing, foreign companies can keep costs low for customers and access the PEV market in China. In addition, to counter possible regional protectionisms, the central government is planning to limit the direct monetary incentives from regional governments to their local automakers (Tian et al. 2017).

In a fiercely competitive environment, learning how to meet the consumer demands is very important for the automakers to win substantial shares of a growing market. So far by 2020, the automakers involve into the PEV market under financial support and policy compliance by the government, while ultimately the consumers are the most important factor that shapes the market (Ou et al. 2019a). On the one hand, the government subsidizes more BEVs with longer electric range than those with shorter electric range, and the automakers also tend to emphasize their efforts on extending the electric range in their BEV products so that the demands of the consumer can be met. However, on the other hand, the low speed electric vehicles, usually with an electric range shorter than 150 km and therefore ineligible for any government subsidies, are ignored by the mainstream automakers but surprisingly well accepted by consumers in China—annual sales of low-speed electric vehicles reached over 1.7 million units in 2017 (ResearchAndMarkets.com 2018). According to the study by Ou et al., for the privately-owned vehicle drivers in China, 99% probability of the daily driving distance is no more than 88.0–112.0 km (Ou et al. 2019b). It might explain why low-speed electric vehicles are so popular in China might be complex, but the sales record indicates that an electric range below 150 km can, more or less, meet the travel demands of many drivers.

Conclusions

From the discussions laid out in this chapter, it is safe to conclude that PEV policies in China have been shaping the vehicle sales, the urban infrastructure, and the industrial supply chain. Even during the period of low oil price, the PEV market has continued to grow rapidly, driven by strong government incentives. Uptake of electric vehicles in China has been favored by several factors: availability of several electric vehicle models, including affordable ones in the compact city car segment, passenger car driving patterns that are compatible with electric vehicle range limitations, environmental awareness of the passenger car buyers, and more. However, currently, the single most important factor is the presence of strong monetary and non-monetary incentives from the government. Electric vehicles are a pillar of the Chinese government's industrial policy and, in order to support the development of PEV automakers and of the related supply chain, huge direct subsidies from both central and local governments for PEV purchase have been provided which have allowed the market to grow rapidly. The subsidies have made PEVs highly price competitive in China and have minimized potential risks for PEV industrial investors. However, monetary subsidies in particular have proved controversial: while supporting the market, they have also brought problems such as subsidy frauds, regional protectionism, financial fiscal waste, and market distortion. Partly for this reason, direct subsidies to PEV purchase are now being phased out and government support is gradually transitioning toward supply-side policy measures such as fuel economy standards and PEV quotas.

This chapter has addressed the question whether the recent rapid growth of the Chinese PEV market can be sustained. In addition to reviewing government incentives, this chapter has also analyzed consumer concerns over PEVs, electric vehicle market development, and industrial competition. The main insights gained from the analysis of the Chinese PEV market are summarized below:

- In the passenger vehicle segment, more PEVs are sold in large cities which are able to offer favorable subsidies, better charging infrastructure availability, and more driving friendly rules to PEVs. In particular, vehicle license plate privilege is deemed as a major motivation to the consumers who purchase PEVs in megacities such as Beijing and Shanghai.
- Meanwhile, the home parking spaces are becoming a severe barrier to the consumer acceptance of the electrified technology in the crowded cities. This study quantified the ownership cost impacted by the residential parking space: For a BEV with an electric range at 150 km, cost might reach $10,687 USD.
- The BEV range anxiety and potential electricity cost are also quantified. For example, in the average Chinese vehicle driver scenario, a BEV-150 km driver with annual VKT at 19,681 km and daily commute distance at 16.22 km, the range anxiety cost can reach about $9,847 USD, and electricity cost is about $1,464 USD. However, it all depends on the consumer travel patterns.
- Currently domestic, privately-owned automakers are one step ahead in the PEV market compared to the large domestic, state-owned ICEV automakers.

- As the government lifts the restrictions on foreign automakers, lowers tariffs on imported vehicles, transforms the subsidies to supply incentives (i.e. the dual-credit policy), China's PEV market is becoming fiercely competitive. Foreign entrants and local hi-tech start-ups are also flooding into this industry.
- We have calculated the PIP of the automotive industry by simulating the NEOCC model. Compared to a scenario where there are no rules in the market, the NEOCC model estimates that with the dual credit policy the automotive industry will "lose" around $2122 per vehicle by 2020. However, the impacts of PIP might be afforded by the automakers through improving their product structure and adopting better energy efficient technologies.
- The variation of the oil price in a current fluctuation range (year 2016) has hardly any impact on PEV sales in China. For small increases in oil prices, consumers prefer to purchase fuel-efficient ICEVs rather than PEVs, but larger increases in oil price would, ultimately, push more consumers to buy PEVs.
- There are still obstacles for mass market adoption of PEVs, such as vehicle cost, home charging availability, and charging infrastructure. As government incentives are phased out, further PEV adoption will increasingly depend on technology and the extent to which it will continue to improve.

So, will the prosperous PEV market in China be sustainable? In the short run, the PEV market is expected to continue booming. Although direct government subsidies for PEV purchases are gradually waning, more investments are being transferred to infrastructure support in order to increase its availability and convenience. At the same time, supply-side policies are being implemented to compel automakers to invest more in the PEV segment: in 2017, China issued the dual-credit policy that requires auto firms to comply with mandatory PEV requirements; non-compliance will be punished by harsh penalties. Other nonmonetary incentives, such as the vehicle license plate registration privilege, are effective on the market, especially in megacities, and are expected to remain in place in the short term. As more consumers in megacities buy PEVs, incentivized by the vehicle license plate registration privilege, the waiting line for obtaining the free vehicle license plates are becoming long, which dampens the effectiveness of this policy.

In the long run, however, the technology attributes and individual consumer acceptance will determine the PEV's competitiveness in the vehicle market. Currently, unlike the individual consumers in the ICEV segment, individuals are not the major PEV buyers. Most PEVs are purchased by the government for public transportation and by fleet companies for ridesharing or rental in China. This seems to be a compromised business model because of the short electric ranges of the PEVs and the inconvenience of charging them. When the government's direct incentives are gone, however, only safe, affordable, clean, and energy-saving PEVs that meet individual consumer travel demands can be competitive in the vehicle market.

Acknowledgements The authors would like to thank Aramco Services Company for their generous financial support for this study, and the China Automotive Technology and Research Center for their data support. The authors are solely responsible for the views expressed in this book chapter.

References

Beijing Municipality Government. (2018). *Provisions of Beijing Municipality on motorized vehicle parking management*. http://zhengce.beijing.gov.cn/library/192/33/50/438650/1543100/index.html. Accessed November 2, 2018.

Beijing Transportation Research Center. (2017). *Beijing Transportation Annual Report 2017, Beijing, China*. www.bjtrc.org.cn.

Bloomberg. (2020). How elon musk built a tesla factory in China in less than a year. https://fortune.com/2020/01/07/elon-musk-tesla-gigafactory-shanghai-china-ceremony. Accessed February 4, 2020.

Castle, S. (2017). Britain to ban new diesel and gas cars by 2040. *New York Times* (2017). https://www.nytimes.com/2017/07/26/world/europe/uk-diesel-petrol-emissions.html.

CATARC. (2014). Annual report on automotive energy-saving in China (1st ed.). Beijing, China: Posts & Telecom Press.

CATARC. (2019). *Automotive data center database of China Automotive Technology and Research Center*. Tianjin, China.

Chai, J., Yang, Y., Lu, Q., Xing, L., Lai, K. K., & Wang, S. (2017). *Green transportation and energy consumption in China* (1st ed.). New York, NY: Routledge Taylor & Francis Group.

Chen, S. (2016). The best-selling electric vehicles in year 2015 (in Chinese). *Micro Electric Vehicles*. http://www.aooev.com/html/201601/59439.html. Accessed July 18, 2016.

Chen, Y. (2018). New energy vehicle after market service: Real electric range is complained most. *D1EV Website*, 1. https://www.d1ev.com/news/shichang/64867. Accessed May 29, 2018.

Chi, D. (2018). Chances of getting a Beijing license plate hit record low of one in 1,907. *Global Times*. https://gbtimes.com/chances-of-getting-a-beijing-license-plate-hit-record-low-of-one-in-1907.

China Automotive Technology and Research Center, Nissan Group of China, Dongfeng Motor Company. (2018). *Annual report on new energy vehicle industry in China*. Beijing, China: Social Sciences Academic Press. http://www.ssapchina.com/ssapzx/c_00000009000200010007/d_1202.htm.

Cui, H., Jin, L., Zhou, H., Zhang, J., Zhang, B., Liu, X., & He, H. (2018a). *The assessment of the new energy vehicle incentive policies in the Chinese cities, Beijing, China*. https://www.theicct.org/publications/evaluation-incentive-policies-China-urban-NEVs.

Cui, H., Jin, L., Zhou, H., Zhang, J., Zhang, B., Liu, X., He, H. (2018b). *Evaluation of incentive policies for new energy passenger vehicles in Chinese cities, Beijing China*. https://www.theicct.org/publications/evaluation-incentive-policies-China-urban-NEVs.

D1EV Web. (2016). *Electric vehicles information database* (in Chinese). http://www.d1ev.com/salecar. Accessed July 18, 2016.

Davis, S., Williams, S.E., Boundy, R.G., & Moore, S. (2016). *Vehicle Technologies Market Report, Oak Ridge, TN*. http://cta.ornl.gov/vtmarketreport/index.shtml.

Hao, Y. (2018). Investments helps internet startup NextEV kick off mass production. *China Daily*, 1. http://www.chinadaily.com.cn/cndy/2017-11/13/content_34464531.htm. Accessed May 15, 2018.

Hao, H., Ou, X., Du, J., Wang, H., & Ouyang, M. (2014). China's electric vehicle subsidy scheme: Rationale and impacts. *Energy Policy, 73*, 722–732. https://doi.org/10.1016/j.enpol.2014.05.022.

Helveston, J. P., Liu, Y. E., Feit, M., Fuchs, E., Klampfl, E., & Michalek, J. J. (2015). Will subsidies drive electric vehicle adoption? Measuring consumer preferences in the U.S. and China. *Transportation Research Part A: Policy Practice, 73*, 96–112. https://doi.org/10.1016/j.tra.2015.01.002.

Howell, S., Lee, H., & Heal, A. (2014). Leapfrogging or stalling out? Electric vehicles in China. *HKS Working Paper*. http://dx.doi.org/10.2139/ssrn.2493131.

Huo, H., Cai, H., Zhang, Q., Liu, F., & He, K. (2015). Life-cycle assessment of greenhouse gas and air emissions of electric vehicles: A comparison between China and the U.S. *Atmospheric Environment, 108*, 107–116. https://doi.org/10.1016/j.atmosenv.2015.02.073.

Huo, H., Zhang, Q., Wang, M. Q., Streets, D. G., & He, K. (2010). Environmental implication of electric vehicles in China. *Environmental Science and Technology, 44,* 4856–4861. https://doi.org/10.1021/es100520c.

International Energy Agency. (2017). *Global EV outlook 2017, Paris, France*. https://www.iea.org/publications/freepublications/publication/GlobalEVOutlook2017.pdf.

Jiang, C. (2016). *MIIT changes the subsidy strategies (Chinese)*. http://www.gov.cn/xinwen/2016-12/21/content_5150850.htm#1. Accessed August 28, 2018.

Jiao, J., Zuo, F., Li, L., Yuan, H., & Li, J. (2017). Estimation of China's alternative policies of automotive fuels—A perspective of oil dependence. *Journal of Cleaner Production*. https://doi.org/10.1016/j.jclepro.2017.05.107.

Li, Z., & Sun, L. (2011). *The impact of the government policy on the Chinese electric vehicle industry and business strategy management: Case of FAW*. Linköpings Universitet. http://www.diva-portal.org/smash/get/diva2:445387/fulltext01.pdf.

Lin, Z. (2014). Optimizing and diversifying electric vehicle driving range for US drivers. *Transportation Science, 48,* 635–650. https://doi.org/10.1287/trsc.2013.0516.

Lin, Z., & Greene, D. L. (2011). Promoting the market for plug-in hybrid and battery electric vehicles: Role of recharge availability. *Transportation Research Record, 2252,* 49–56. https://doi.org/10.3141/2252-07.

Liu, H., & Yu, Q. (2018). *Forbidden regional protectionism on new energy vehicles (Chinese)*. Xinhua News Agency. http://www.gov.cn/xinwen/2018-02/15/content_5266998.htm. Accessed February 15, 2018.

Ma, S., Fan, Y., & Feng, L. (2017). An evaluation of government incentives for new energy vehicles in China focusing on vehicle purchasing restrictions. *Energy Policy, 110,* 609–618. https://doi.org/10.1016/j.enpol.2017.07.057.

Marquis, C., Zhang, H., & Zhou, L. (2013). *China's quest to adopt electric vehicles*. http://hdl.handle.net/1813/36438.

MIIT. (2015). *Notice on financial support policies on the promotion and application of new energy vehicles (2016–2020)*. http://www.miit.gov.cn/n1146290/n4388791/c4214459/content.html. Accessed August 28, 2018.

MIIT. (2013). Notice of continuing on the promotion and application of new energy automobiles. http://www.gov.cn/zwgk/2013-09/17/content_2490108.htm. Accessed August 27, 2018.

MIIT. (2017). *Measures on parallel administration of passenger car enterprise average fuel consumption and new-energy vehicle credits*. http://www.miit.gov.cn/n1146290/n4388791/c5826378/content.html. Accessed October 30, 2018.

MIIT. (2018). *Notice on adjusting and improving the policy of financial subsidy for promoting and applying new energy vehicles* (Chinese). http://www.miit.gov.cn/n1146295/n1652858/n1652930/n3757018/c6064786/content.html. Accessed August 28, 2018.

Ministry of Science and Technology of P.R.C. (2014). Notice of further promotion and application of new energy vehicles (in Chinese). http://www.most.gov.cn/tztg/201402/t20140212_111801.htm. Accessed August 27, 2018.

Moss, T. (2018). Tesla to build auto factory in Shanghai. *Wall Street Journal* (2018). https://www.wsj.com/articles/tesla-to-build-auto-factory-in-shanghai-1531230784. Accessed October 7, 2018.

Moss, T., & Colias, M. (2018). China to ease rules on foreign auto makers. *Wall Street Journal,* 1. https://www.wsj.com/articles/china-to-ease-rules-on-foreign-auto-makers-1523963345. Accessed May 25, 2018.

NDRC. (2018). *Notice of the State Council on implementing the price and tax reform of refined oil, China*. http://www.chinatax.gov.cn/n810341/n810765/n812171/n812675/c1190752/content.html.

Office of Energy Efficiency & Renewable Energy. (2016). *Global plug-in light vehicle sales increased by about 80% in 2015*. U.S. Department of Energy. http://energy.gov/eere/vehicles/fact-918-march-28-2016-global-plug-light-vehicle-sales-increased-about-80-2015. Accessed July 18, 2016.

Office of Energy Efficiency & Renewable Energy. (2017). *China has the highest number of sales of plug-in vehicles in the world*. U.S. Department of Energy. (2017). https://www.energy.gov/eere/vehicles/fact-976-may-8-2017-china-has-highest-number-sales-plug-vehicles-world. Accessed May 14, 2018.

OFweek. (2020). NIO Auto's sales grew more than 80% in 2019 and it still faces more difficulties in 2020 (Chinese). https://nev.ofweek.com/2020-01/ART-71008-8440-30424276.html. Accessed February 4, 2020.

Ou, S., Hao, X., Lin, Z., Wang, H., Bouchard, J., He, X., et al. (2019). Light-duty plug-in electric vehicles in China: An overview on the market and its comparisons to the United States. *Renewable and Sustainable Energy Reviews, 112,* 747–761. https://doi.org/10.1016/j.rser.2019.06.021.

Ou, S., Yu, R., Lin, Z., Ren, H., He, X., Przesmitzki, S., & Bouchard, J. (2019). Intensity and daily pattern of passenger vehicle use by region and class in China: estimation and implications for energy use and electrification. *Mitigation and Adaptation Strategies for Global Change,* 1–21.

Ou, S., Lin, Z., He, X., & Przesmitzki, S. (2018a). Estimation of vehicle home parking availability in China and quantification of its potential impacts on plug-in electric vehicle ownership cost. *Transport Policy, 68,* 107–117. https://doi.org/10.1016/j.tranpol.2018.04.014.

Ou, S., Lin, Z., Qi, L., Li, J., He, X., & Przesmitzki, S. (2018b). The dual-credit policy: Quantifying the policy impact on plug-in electric vehicle sales and industry profits in China. *Energy Policy, 121,* 597–610. https://doi.org/10.1016/j.enpol.2018.06.017.

Ou, S., Lin, Z., Wu, Z., Zheng, J., Lyu, R., Przesmitzki, S., et al. (2017). A study of China's explosive growth in the plug-in electric vehicle market. *Knoxville TN*. https://doi.org/10.2172/1341568.

Panasonic News Bureau. (2015). Panasonic Factory in Dalian, China begins mass production and shipments of automotive lithium-ion batteries. *Panasonic*, 1. https://news.panasonic.com/global/press/data/2018/03/en180313-3/en180313-3.html. Accessed May 15, 2015.

ResearchAndMarkets.com. (2018). *China Low-speed Electric Vehicle Industry Report, 2018–2022, Dublin*. https://www.prnewswire.com/news-releases/china-low-speed-electric-vehicle-industry-report-2018-byvin-yogomo-and-shifeng-group-are-the-top-3-players-300657214.html.

Schmidt, E. (2018). Electric vehicle sales in Canada, 2017. *Fleetcama*. https://www.fleetcarma.com/electric-vehicle-sales-canada-2017/. Accessed May 14, 2018.

Schmitt, B. (2016). Large number of Chinese EV sales fake, investigators say. *Forbes*. http://www.forbes.com/sites/bertelschmitt/2016/02/01/large-number-of-chinese-ev-sales-fake-investigators-say/#5c0fb25f4dc3.

Shenzhen Municipal Finance Commission, Shenzhen Development and Reform Commission. (2017). *Financial support policy for the promotion and application of new energy vehicles in 2017 in Shenzhen*. Shenzhen Municipality Government, 1. http://www.sz.gov.cn/zfgb/2017/gb1016/201708/t20170816_8142020.htm. Accessed May 28, 2018.

Shenzhen Municipal Government. (2018).*Passenger car ownership control in Shenzhen*. http://www.szxqcjj.com/jjjg.shtml. Accessed October 31, 2018.

SHEVDC. (2017). *Shanghai New Energy Vehicle Market Characteristics and User Behavior Research Report 2017, Shanghai, China*. http://www.shevdc.org/report/.

Sohu Auto. (2018). *Trend analysis of government subsidies in 2018*. http://www.sohu.com/a/238423857_383324. Accessed October 31, 2018.

Sohu, Former Director of the Planning Division in the Ministry of Industry and Information Technology blamed the New Energy Vehicle Planning in Beijing (in Chinese) (2016). http://auto.sohu.com/20160217/n437613315.shtml. Accessed July 19, 2016.

Tamor, M. A., Gearhart, C., & Soto, C. (2013). A statistical approach to estimating acceptance of electric vehicles and electrification of personal transportation. *Transportation Research Part C: Emerging Technologies, 26,* 125–134. https://doi.org/10.1016/j.trc.2012.07.007.

Temple, J. (2018). China's ambition to power the world's electric cars took a huge leap forward this week. *MIT Technology Review*, 1. https://www.technologyreview.com/s/611448/chinas-ambition-to-power-the-worlds-electric-cars-took-a-huge-leap-forward-this-week/.

The State Council of P.R China. (2012). Planning for the development of the energy-saving and new energy automobile industry (2012–2020) (p. 16). http://www.scio.gov.cn/32344/32345/32347/33535/xgzc33541/Document/1450371/1450371.htm.

The Strategic Advisory Committee, SAE-China. (2016). *Technology roadmap for energy saving and new energy vehicles* (1st ed.). Beijing, China: China Machine Press.

The Tariff Commission. (2018). *Notice on reducing vehicles and components import tariffs by the Tariff Commission*. Minist of Finance of the P.R.C, 1. http://gss.mof.gov.cn/zhengwuxinxi/zhengcefabu/201805/t20180522_2903728.html. Accessed May 29, 2018.

Tian, Y., Xie, H., Zhao, Y. (2017). China plans to kill local subsidies for electric cars. *Bloomberg News*. https://www.bloomberg.com/news/articles/2017-12-18/china-is-said-to-plan-killing-local-subsidies-for-electric-cars-jbbq58bn. Accessed May 18, 2018.

U.S. Department of Energy. (2020). *Alternative fuels and advanced vehicles data center*. http://www.afdc.energy.gov/. Accessed Feburary 5, 2020.

Xie, F., & Lin, Z. (2017). Market-driven automotive industry compliance with fuel economy and greenhouse gas standards: Analysis based on consumer choice. *Energy Policy., 108*, 299–311. https://doi.org/10.1016/j.enpol.2017.05.060.

Xinhua News Agency. (2016). *China's unveils measures to promote the private investment*. State Council Information Office. http://www.scio.gov.cn/32344/32345/33969/35257/35264/Document/1493927/1493927.htm. Accessed January 1, 2016.

Yamazaki, M. (2018). Japan's Panasonic says it could make batteries with Tesla in China. *Reuters*, 1. https://www.reuters.com/article/us-panasonic-results/japans-panasonic-says-it-could-make-batteries-with-tesla-in-china-idUSKBN1IB0X2. Accessed May 15, 2018.

Zheng, J., Mehndiratta, S., Guo, J. Y., & Liu, Z. (2012). Strategic policies and demonstration program of electric vehicle in China. *Transport Policy, 19*, 17–25. https://doi.org/10.1016/j.tranpol.2011.07.006.

Plug-in Electric Vehicle Market Research on Consumer in Six Cities in China

Yan Xing, Gil Tal and Yunshi Wang

Abstract China's Plug-in Hybrid Electric Vehicle (PEV) market's recent surge makes it the world's largest PEV stock. A series of supportive policies in China contributed greatly to the rapid PEV adoption. However, what role these policies play in encouraging consumers to purchase PEVs rather than conventional vehicles is unknown. Additionally, factors other than incentives that may help maintain the current adoption trend are still unclear. The latter are especially critical in understanding how the market reacts to phasing out subsidies while aiming to achieve the next goal of 5 million PEVs on Chinese roads by 2020. Therefore, in this study we employed both quantitative and qualitative methods to explore these research questions through a cross-sectional study of the current PEV consumer market in six cities in China. The sample size of this study is by far the largest survey of PEV users ever conducted in China. Additionally, to the best of our knowledge, it is the first study focusing on PEV adoption in third-tier cities in China, where there is much room for PEV sale improvement. Further, collective attitudes toward micro-scale electric vehicles, which have not received much attention due to their smaller market share in big cities, were captured in this study for the first time. The results altogether provide meaningful insights into ways to increase PEV adoption in China and internationally.

Keywords PEV adoption · Six cities in China · Quantitative and qualitative methods · Cross-sectional study

Introduction

Recently, Plug-in Electric Vehicles (PEVs), including Plug-in Hybrid Electric Vehicles (PHEVs) and Battery Electric Vehicles (BEVs), have been garnering more global attention and are emerging as an important element of the transportation sector in reducing local emissions and greenhouse gases (GHGs). To achieve a large-scale adoption of PEVs, which are currently costly and unfamiliar to the general public, considerable stimulus incentives, including monetary subsidies and non-monetary

Y. Xing · G. Tal (✉) · Y. Wang
University of California, Davis, USA
e-mail: gtal@ucdavis.edu

M. Contestabile et al. (eds.), *Who's Driving Electric Cars*, Lecture Notes in Mobility,
https://doi.org/10.1007/978-3-030-38382-4_12

strategies, are provided in China and globally. As a result, some countries have achieved great increases of PEV sales. For instance, China experienced a great surge in PEV sales: 331,092 PEVs were sold in 2015 alone, a 340% increase over 2014. The year 2016 saw a sustained market growth, with total PEV sales reaching 322,271 by the end of the year. An accumulated total number of 645,708 light-duty plug-in electric passenger cars through December 2016 makes China the world's largest PEV stock (Cobb 2017). Despite the rapid global growth of PEVs in some countries, PEVs are still marginal in number compared to conventional fuel vehicles; current data shows the market share of passenger PEVs in China, most of which were sold in the first-tier cities, is only close to 1.5% (International Energy Agency 2017). China's experience, therefore, deserves consideration globally for its dramatic expansion of the PEV market, which may suggest great possibilities for increasing PEV adoptions in other countries.

The factors that drove the recent surge of PEV sales in China stem from a series of supportive public policies that create demand for the vehicle supply and a set of increased incentives to purchase them. For example, in addition to subsidies and tax exemptions, PEV consumers can take advantages of getting a free license plate (i.e. permission to purchase a vehicle) and/or no driving restrictions on PEVs during weekdays in big cities with heavy traffic. However, what role these policies play in encouraging consumers to purchase PEVs rather than conventional vehicles is unknown. Further, the Chinese government has already begun to gradually remove the monetary incentives from PEV sales, which will be completely phased out by the end of 2020. The key question then will be where and how to sell these PEVs from 2021 onwards. Therefore, it is especially critical to understand how the market will react to lower incentive levels. Factors other than incentives that may help maintain the current adoption trend to achieve the next goal of 5 million PEVs on the road by 2020 in China are still unclear. Therefore, this study aims to explore these research questions by studying the current PEV market on consumers in Beijing, Shanghai, Chengdu, Wuhan, Shijiazhuang, and Linyi.

Conceptual Model and Literature Review

Conceptual Model

The theory of diffusion of innovations provides a useful conceptual basis for understanding vehicle choice behavior and the adoption of electric mobility as a new technology that spreads over time and space in a society. Diffusion of innovations theory (Rogers 2003) focuses on explaining how and why an innovation spreads through certain channels among adopters over time. The main elements in diffusion of innovations are the innovation, adopters, communication channels, time, and social system, which play important roles in the adoption of an innovation. Innovation indicates the new idea, practice, or object of interest; adopters are the analysis unit such as individuals, organizations, etc.; communication channels are the patterns by which the information on innovation transfers from one unit to another; time indicates the necessary period of time for innovations to be adopted; and the social system includes all external and internal influences in a society. With this basic theory as our foundation, we constructed a conceptual model for BEV adoption as affected by multiple levels of factors which are grouped as (1) vehicle attributes, (2) individual factors, and (3) the social system. Vehicle attributes include characteristics of BEVs such as their benefits and costs, etc. Individual factors consist of socio-demographics and attitudes including preferences, beliefs, and life styles etc. Within social system, several types of influences are categorized and defined as external influences, i.e. infrastructure settings, political conditions, societal norms and culture in a community, and internal influences measuring interpersonal networks with near-peers in a society (in the theory of diffusion of innovation, the analogous counterparts of external and internal influences are social structure and communication structure). It is notable that this conceptual framework does not include the time effect and diffusion channel due to the limit of the cross-sectional design of our study.

Policy Background

PEV marketing started in 2009: the program known as "ten cities, ten thousand PEVs" was carried out, and ten cities were selected as pilot cities for promoting public fleet PEVs. Passenger PEV markets had not been financially subsidized until 2013 and subsequently all PEV buyers were exempted from vehicle purchase tax in 2014. In 2015, however, the Chinese government implemented a policy to phase out the monetary incentives on PEVs over time. A more detailed discussion of China PEV policies is provided in Chap. 11.

Literature on Factors Associated with PEV Adoption

Qualitative methods were employed to explore factors influencing PEV adoption. Selected previous studies based on qualitative methods are briefly discussed here. One study (Vergis et al. 2014) used case studies to examine potential determinants of PEV adoption by comparing PEV market development in seven nations. Studies found that PEV adoption is supported through government incentives, government resources, and other legitimation activities. Some qualitative studies used pre-post methods to examine the change of attitudes toward PEVs that influenced later decisions about PEV adoption. One study (Carroll 2011) found the attitudes of fleet users and managers were more positive toward PEVs after they had experiences with them. Kurani et al. (2014) designed workshops which included both PEV and conventional gasoline vehicle (ICE) owners from 3 regions in California and found ICE owners showed a more positive interest in PEVs after hearing accounts of PEV owners responding to their questions about PEVs; these were discussed in Chap. 3 of this book. Another study (Bühler et al. 2014) employed qualitative, combined with quantitative methods and showed that the perception of many PEV features was positively influenced shortly after experiences with a PEV. Qualitative methods were also used in some studies (Bakker and Trip 2013; Graham-Rowe et al. 2012) to reveal inherent understanding or explanation mechanisms underlying actual experiences based on limited sample sizes.

However, quantitative methods, including descriptive and explanatory analysis, are more commonly used in previous literature on PEVs. Descriptive analysis focuses on describing the characteristics of the variables of interest, i.e. PEV adoption. For instance, the report from the National Renewable Energy Laboratory (Singer 2016) describes the national level household vehicle fuel type shares, shares of households willing to consider purchasing PEVs, etc. However, the nature of PEV adoption cannot be explained by this method. The method of explanatory analysis was, therefore, used in many studies to infer factors related to PEV adoption. Specifically, explanatory analysis employed in previous literature includes chi-square test (Bakker and Trip 2013), ordinary least squares (OLS) regression (Carley et al. 2013; Sierzchula et al. 2014; Vergis and Chen 2015), cluster analysis (Campbell et al. 2012), binary logit model (Zhang et al. 2011), multiple logistic regression (Javid and Nejat 2017), or a mixed logit (MXL) model (Hackbarth and Madlener 2013), etc. The data of these studies can be categorized into two types: aggregate data at a national level (Sierzchula et al. 2014) or at state level (Vergis and Chen 2015), and disaggregate data at an individual level (e.g. (Carley et al. 2013; Zhang et al. 2011)); most of the latter are collected from stated preference surveys.

Various potential factors associated with vehicle choice for PEVs have been tested in previous studies. Guided by our conceptual model, factors associated with PEV adoption based on previous literature reviews can be grouped into three major explanatory categories rather than being exhaustively enumerated individually. (1) Vehicle attributes correlate to choice of PEVs: it is found that PEVs' high purchase price, limited driving range, and long charging times deter people from owning PEVs (Carley et al. 2013; Graham-Rowe et al. 2012). (2) Individual factors play important roles in PEV adoption. Individual socio-demographics which include gender, age (Carley et al. 2013; Hackbarth and Madlener 2013), income, education level (Hackbarth and Madlener 2013; Sierzchula et al. 2014), number of driving license holders in a family, and number of vehicles (Zhang et al. 2011) are found to significantly influence consumers' choice of PEVs. Attitudes including preference, beliefs, and life style etc. are also important in explaining PEV adoption. For example, people's environmental awareness and interest in PEVs (Egbue and Long 2012), beliefs, pro-environmental identity and lifestyle, knowledge of environmental problems, concern for environment (Lane and Potter 2007), concern for energy independence and climate change (Zheng Krupa et al. 2014), vehicle confidence, environmental beliefs, and perception of electric vehicles (Graham-Rowe et al. 2012) etc. all affect PEV market penetration. (3) The social system, especially external influences, including charging infrastructure, are significantly associated with PEV adoption (Egbue and Long 2012; Graham-Rowe et al. 2012; Jensen et al. 2013; Sierzchula et al. 2014;

Skippon and Garwood 2011). Additionally, awareness of electric vehicle charging stations in a community also correlate to the intention of purchasing a PEV (Carley et al. 2013). Public policies pertaining to PEVs are found to be important in previous studies on the adoption of PEVs: Consumer adoption of PEVs is encouraged by governments' environmental regulations (Lane and Potter 2007), tax incentives or manufacturer rebates (Zheng Krupa et al. 2014), or policy measures to decrease the purchase costs (Peters and Dutschke 2014). However, a recent study found that consumers in China are more receptive to BEVs than consumers in the U.S. regardless of financial subsidies (Helveston et al. 2015). Although the importance of internal influences on PEV adoption is suggested by the conceptual model, to the best of our knowledge, few studies have explored its impact on PEV adoption empirically.

Overall, previous research provides important insights into factors associated with PEV adoption. However, empirical knowledge is still limited about the importance of some key variables such as PEV affection. Additionally, internal influences represented by interpersonal networks with near-peers in a society where PEVs spread have not been measured in previous studies, and therefore empirical testing of their impacts on PEV adoption is limited. Furthermore, factors facilitating the rapidly increasing PEV sales in China have not been effectively explored empirically. More importantly, few studies have examined the decision process of consumers to understand the formation of preferences for PEVs as well as the reasons for rejection. The issues mentioned above are all research questions in this study, and are expected to be addressed by applying both quantitative and qualitative analyses of original surveys conducted in six cities in China.

Methodology

Survey Sampling and Administration

From December 2016 to March 2017, with support from Energy Foundation, Beijing Transportation Research Institute, Shanghai PEV Public Data Collect and Monitor Research Center, and Institute of Transportation Studies, UC Davis jointly conducted a large-scale survey on PEV consumers in Beijing, Shanghai, Chengdu, Wuhan, Shijiazhuang, and Linyi in China. The study employs a cross-sectional research design to explore influences of policy incentives, individual factors including socio-demographics and attitudes, driving behaviors, and charging behaviors on PEV adoptions.

Six cities were selected for this study based on several factors. First, we looked for stand-out cities in 88 pilot cities which lead PEV sales in China. Our hope was to then find cities that differed in economic status, population, and area. More importantly, these cities have implemented different incentives for promoting PEV sales. The purpose of this design was to ensure variation in potential explanatory factors. Two first-tier cities, Beijing and Shanghai; two second-tier cities, Chengdu and Wuhan;

and two third-tier cities, Shijiazhuang and Linyi were chosen as comparison cities (Table 1).

Shanghai PEV Public Data Collect and Monitor Research Center used a list with a random sample of PEV buyers from PEV dealers, previous PEV users surveyed, PEV users recommended by the respondents, etc. to find PEV users in Shanghai, Chengdu, and Linyi. The highest response rate was 60% in Shanghai and the lowest was 45% in Linyi. Beijing Transportation Research Institute used the list of PEV buyers from the license plate registration bureau, and randomly chosen PEV buyers were contacted through email in Beijing; the Beijing institute used a similar method to that of the Shanghai research group to find PEV buyers in Wuhan and Shijiazhuang. The highest response rate is 80% in Wuhan and the lowest is 38% in Beijing. The survey questions include the socio-demographics of PEV buyers and households, attitudes, driving and charging behaviors, and public policies pertaining to PEVs. Most parts of the questionnaires are the same in the six cities but differ when considering incentives for PEVs. An online or Wechat survey was conducted in the six cities to collect responses from these PEV users. In addition, we offered to send a hard copy of the survey upon request. In total, we achieved valid sample sizes of 1008 in Beijing, 657 in Shanghai, 63 in Chengdu, 60 in Wuhan, 51 in Shijiazhuang, and 63 in Linyi.

Although we designed the survey to be available to all PEV owners, it is possible that individuals who like their PEVs were more inclined to complete the survey, leading to a self-selection bias. Because our survey had the added barrier of being online, non-response bias is also a concern in this study. For example, the survey results show that 11.2% of people included in the survey who have bought new PEVs in 2016 are Denza owners, in comparison to only 2.4% of the 2016 market share of Denza in Beijing.[1] Because the focus of our study is on comparison of the PEV markets by employing bivariate analyses, or explaining mode choice as a function of other variables, and not on describing the simple univariate analyses, these differences are not expected to materially affect the results (Babbie 1998).

Analytical Methods

Quantitative Analysis: Descriptive Analysis

Socio-demographics of early PEV adopters and characteristics of their households, such as gender, education, household income, their attitudes toward PEVs, driving and charging behavior, and effects of different incentives on their PEV adoptions were analyzed by employing descriptive methods to provide an overview of the survey data. The descriptive analysis provides an overview of early adopters in the six cities, though limited in explaining the nature of the PEV adoption.

---

[1]From China Automotive Technology and Research Center (CATARC).

Table 1 Survey cities

Category	City	GDP (billion)	Area (1000 km^2)	Population (1000)	Annual income/capita (¥)[a]	Non-monetary incentive (¥)	Monetary incentive (¥)
First tier	Beijing	0.25	16	21,448	52,530	Free license plate; no traffic restriction	National: local subsidy: 1:1
	Shanghai	0.27	6	23,019	54,305	Free license plate	BEV 40,000;PHEV 30,000
Second tier	Chengdu	0.12	14	14,048	35,902	2-hour free public parking; no traffic restriction	National: local subsidy: 1:0.6
	Wuhan	0.12	8	10,766	35,383	no traffic restriction	National: local subsidy: 1:1
third tier	Shijiazhuang	0.05	16	10,160	30,459	Households no purchase restriction for the third vehicle; no traffic restriction	National: local subsidy: 1:0.5
	Linyi	0.04	17	10,039	19,912	None	National: local subsidy: 1:0.6

[a]2016 National and social development statistical report

Quantitative Analysis: Explanatory Analysis

To explore the effect of potential factors associated with PEV adoption specifically in the China market, we focused on the Beijing PEV market due to its large market size and strong multi-incentives for PEV buyers. Based on data collected in Beijing, an experimental choice set was designed for each respondent: respondents were asked to make the choice from the following: "What would you choose if the free license plate was no longer offered in return for purchasing a PEV, but other incentives were kept the same?" Four choices were offered: 1. Would buy a PEV. 2. Would buy a conventional fuel vehicle. 3. Would not buy any vehicle. 4. Other. In this sample, almost a half (about 44%) of the total would no longer choose a BEV: including 35% of the total would turn to purchase a conventional car; 71 people would not buy any vehicle without the incentive of a free license plate; furthermore, 2% of the respondents were undecided and would turn to other choices, probably renting or borrowing a car. In total, only about 56% PEV owners would again choose a BEV. Because the latter two categories ("Not to buy" and "Other") of the four choices share common characteristics, i.e. respondents would not buy any vehicle, we combined them as one group. The dependent variable was thus derived from this survey question as a set of 3 alternatives consisting of nominal categories: 1—Would buy a PEV; 2—Would buy an ICE; 3—Would not buy or Other.

Because the dependent variable is a set of 3 alternatives consisting of nominal categories: 1—Would buy a PEV; 2—Would buy an ICE; 3—Would not buy or Other, first, a multinomial logit model (MNL) was used, which is appropriate if the IIA (the Independence of Irrelevant Alternatives) holds for the three nominal categories. The Hausman-McFadden method was employed to test the IIA assumption. The result of the Hausman-McFadden test is inconclusive, but another test approach Nested Logit model test fails to reject the MNL structure for this study. Therefore, a MNL model was applied to explore the factors associated with PEV adoption in an experimental situation.

Qualitative Analysis

Although both descriptive and explanatory studies have explored potential factors influencing PEV adoption through collecting data via a large-scale survey, the cross-sectional data they relied upon weakened the persuasive power to support causal relationships between the factors and PEV adoption. Second, the analyses of PEV adoption were based on customers' stated intention to buy a PEV under certain hypothesized circumstances. Here, stated preference bias may be a big concern, as individuals may misjudge their pretentions regarding new and unfamiliar technology. Third, PEV consumers in second- and third-tier cities, which are supported by fewer government incentives but have room for PEV sales improvements, have not received equal attention to first-tier cities in China. Ideally, time series studies are needed to examine before-and-after evaluations of the impacts of influencing factors on individuals through the whole decision process. Due to time and budget limits, in-depth qualitative studies aiming to understand individuals' underlying reasons, opinions, and motivations can be a more feasible way to explore the decision process and formation of individuals' preference for PEVs. Therefore, we try to fill the gap by employing a qualitative method to explore the mechanism of decision-making about PEV adoption in a second-tier and a third-tier city in China.

Considering that the number of PEV adopters in some cities, especially third-tier cities, is relatively small, and our purpose is to examine what and how people think during their decision process, and why they come to their conclusions on whether or not to adopt, the method of using focus groups was employed. Focus groups are a widely used qualitative research method and generally take the form of 6 to 12 people with a mediator asking questions of participants about their characteristics, ideas, opinions, attitudes, etc. toward an object. The participants can talk freely with other participants. Focus groups are particularly useful for exploring people's knowledge, experiences, and attitudes, and more importantly, for exploring how they construct and operate within a given cultural context (Kitzinger 1995).

Chengdu and Linyi, a second-tier and a third-tier city in China, were chosen for their reasonable amount of PEV sales, supported by less aggressive incentives compared to first-tier cities. According to the market shares of PEV models and helped by local PEV dealers, PEV adopters were contacted and invited to participate in the focus groups voluntarily. In this study we neglected those who did not consider PEVs at all. We pre-defined PEV rejecters as those who took the initiative to seek information, but ultimately rejected and chose a gasoline car instead. This group of people was screened by whether participants had visited local PEV dealers. Potential adopters who left their personal information at the dealerships were randomly selected and then contacted by local PEV dealers directly. In total, 24 individuals including 6 PEV owners and 6 PEV rejecters were recruited for 2 separate in-depth focus groups in each city in December 2016. The conversations were designed to be exploratory in nature and started with recalling their initial motivations for seeking a PEV. Their knowledge, perceptions, observations, decisions, and expectations related to PEVs were discussed in detail based on the theory of innovation-decision process (Rogers 2003), which provides a useful conceptual basis for understanding the decision-making process (Fig. 1). Further, PEV adopters talked about their experiences with driving and charging as well as their feelings; similarly, PEV rejecters reported how they used their current gasoline cars. All questions were intentionally open-ended and led by an experienced moderator. Additionally, another three researchers observed all the focus groups simultaneously without being seen by the participants.

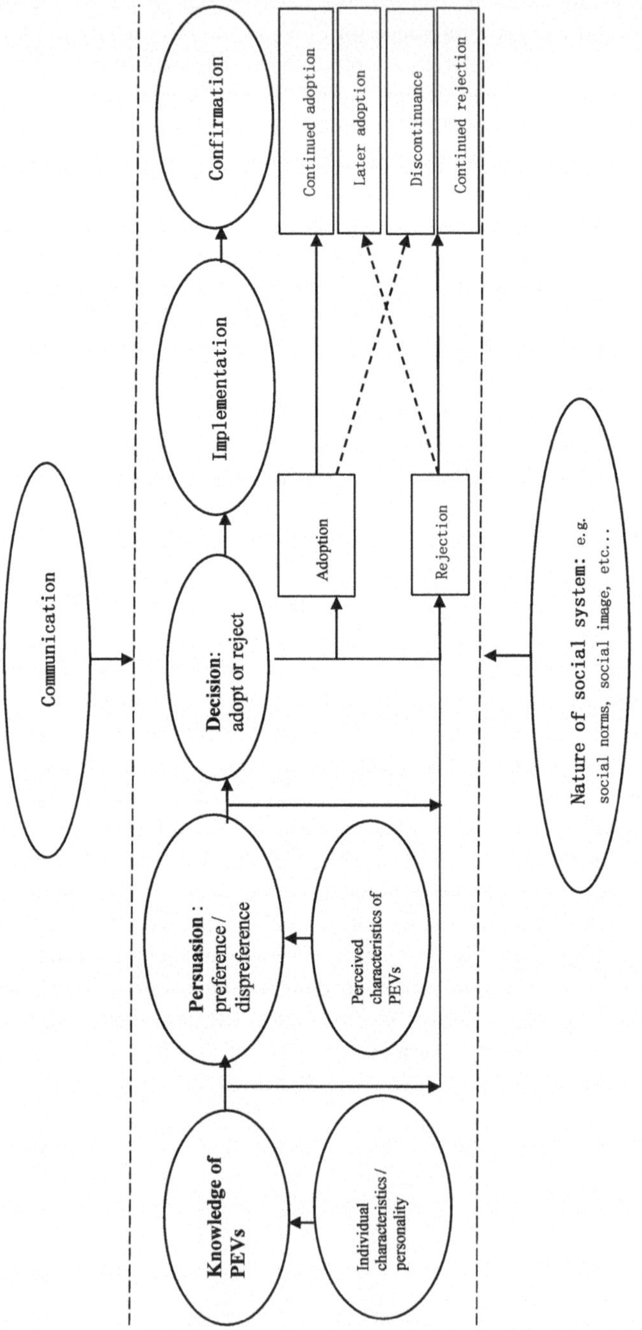

Fig. 1 Five stages in decision process *Source* The theory of innovation-decision process (Rogers, 2003)

Results

Summary of the Survey from Six Cities

Based on the surveys conducted in 6 cities in China, socio-demographics of PEV users, characteristics of the households owning PEVs, users' driving and charging behaviors, as well as policy influences on PEVs were comparably analyzed.

Many PEV users in China are males younger than 40 with high education levels and most of them have children. PEV consumers in the first-tier cities own predominantly mid to high-end PEVs as the first vehicle in their households, while more PEV consumers in the second or third-tier cities have mid or low-end vehicles as a second vehicle in their households. The descriptive comparisons show that most high-end PEV users like their vehicles more and more with time. However, low-end BEV users do not like to talk about their vehicles despite liking them, which may be due to a negative social image for low-end vehicles.

In the six cities, the average daily Vehicle Kilometer Traveled (VKT) of PHEV and high-end BEVs is about 30–40 km; the average daily VKT of mid-end BEVs is about 30 km; the average daily VKT of low-end BEVs is about 20 km. Comparing VKT shows that PEV driving distances are within the vehicle ranges, however they are shorter than those of conventional Internal Combustion Engine (ICE) vehicles. The usage of PEVs is mostly for commuting or picking up family members. Additionally, the average daily vehicle kilometer traveled (VKT) by ICE vehicles in households that own PEVs decreases, implying that some trips by ICE vehicles are partly substituted by PEVs.

Survey data shows home charging is the most important charging method for PEV owners followed by workplace and public fast charging locations. Charging frequency is closely associated with charging infrastructure and vehicle range; in a city with more charging infrastructure, PEV owners charge more frequently than owners in the less infrastructure-dense cities. Shorter vehicle range also correlates to more frequent charging behavior and PHEV users whose cars have small ranges charge their vehicles very frequently.

Non-monetary incentives play a very important role in promoting PEV sales in China. The effects of some of them, e.g. the license plate privilege for PEVs, are greater than effects from monetary incentives. Non-monetary incentives may even induce some transit riders, bicyclists, or pedestrians to buy PEVs. The license plate privilege for PEVs is the number one motivation in the first-tier cities, while PEV adopters in the second- and third-tier cities are motivated by low operating and maintenance costs of PEVs. BEV adopters are more sensitive to vehicle price while PHEV adopters focus more on vehicle performance and interiors.

Model Results of Explanatory Analysis

Multinomial Logit (MNL) was employed to explore the factors associated with PEV adoption in this study. The final data set included in the model comes from 968 individuals and consists of 543 BEV owners who would still choose BEVs, 339 individuals who would choose an ICE instead, and 86 individuals who would choose not to buy a car or Other, under the scenario of removing the incentive of a free license plate but keeping other incentives the same. The best-fitting MNL model for vehicle choice is shown in Table 2. The McFadden ρ^2 measure is based on the Market Share

Table 2 Results of the MNL model for PEV choice

Explanatory	Would buy a PEV			Would not buy or other		
Variable	Coefficient		*p*-value	Coefficient		*p*-value
Constant	−0.351		0.370	−1.851	**	0.006
Individual Factors: Socio-demographics						
Income	−0.159	**	0.004	−0.262	**	0.006
Household Size				0.332	***	0.000
Female				0.435	*	0.067
Individual Factors: Attitudes						
Important license plate	−0.181	***	0.000	−0.133	*	0.076
Important subsidy	−0.152	**	0.002			
Important Environment & energy	0.194	***	0.000			
Price of license plate	0.074	**	0.003			
Like PEV	0.238	*	0.050	0.414	*	0.014
PEV better	0.477	***	0.000			
Maintenance less	−0.562	***	0.000	−0.281	*	0.097
External influences: Charging Infrastructure						
Charger available	0.269	***	0.000			
Charge home				−0.641	**	0.007
Experience: *Charging behavior*						
Charging frequency	0.027	*	0.012	0.030	*	0.053
Number of observations				968		
LL(MS)				−877.805		
LL ($\hat{\beta}$)				−763.03		
ρ^2 MS base				0.131		
$\bar{\rho}^2$ MS base				0.110		

Note Base category: Would buy an ICE
*10% significance level, ** 5% significance level, *** 1% significance level
Blank means the alternative specific variable was allowed to enter the model but was excluded for its insignificance

model, i.e. the model contains constant terms (also labeled the intercepts) only and no any explanatory variables, for this model are 0.131, which indicates that about 13.1% of the information contained in the data has been explained by this model relative to the Market Share model. Although this value is small, which may imply important explanatory variables are still lacking, it is still fair for a disaggregate model with 3 alternatives and a relatively large sample size. Analogous to the adjusted R-square of linear regression models, the adjusted ρ^2 ($\bar{\rho}_2$) is 0.110, which corrects for the number of estimated parameters.

Individual Factors: Socio-Demographics

Socio-demographic factors are associated with vehicle choice. Household annual income level correlates to PEV adoption: people with lower income levels are more likely to still buy PEVs or not to buy vehicles, whereas people with higher income levels intend to buy a conventional vehicle if no free license plate is offered for purchasing a PEV. We also found that people with larger household size are more likely to not buy a vehicle or choose "Other" instead of buying a car, which is likely due to the association between the variable "Household Size" and vehicle ownership ($r = 0.208$, $p = 0.000$). If no free license plate is offered for PEVs, the high cost and difficulty to get a license plate discourages households from purchasing vehicles, especially larger sized families, which own more vehicles thus have no urgent mobility demands. An interesting finding is that female decision-makers are more likely to be conservative and not buy any vehicle. In contrast, males still have the propensity to purchase PEVs or conventional vehicles, even if no free license plate is offered.

Individual Factors: Attitudes

Accounting for socio-demographics, attitude factors show great influence on explaining pro-PEV choice, even without the free license plate incentive. The more important the free license plate was for the decision to buy a PEV, the more likely that people would choose conventional vehicles if this incentive was removed. This finding further emphasizes the important role the free license plate incentive plays in Chinese PEV adoption. PEV owners who were motivated more by subsidies are associated with a lower likelihood of purchasing a PEV without the free license plate, which may indicate that a free license plate is currently viewed as a big subsidy in the Beijing vehicle market. The belief that PEVs are important for environmental protection and energy conservation, which was a reason for purchasing a PEV originally, positively correlates to choosing PEVs under this experimental scenario. The model results also show that people who would pay a higher price for a license plate, which is associated with a higher urgency level for mobility, are more likely to buy PEVs under the scenario that license plates are not free but are available in the market. This finding indirectly reflects the extraordinary difficulty of getting a license plate in Beijing now that people with more urgent mobility demands would rather pay for them if they are available on the market. Affection for PEVs is positively associated with choosing PEVs or choosing not to buy a vehicle. In other words, people who grow to like PEVs more through driving experience are less likely to buy conventional vehicles. PEV owners who agree that PEVs with the subsidies are better than conventional fuel vehicles at similar prices are more likely to still choose PEVs. On the one hand, this finding indicates the important role the configuration, performance, and quality of PEVs play in PEV adoption; on the other hand, it shows the critical role of monetary and non-monetary incentives for the current market. A counter-intuitive result shows that the more people agree the cost of PEV maintenance (excluding batteries) is less than that of a conventional vehicle, the more likely it is that the PEV owners would not choose PEVs. It is possible that although some long-time PEV users may experience lower maintenance costs, this statement may raise their concern for the high cost of replacing the battery. Additionally, based on the feedback of further in-depth interviews, some PEV users in Beijing worried that reduced regular maintenance requirements as compared to conventional vehicles would impact vehicle safety.

External Influences: Charging Infrastructure

The accessibility of charging infrastructure, indirectly measured by the perception of the charging infrastructure around the community, is significantly associated with PEV adoption; PEV owners who feel they can find a charging place in their community are more likely to choose a PEV. One assumption was that charging availability at home would be a key factor in purchasing a PEV. However, the results show that people who can plug in their PEV at home prefer to buy both PEV and conventional vehicles rather than not to buy vehicles. This may be due to a design flaw in the survey; the survey question measuring availability of home charging fails to identify respondents owning private charging poles from those having shared charging poles or using a fly line, i.e. a long charging cable through windows for home charging. The inconvenience of home charging with the two latter charging methods may lead to the choice of conventional vehicles.

Experience: Charging Behavior

Charging behavior is also found to be associated with PEV adoption. People with higher charging frequencies are less likely to choose a conventional vehicle. They may realize the economic benefits of PEVs by using theirs more frequently, or alternatively, the frequent charging behavior indirectly indicates the accessibility and availability of nearby charging infrastructure.

Findings from Qualitative Analysis

The focus group conversations were designed to simulate the individual's decision process. All participants reported their experiences with PEVs, from gaining knowledge of them, their original motivations for purchase, perceived relative advantages compared with other choices, to taking a trial drive and making the decision. Differences were widely recognized between adopters and rejecters, as well as between the two cities (Table 3).

Knowledge Pertaining to PEVs

Although all participants had mobility needs, they presented varied interests in vehicle features. Adopters were more interested in some specific vehicle features such as size, cost savings, or low purchase prices. Rejecters were more attracted to general advantages of PEVs, such as representing new technology and social trends, in addition to their low operation and maintenance cost.

Chengdu participants were more aware of energy saving and environmental friendliness aspects, and they were more knowledgeable about public incentives for PEVs. It is likely that higher education levels led Chengdu participants to be more aware of PEVs through broader methods and access to communication channels. On the other hand, Linyi participants were more interested in the lower prices and operating costs, which may be due to their lower household income levels compared to participants in Chengdu.

Socio-demographic Characteristics

We captured more female PEV owners in Linyi. Chengdu PEV adopters chose more subsidized and higher-end PEVs than those in Linyi. Most BEVs are an additional vehicle, whereas PHEVs are the first vehicle in these households.

All PEV rejecters in both focus groups are males. According to their reported socio-demographic information, PEV rejecters in both cities present relatively higher education level, and correspondingly higher household income levels than the PEV adopters in each city. Further, more PEV rejecters are first-time vehicle buyers than the PEV adopters in each city. Particularly, most participants are families with children, but most Chengdu rejecters are single, or families without children.

Awareness of PEVs

All PEV adopters were driven by their mobility needs, e.g. many of them have children. Thus they took the initiative to seek vehicle information through recommendations from peers, and/or mass media such as the internet and advertising. All

Table 3 Responses of participants by focus group in different decision stages

Decision stage	Influence factors	Chengdu adopters	Linyi adopters	Chengdu rejecters	Linyi rejecters
Knowledge	Female (%)	16.67	50	0	0
	Age	41.17	32.67	26.5	35.67
	Education[a]	0%:66.7%:33.3%	33.3%:66.7%:0%	0%:33.3%:66.7%	16.7%:83.3%:0%
	Marriage Status[b]	0%:0%:100%	16.7%:0%:83.3%	50%:33.3%:16.7%	0%:0%:100%
	Vehicle ownership[c]	33.3% (PHEV):50%:16.7%	0%:16.7%:83.3%	83.3%:0%:16.7%	50%:16.7%:33.3%
	HH income/month	15,500 Yuan	11,800 Yuan	22,500 Yuan	12,300 Yuan
	Awareness of PEVs	Originally attracted by the low operating cost, bigger vehicle size, and good acceleration speed. Later knew battery range; charging time; energy saving and environmental friendliness; performance of speed accelerating; maneuverability; BYD is a domestic manufacturer with good reputation; BYD Tang is a SUV	Originally attracted by the low price; low operating cost; parking and driving easily with its tiny size. Later familiar with battery range, charging time, fake range; maintenance, battery replacement	Originally impressed by the new technology and trends; energy saving and environmental friendliness; then later familiar with the battery range; charging time; acceleration; maneuverability; energy saving, environmental friendliness. BYD is the best domestic PEV manufacturer	Originally impressed by the new technology and trends; low operating cost; good for commuting. Then later familiar with its social image: for old people and women only; battery range; charging time; charging infrastructure around; energy saving; environmental friendliness;

(continued)

Table 3 (continued)

Decision stage	Influence factors	Chengdu adopters	Linyi adopters	Chengdu rejecters	Linyi rejecters
	Awareness of public incentives	Knew all incentives including national and local subsidies, purchase tax exemption; subsidies on installation of charging pole, 2-h free public parking, etc. They also knew the differences between subsidies in Chengdu and nearby cities	Not clear; nowhere to know except dealers	They knew total subsidies but were not clear about the exact amount from national and local. They also knew the differences of subsidies between local and nearby cities	They knew total subsidies but were not clear about the exact amount from national and local
Persuasion	Relative advantage	Low operating and maintenance cost; good acceleration; energy saving and environment friendliness; 2-h free public parking; big space. Short range; long charging time; insufficient charging infrastructure around	Low price; Quiet driving; low operating and maintenance cost; parking and driving easily with its tiny size Short range; long charging time; insufficient charging infrastructure around	Low operating and maintenance cost; good acceleration; energy saving and environment friendliness Short range; long charging time; insufficient charging infrastructure around; Vehicle safety concern; immature battery technology; high battery replacing cost	Low operating and maintenance cost; good acceleration; energy saving and environment friendliness Short range; long charging time; insufficient charging infrastructure around; Tiny size; unfavorable social image; vehicle safety concern; immature battery technology; high battery replacing cost

(continued)

Table 3 (continued)

Decision stage	Influence factors	Chengdu adopters	Linyi adopters	Chengdu rejecters	Linyi rejecters
	Complexity			Fuel efficiency complexity for PHEV users: If the electricity runs out, it is for sure less fuel efficient than a conventional gasoline car	
	Compatibility	It fits daily needs: most trips are for daily commuting or within city. Positive social profile: PEV adopters have median income level, are creative and easy to accept new things	It fits daily needs: most trips are for daily commuting or within city Negative social profile: the older- and women-only vehicle	Daily commuting, recreational or social trips, sometimes long trip traveling needed Positive social profile: PEV adopters have median income level, are creative and easy to accept new things	Business trips, daily commuting, recreational trips with family. Sometimes long trip traveling needed Negative social profile: the older- and women-only vehicle
	Trialability	Took a driving trial	Took a driving trial	Took a driving trial, found the salesman was not familiar with PEVs	Took a driving trial, but found very limited vehicle models in local market
	Observability	Low operating cost; big size; free public parking; insufficient public charging. Some chargers are broken or destroyed	Low operating cost; flexibility with tiny size; insufficient public charging; some chargers are broken or destroyed	Low operating cost; insufficient charging infrastructure; insufficient local government support	Low operating cost; insufficient charging infrastructure; insufficient local government support

(continued)

Table 3 (continued)

Decision stage	Influence factors	Chengdu adopters	Linyi adopters	Chengdu rejecters	Linyi rejecters
Decision	Taking a trial	Big space; good interior features; good acceleration; quiet driving; good maneuverability	Tiny size; acceleration is similar or better than gasoline cars; quiet driving	The appearance and quality are not worth its value. Questions were not solved with unprofessional salesperson	Poor quality; shabby interior features; appearance doesn't look fine; very crudely made; very tiny size not suitable for family; after-sale service concern
Implementation	Driving	Very low operating cost, about 100 Yuan/month. 90% of trips of PHEV users driven by electricity	Very low operating cost, about 100 Yuan/month. Most of them think it has very poor quality and the car had been repaired for several times already. Some of them experienced "fake range" Not like a real car, for the older or women only	NA	NA
	Charging	Most of them charged at home. One user used fly line. They had experienced that they couldn't find chargers or couldn't charge because Public chargers were limited or blocked or broken or the charger was incompatible	Most of them used fly line or installed an outlet for BEV specifically at home or in community parking lot but with big safety concern. They had experiences that Public chargers were limited or blocked or broken or the charger was incompatible	NA	NA
Confirmation	Experience	Sustainable adoption: would recommend to others. It's worth it	Would not recommend to others	NA	NA

[a]Percentage of high school, 3-year college, Bachelor respectively
[b]Percentage of single, married and having no kids, married and having kids
[c]Percentage of first purchase, replacing old car, an additional car

PEV adopters in both cities are familiar with the purchasing prices of PEVs and the low operating cost of PEVs. They clearly knew where or how to charge a PEV based on their experiences before and after their purchase. Some PEV adopters in Linyi, especially the low-end PEV adopters, were attracted by low PEV prices. However, PEV adopters in Chengdu were originally attracted by the bigger vehicle size, e.g. BYD Tang, and its low operating cost, and were more aware of the energy saving advantages and environmental friendliness. Moreover, they had prior information regarding PEV performance including speed, acceleration, and maneuverability.

All rejecters had mobility needs and sought vehicles that met those requirements. Compared with adopters, rejecters were more attracted by the social trends and new technology associated with PEVs. They also favored energy saving advantages, environmental friendliness and were impressed by low operating costs and commuting abilities. All rejecters, like adopters, were aware of the limitations of PEVs, such as short range, long charging time, immature technology, insufficient charging infrastructure, etc.

Awareness of Public Incentives

PEV adopters in Linyi were not clear about subsidies. They were told the final price by dealers when they bought the PEV. Most of them said they did not know how to go about determining that information. PEV adopters in Chengdu were more aware about government incentives, including national and local subsidies, purchase tax exemption, subsidies on installation of charging poles, and 2-h free public parking etc. They even made comparisons of local subsidies between cities and bought cars in adjacent cities with higher subsidies. However, some of them still misunderstood that they needed to apply for the local subsidy after they bought the car.

All the rejecters knew of total subsidies, but were not clear about national and local amounts. Most of the information was from local dealers. Some Chengdu rejecters also compared the subsidies between cities.

Persuasion

In the persuasion stage, an attitude of PEV preference is formed. Pro-PEV attitude formation is greatly influenced by individuals' perceptions of relative advantages, complexity, compatibility, trialability, and observability. Additionally, the attitudes toward PEVs from peers and society provide social reinforcement on individuals' attitude formation.

This stage shows that the adopters and rejecters have similar perceptions of advantages and disadvantages of PEVs. A key difference is that the rejecters have more technology and safety concerns than the adopters. Further, technology complexity, incompatibility of their daily usage and self-social-identification, lacking reliability, and downsides of observational learning also contributed to their decisions to reject PEVs.

Relative Advantage

Relative advantage is the process in which the individual compares the total possible benefits (perceived advantages) of PEV adoption with the cost (perceived disadvantages), either monetarily or non-monetarily.

Although the adopters in Linyi had concerns and dissatisfactions about short range, long charging time, and insufficient charging infrastructure, they emphasized benefits of PEV adoption with the comments:

- "The price (of the BEV) is cheap",
- "Operating costs are very low",
- "Its range is enough for my commuting needs",
- "The BEV size is small, easy and flexible to park and drive especially in traffic congestion",
- "It's very quiet".

In addition to low running costs and good accelerating speed, Chengdu adopters reported their perceived relative advantages as the following:

- "It is gasoline saving and environment friendly",
- "I like 2-h free public parking",
- "The (BYD Tang) is spacious".

However, the rejecters in Linyi had different comments than the adopters. Although they agreed on the low operating cost and good accelerating speed, they were not satisfied with small sizes of most PEVs insufficient for family usage. Additionally, they were concerned about vehicle safety including immature battery technology, short range, long charging time, insufficient charging infrastructure and vehicle safety. They agreed that both operating and maintenance cost for PEVs is lower than conventional gasoline cars, however, they believed the cost of replacing the battery is too high, which may lead to a total cost that is less economical than gasoline cars. Chengdu rejecters also made similar comments on short ranges, long charging times, insufficient charging infrastructure, and the high battery replacement cost. Furthermore, they expressed different vehicle safety concerns:

- "The battery is installed inside the chassis which is on the bottom of the vehicle and will be very dangerous if crashed or driven in water. "

Complexity

Complexity relates to how difficult it is for individuals to understand and use PEVs. Differing opinions and knowledge about batteries, and a lack of sufficient knowledge of PEV technology adds to misunderstanding benefits of PEV adoption. For example, almost all of the Chengdu rejecters who had been interested in PHEVs questioned the fuel efficiency of PHEVs. A typical comment is as follows:

- "BYD Tang is very fuel efficient if driven with full charge of electricity. However, the range is only 80 km (50 miles) at most, actually 60 km (about 40 miles), which is too short for daily travel. If the electricity runs out, I am driving a vehicle with a very heavy battery, which is for sure less fuel efficient than a conventional gasoline car."

Compatibility

Compatibility indicates to which degree a PEV fits an individual's social value, lifestyle, and daily needs. All Chengdu PEV adopters and even rejecters have a common positive opinion that early PEV adopters generally have median income level, are creative and willing to accept new technologies. Whereas both Linyi adopters and rejecters, especially men, think PEVs are incompatible with their self-social-identity because they perceived the social profiles of PEV drivers as older people or women.

Although most adopters worried about battery range of PEVs, they agreed that PEVs satisfied most of their daily needs if they carefully planned beforehand based on their experience. Most trips were short, within the city, and vehicles were used primarily for commuting. Many rejecters were concerned about short ranges of BEVs for their long trips, e.g. business trip, traveling etc. or lower fuel efficiency of PHEVs after electricity running out.

Trialability

Trialability deals with the extent to which a PEV can be experienced before purchase. Potential PEV buyers can dispel uncertainty about a PEV after testing it at dealer shops. All the participants in the focus groups had experimented driving a PEV in their local dealer shops. However, Linyi rejecters were not satisfied with limited PEV models available in the local market.

"The model choices are not as many as those in big cities", one rejecter commented.

Observability

Observability indicates the results (benefits or disadvantages) of PEVs observed by people. Some benefits are easily observed, e.g. all participants agreed on fuel cost savings. Chengdu adopters observed free public parking lots for PEVs. Linyi adopters noticed the flexibility of small vehicle size in traffic. There are also observed downsides of PEVs: in the city Linyi, older people and women are the primary cohort observed driving BEVs. Further, some charging poles are found broken or destroyed. Based on their observations, all rejecters felt that support by local governments for PEVs was insufficient compared to other cities in China.

Decision

In the decision stage, an individual actively chooses between PEV adoption and rejection. A critical step in this stage is a trial of whichever PEV in which the individual is interested. It is also possible that an individual purchased a PEV without trying it out if a close acquaintance had taken a trial already. However, because vehicle purchase is viewed as a significant expenditure in a household, most consumers prefer trialing a PEV themselves over purchasing it blindly.

The results show that the trials reinforced the perceived advantages and disadvantages of the adopters and rejecters; the rejecters confirmed their perceived low quality of PEVs, and as such did not end up purchasing a PEV.

Most PEV adopters confirmed their perceived characteristics of PEVs such as good accelerating speed, quietness (Linyi adopters), good performance, good maneuvering, larger space and good interior features (Chengdu adopters). Some rejecters declined PEVs during this stage because of small space, fewer interior features (Linyi rejecters), and low quality (Chengdu adopter). Linyi rejecters commented as following:

- "The car is too small, not designed for tall and strong man",
- "There are only 2 seats. It does not fit families with kids. And the appearance of car's exterior is less attractive",
- "There are only few dealers in the city, so I worry about the after-sale services".

One Chengdu rejecter said:

- "It looks like a 100,000 Yuan cheap gasoline car",
- "It seems the salesman knows very little about PEVs".

Implementation: The Decision of Sustainable Use Through Driving and Charging Experiences

Sustainable use is defined as the degree to which a PEV continues to be used over time after the end of a PEV adoption decision. Most innovation diffusion studies ended with the decision to adopt or with implementation of the new innovation; however, innovations are important only if they continue to be used. This is especially true for PEVs: although some customers have decided to adopt a PEV and put it to use in reality, i.e. implementation, this does not guarantee a sustainable acceptance of PEVs. The early PEV adopters may experience difficulties in their driving or charging behaviors, and thus are discouraged to re-invest in their future purchase of PEVs or provide positive messages about the PEVs to their peer networks, which is an important interpersonal communication channel at the persuasion stage in the adoption decision process. The opinions and attitudes developed during this stage are critical under the diminishing incentives offered by the governments of China. We therefore look further into the driving and charging experiences of the PEV adopters in both cities. In summary, both adopters complained about short ranges of PEVs as well as insufficient charging infrastructure. Specifically, most Linyi adopters were frustrated with the quality and social image of their PEVs, leading to unsustainable adoption in the long run. Chengdu adopters who purchased higher-end PEVs were more satisfied with their PEVs, leading to sustainable adoption of PEVs.

Most Linyi adopters were disappointed with their PEVs and would not recommend one to others in the future, although they acknowledged some advantages including cost saving, quietness, good accelerating speed. Their complaints include the following:

- "I am so disappointed with the vehicle quality that I had my car repaired several times",
- "Its range is only about 100 km (about 60 miles), so I have range anxiety, usually I would worry when I had 30% range left,"
- "I charge my car by using a fly line, but I have safety concerns about it",
- "I could not find charging poles outside or the charging lot was occupied by gasoline cars, or I couldn't charge my car because of incompatibility of the charging interface".
- "I don't want to talk about my electric car with other people because it does not look like a real car ... Most people think only older people or women drive this kind of car".

However, Chengdu adopters who purchased higher-end PEVs were generally more satisfied with their PEVs. Their comments include:

- "I usually drive within the city and the range of my car (BYD EV 300) is enough for most of my trips", "I have nothing to complain about my car",
- "I experienced situations where I couldn't find chargers outside or I could not charge my car either because it's occupied by others or the charging interface was not compatible",
- "I'd like to recommend to others because it has high cost-performance ratio and good maneuverability".

Additionally, all PHEV (BYD Tang) adopters stated that ninety percent of their trips were driven by electricity and they always started their trips with full electricity. Their complaint is: "(BYD Tang) only has a range of 80 km (50 miles), I have a range anxiety if my round trip is longer than 80 km".

Confirmation

Through the implementation stage, even an individual who has already adopted a PEV may reverse their original decision based on their experiences, e.g. most Linyi adopters formed unfavorable attitudes toward BEVs and would not recommend their BEVs to others. This group of adopters may make a decision of discontinuance, i.e. rejecting PEVs in the future unless their concerns were alleviated. By contrast, Chengdu adopters satisfying experiences reinforced their prior decisions to adopt PEVs. Their confirmed favorable attitudes toward PEVs are illustrated by their recommendations of PEVs to others.

Summary and Conclusions

The empirical analysis of individuals' vehicle choice based on the stated preference data provides new and potentially important insights into factors associated with PEV adoption in China. The findings reveal that the supportive policies, free license plates and subsidies on PEV purchase, significantly facilitate PEV adoptions. About 44% of respondents in Beijing stated that they would not buy BEVs if no free license plate was offered for the purchase, however this statistic suggests some people are still interested in PEVs as a suitable option for their transportation needs even without this incentive. Considering a stated preference bias may exist, the actual percentage of people who would reject PEVs if there were no free license plate incentive should be even higher. The model results also reveal the influences of the free license plate on specific groups of consumers: if free license plates were not offered, people with higher income and people who viewed the free license plate as the most important factor in their decision to purchase the BEV would be more likely to purchase conventional vehicles instead. Additionally, controlling for driving and charging experiences of these PEV owners, the significance of the variable "PEVs are better with subsidies" indirectly suggests the importance of subsidies on PEVs, despite the subsidy impact on the decision to adopt PEVs not showing in the model results. The Multinomial Logit model, when controlling for the influence of this incentive, shows that other factors such as individual socio-demographics, attitudes, charging infrastructure, and charging experiences play important roles in influencing PEV adoptions. The attitude of affection for BEVs strongly encourages people to adopt a BEV, and so does the attitude of concern for the environment and energy security. The results also indicate BEV adoption in China could be driven by mobility needs. Under the experimental scenario that the license plate was no longer free but could be purchased via the market, the price people were willing to pay for it (reflecting the extent to which they need a vehicle) was positively associated with BEV adoption. Among external influences, accessibility of charging infrastructure, including both private charging poles and outdoor charging infrastructure, plays an important role in encouraging the choice of a BEV. Conversely, internal influences do not show significant direct effects on PEV adoption in this model. It is possible that the interpersonal network may work as a mediator indirectly through the attitude of affection, e.g. being encouraged by behaviors of near-peers favoring PEVs. People living in a PEV supportive interpersonal network may be more inclined to like PEVs and feel positively towards them over time, which positively correlates to PEV adoption.

The results of qualitative methods focusing on the individual decision process could be beneficial in improving the understanding of attitude and preference formation for PEVs. We found that in the knowledge stage, high-end PEV adopters are more aware of government incentives (including monetary and non-monetary policies) than the rejecters. In the persuasion stage, the attitude of PEV preference is weakened by the perception of the relative disadvantage of PEVs due to their technological complexity, incompatibility for daily usage, vehicle safety concerns, limited model availability, and downsides of observational learning. Additionally, the negative attitudes toward low-end PEVs from peers and society reinforce formation of an attitude of disliking PEVs. In the decision stage, the perceived low quality of PEVs and discouraging actions from local dealers (such as little after-sale service and dealers' lack of knowledge of PEVs) decreased the likelihood of purchasing a PEV. Further, in the implementation and confirmation stage, though both high and low-end PEV owners reported dissatisfying PEV ranges and insufficient charging infrastructure after they used the vehicle, low-end PEV owners were more concerned with home-charging safety and more likely to have a perception of poor quality as well as negative social image of their PEVs, which lead to their confirmation of unsustainable adoption of PEVs in the long run. Generally speaking, ***consumer awareness of PEVs,*** including knowledge of related supportive policies and familiarity with the new technology, range compatibility of daily travel, vehicle safety and quality, local market model variety, and awareness of the social image of PEVs are important factors affecting the decision-making process of PEV adoption. It also shows the ***social image of PEVs, vehicle quality, and home charging safety*** play critical roles in the long-term sustainable adoption of PEVs.

The results altogether offer meaningful insights into ways of increasing PEV adoption. Planners usually focus on tangible strategies, such as improving charging infrastructure, to promote PEV sales. Changing attitudes has not traditionally fallen within the realm of PEV promotional strategies. This study, however, points to the importance of the ***attitude toward PEVs*** and ***vehicle social image*** on PEV adoption, which may lead to other potential ways of influencing more people to choose PEVs. Although limited, the available evidence of China's PEV market suggests that "soft" strategies may have a measurable impact on PEV adoption, given a certain level of initial charging infrastructure. The results imply that planners need to consider comprehensive programs that affect PEV adoption on all three levels—individual attitudes, charging infrastructure, and incentives. One concern under China's current plan to reduce incentives gradually is that PEV demand will decrease before a self-sustained market can be achieved. In this situation, the findings of this study suggest that the increasing trend may be maintained through programs such as interventions by media advocacy programs designed to provide helpful information about the benefits and costs of PEVs. Practically, governments, local communities, and scientific institutions can play synergetic roles in advocating PEVs as an environmentally and economically beneficial mode of transport to build a positive image of PEVs that may help attract adopters who seek to be more protective of the environment and more energy efficient, or help foster the attitude of affection for PEVs. Additionally, given the importance of ***consumer awareness of PEVs***, efforts are needed to provide more information about PEV policies and technology through social media marketing activities and events by governments, automakers and dealers that could help remove barriers to awareness of PEVs. For instance, both the government and automakers may provide detailed overall performance reports for PEVs, especially PHEVs, e.g. fuel efficiency performance based on real operating data under different scenarios. Meanwhile, dealers are an especially important information channel due to their direct interaction with consumers. More PEV awareness and education programs for auto dealers are necessary before they can effectively/persuasively convey PEV knowledge to consumers. Further, governments need to take high responsibility in both regulating and administering ***vehicle quality and safety for PEVs***. While it is the responsibility of the Chinese government to place stricter regulations on vehicle quality and safety control, those regulations may not be entirely sufficient for enacting change to vehicle quality and safety. To achieve sustainable PEV adoption, it is also important to increase efforts to hasten the development of both home and outdoor ***charging infrastructure*** especially in 3rd-tier cities. Local governments need to coordinate with PEV dealers, utility companies, and homeowners associations to ensure successful home charging installation. To provide a friendly charging environment for PEV adopters, strategies may also include standardizing PEV charging standards nationwide. These strategies, together with incentives, should help expand the PEV market in China. These strategies may also be effective for other countries, if only ones with a similar economic status.

The sample size of this study is by far the largest of the most recent surveys on PEV users to have ever been conducted in China. To our best knowledge, this is the first study focusing on PEV adoption in third-tier cities in China, which still have much room for PEV sales improvements. Collective attitudes toward micro-scale electric vehicles, which have not received much attention due to their smaller market share in big cities, were for the first time captured in this study. Although we designed the survey to be relevant to all PEV owners, this is a marginal population that can be hard to reach. Out of the PEV owners recruited for the survey, it is possible that individuals who do not like PEVs were less inclined to complete the survey. Because our survey was online, non-response bias may be another concern. Moreover, our empirical model is limited by the experimental design in which a stated preference bias may exist. Some issues cannot be resolved without further research, such as improvements in survey methodology to achieve time series rather than cross sectional data. Nevertheless, this study offers an initial understanding of potential determinants of PEV engagement, and helps support the formation of policies directed at getting more people to buy and love PEVs.

References

Babbie, E. R. (1998). *The practice of social research*. Belmont, CA: Wadsworth Publishing Company.

Bakker, S., & Trip, J. J. (2013). Policy options to support the adoption of electric vehicles in the urban environment. *Transportation Research Part D, 25,* 18–23.

Bühler, F., Cocron, P., Neumann, I., Franke, T., & Krems, J. F. (2014). Is EV experience related to EV acceptance? Results from a German field study. *Transportation Research Part F, 25,* 34–49.

Campbell, A. R., Ryley, T., & Thring, R. (2012). identifying the early adopters of alternative fuel vehicles: A case study of Birmingham, United Kingdom. *Transportation Research Part A, 46,* 1318–1327.

Carley, S., Krause, R. M., Lane, B. W., & Graham, J. D. (2013). Intent to purchase a plug-in electric vehicle: A survey of early impressions in large US cites. *Transportation Research Part D-Transport and Environment, 18,* 39–45. https://doi.org/10.1016/j.trd.2012.09.007.

Carroll, S. (2011). The smart move case studies (Smartmove-10-010). Retrieved from https://www.cenex.co.uk/wp-content/uploads/2013/06/2011-11-21-Smart-Move-Case-Studies.pdf.

Cobb, J. (2017, January 16). The World Just Bought Its Two-Millionth Plug-in Car. *Hybrid-Cars.com*. Retrieved from https://www.hybridcars.com/the-world-just-bought-its-two-millionth-plug-in-car/.

Egbue, O., & Long, S. (2012). Barriers to widespread adoption of electric vehicles: An analysis of consumer attitudes and perceptions. *Energy Policy, 48,* 717–729.

Graham-Rowe, E., Gardner, B., Abraham, C., Skippon, S., Dittmar, H., Hutchins, R., et al. (2012). Mainstream consumers driving plug-in battery-electric and plug-in hybrid electric cars: A qualitative analysis of responses and evaluations. *Transportation Research Part A, 46*(1), 140–153.

Hackbarth, A., & Madlener, R. (2013). Consumer preferences for alternative fuel vehicles: A discrete choice analysis. *Transportation Research Part D, 25,* 5–17.

Helveston, J. P., Liu, Y., Feit, E. M., Fuchs, E., Klampfl, E., & Michalek, J. J. (2015). Will subsidies drive electric vehicle adoption? Measuring consumer preferences in the U.S. and China. *Transportation Research Part A, 73,* 96–112.

International Energy Agency. (2017). Global EV Outlook 2017: Two million and counting. Retrieved from International Energy Agency: https://www.iea.org/publications/freepublications/publication/GlobalEVOutlook2017.pdf.

Javid, R. J., & Nejat, A. (2017). A comprehensive model of regional electric vehicle adoption and penetration. *Transport Policy, 54,* 30–42.

Jensen, A. F., Cherchi, E., & Mabit, S. L. (2013). On the stability of preferences and attitudes before and after experiencing an electric vehicle. *Transportation Research Part D, 25,* 24–32.

Kitzinger, J. (1995). *Introducing focus groups. BMJ, 311.*

Kurani, K. S., Caperello, N., TyreeHagenam, J., & Davies-Shawhyde, J. (2014). I am not an environmental wacko! Getting from early Plug-in Vehicle owners to potential later buyers (Working Paper—UCD-ITS-WP-14-05).

Lane, B., & Potter, S. (2007). The adoption of cleaner vehicles in the UK: exploring the consumer attitude-action gap. *Journal of Cleaner Production, 15*(11–12), 1085–1092.

Peters, A., & Dutschke, E. (2014). How do consumers perceive electric vehicles? A comparison of German consumer groups. *Journal of Environmental Policy and Planning, 16*(3), 359–377.

Rogers, E. (2003). *Diffusion of innovations* (5th ed.). New York: Free Press.

Sierzchula, W., Bakker, S., Maat, K., & van Wee, B. (2014). The influence of financial incentives and other socio-economic factors on electric vehicle adoption. *Energy Policy, 68,* 183–194.

Singer, M. (2016). Consumer Views on Plug-in Electric Vehicles-National Benchmark Report (NREL/TP-5400-65279). Retrieved from www.nrel.gov/publications.

Skippon, S., & Garwood, M. (2011). Response to battery electric vehicles: UK consumer attitudes and attributions of symbolic meaning following direct experience to reduce psychological distance. *Transportation Research Part D, 16*(7), 525–531.

Vergis, S., & Chen, B. (2015). Comparison of plug-in electric vehicle adoption in the United States: A state by state approach. *Research in Transportation Economics, 52,* 56–64.

Vergis, S., Turrentine, T. S., Fulton, L., & Fulton, E. (2014). Plug-In electric vehicles: A case study of seven markets. Research report-UCD-ITS-RR-14-17.

Zhang, Y., Yu, Y., & Zou, B. (2011). Analyzing public awareness and acceptance of alternative fuel vehicles in China: The case of EV. *Energy Policy, 39,* 7015–7024.

Zheng Krupa, J. S., Rizzo, D. M., Eppstein, M. J., Lanute, D. B., Gaalema, D. E., Lakkaraju, K., et al. (2014). Analysis of a consumer survey on plug-in hybrid electric vehicles. *Transportation Research Part A, 64,* 14–31.

Conclusions

Lessons Learned and Future Prospects

Marcello Contestabile

Abstract At the end of this journey around the leading PEV markets in America, Europe and Asia, guided by the scientists who have been studying them over the last several years, what have we learned? Firstly, that much evidence is already available on consumer adoption and use of PEVs which can be used to answer, at least in part, important questions that policymakers and other stakeholders are currently faced with. Secondly, that undoubtedly much work still remains to be done and that, as the markets continue to expand, it is essential that research efforts in this area continue. Technology, policy and consumer behaviour will continue to co-evolve. Some of the answers to current questions may therefore change, and new questions may arise. However the evidence available today can already be a good guide to policymakers and other stakeholders in their efforts to support a rapid and effective transition to electric mobility. The evidence presented in this book also points at new questions that are worth researching and at the methods and new data that would be needed to tackle them. This concluding chapter is organised around key questions currently being asked. For each question in turn, it briefly discusses the high-level messages that have emerged from the market-specific studies presented in the previous chapters and, by comparing and contrasting them, seeks to derive more general insights. The chapter concludes with a brief discussion of future prospects for this field of research and their implications for policy-making.

Keywords PEV adoption · PEV use · PEV policy · PEV consumer research

What Have We Learned?

At the end of this journey around the leading PEV markets in America, Europe and Asia, guided by the scientists who have been studying them over the last several years, what have we learned? Firstly, that much evidence is already available on consumer adoption and use of PEVs which can be used to answer, at least in part, important questions that policymakers and other stakeholders are currently faced

M. Contestabile (✉)
Imperial College London, South Kensington Campus, London SW7 2AZ, UK
e-mail: marcello.contestabile@imperial.ac.uk

M. Contestabile et al. (eds.), *Who's Driving Electric Cars*, Lecture Notes in Mobility,
https://doi.org/10.1007/978-3-030-38382-4_13

with. Secondly, that undoubtedly much work still remains to be done and that, as the markets continue to expand, it is essential that research efforts in this area continue. Technology, policy and consumer behaviour will continue to co-evolve. Some of the answers to current questions may therefore change, and new questions may arise. However the evidence available today can already be a good guide to policymakers and other stakeholders in their efforts to support a rapid and effective transition to electric mobility. The evidence presented in this book also points at new questions that are worth researching and at the methods and new data that would be needed to tackle them. Further research would in turn generate more evidence that policymakers and other stakeholders could use to refine their future strategies.

Crucially, the evidence, both present and future, needs to be provided to the non-experts in a form that they can consume, and we hope that this book has gone some way towards achieving that. It is in this spirit that we conclude the book with a brief discussion of the high-level messages that have emerged from the market-specific studies presented in the previous chapters. In particular, by comparing and contrasting the findings across different markets, characterised as they are by their own socio-economic, policy, geographic and market conditions, we seek to derive more general insights. For ease of reading, we have organised the insights around some of the key questions that are being asked about the PEV market. The questions, which are clearly not exhaustive, are addressed in turn in the remainder of this section. In the following and concluding section of the chapter, and indeed of the book, we will briefly touch upon future prospects for this field of research and the implications that these may have for policymaking.

Who Has Already Adopted PEVs?

Firstly, in keeping with the title of this book, let us reflect on what we have learned so far on the profiles of the adopters of PEVs and their motivations for adopting. The research presented in this book has provided us with a very rich picture: PEVs are being adopted by a mix of private and corporate car users who have different socio-economic attributes and live in different types of dwellings and geographical areas. Moreover, substantial differences can be observed across the markets we have studied, which clearly reflect the particular policy approaches of the governments as well as the broader regulatory, demographic and economic environment.

Let us first focus on BEVs. One high level message that can be derived from the research presented in the book is that, despite the growth experienced by the market in the last few years, many private adopters of BEVs across the world still broadly fit the profile of the innovators and early adopters as defined by Rogers in his theory of diffusion of innovation. BEV adopters tend to be high-income and highly-educated individuals with a strong interest in technology and the environment. In particular, the interview studies conducted by Jonn Axsen and co-workers in Canada (Chapter "The Present and Future Market for PEVs in Canada: Evidence from a Mixed-Method Research Program") show that BEV adopters have high awareness and a sophisticated understanding of the technology, have adopted for a combination of practical, symbolic and societal reasons, and had to adapt their vehicle usage patterns. Interestingly though their socio-economic profiles are much more heterogeneous than Roger's innovators. The study of the evolution of the US market presented in Chapter "Electric Cars in California: Policy and Behavior Perspectives" further shows that the share of BEV adopters broadly corresponding to the innovator and early adopter profiles has been slightly decreasing over time, although still growing in absolute terms, and that the share of early majority adopters has also been growing. If however we consider that the mainstream or majority segment of the population is much larger than the innovators and early adopters, we can see that PEV adoption is still very much an early adopter phenomenon in the US as well as in many other markets. BEVs, at least in Europe and North America, are also predominantly adopted in multi-vehicle households, where their range restrictions can be managed by shifting trips across vehicles. The Chinese market has its peculiarities, which we will discuss later.

Norway is by far the most advanced BEV market in the world. In Norway BEV drivers are for the vast majority private consumers who want to save money on vehicle purchase price, fuel, toll roads and ferries, and want to be able to drive in bus lanes to save commute time. Even in Norway though BEV adopters have a profile that somewhat differs from the average car user. In particular, they tend to be younger, better educated, with higher rate of employment, part of larger, multi-vehicle households, and having longer commutes than ICEV and PHEV buyers. BEVs for them prove more time efficient than public transport and less costly than ICEVs. BEVs are driven similar mileages per year to ICEVs, however they tend to be used predominantly on shorter trips which suggests that there is a degree of trip redistribution between the vehicles. In fact, 4 out of 5 BEV owners in Norway also own another car, which allows them to choose vehicles according to journey types and length. A similar situation can be observed in other markets as well. For example in Sweden BEVs are a small fraction of the total PEV market, dominated as it is by PHEVs, however their adopters are also multi-vehicle families. Similar findings apply to Germany too, where BEV adopters are typically middle aged men with high socio-economic status, part of multi-vehicle households, living in suburban areas and characterised by a strong interest for technology and the environment.

As previously mentioned, the Chinese BEV market shows some peculiar traits. Firstly, unlike other markets where BEVs are predominantly adopted by private car users, in China around half of the BEVs belong to shared mobility fleets, particularly short rentals and ride share companies. A key reason for this is that in China government grants are paid to the BEV manufacturers and not their buyers, and it is therefore not a coincidence that some of these shared mobility companies are owned by the BEV manufacturers themselves and are used as a way of rolling out their vehicles. As for the private owners of BEVs, many of them are males younger than 40 with high education levels and most of them have children, which means that Chinese BEV adopters tend to be younger than their American and European counterparts. Moreover, unlike other markets, many privately owned BEVs have been adopted by first-time car buyers and single vehicle households as well.

Substantial differences also exist in China between BEV adopters in so-called first tier cities such as a Beijing and Shanghai, and BEV adopters in second and third tier cities. In first-tier cities in China, where strong restrictions apply to the registration of new ICEVs and their circulation, non-monetary incentives play a crucial role in promoting adoption of BEVs and PHEVs. In particular, the registration privilege for new PEVs has been identified as the number one driver and its effect, combined with that of other non-monetary incentives such as exemption from traffic restrictions, has even induced some users of public transport or non-motorised transport modes to buy BEVs and PHEVs. Moreover, first-tier city dwellers tend to prefer medium to high end BEVs and PHEVs, and many purchase them as their first or only vehicle even if they don't have access to off-street private parking. The latter is possible thanks to the extensive provision of public charging infrastructure in parking bays where PEVs have priority, hence giving their owners an additional advantage in areas where demand for public parking spaces is high.

BEV adopters in the second- and third-tier Chinese cities however do not benefit from the same non-monetary incentives and are therefore mainly motivated by the low operating and maintenance costs of BEVs. For this reason, BEV buyers in these cities tend to prefer the low-end, short-range BEVs offered by the Chinese OEMs that cater to the needs of households with limited disposable income. The users of these vehicles generally like them and appreciate their practical attributes however they don't like to talk about them due to the negative social image attached to low-end vehicles. In general in China we observe that BEV adopters are more sensitive to vehicle price while PHEV adopters focus more on vehicle performance and interiors, and that high-end BEVs and PHEVs are driven more kilometres per day on average than the low end BEVs but less than ICEVs. BEVs are typically used for commuting or picking up family members.

Let us now shift our attention to PHEVs. One important observation is that in some countries, like the Netherlands and Sweden, adopters of PHEVs are predominantly company car drivers who drive more than the national average and adopt PHEVs thanks to the financial and practical benefits they provide without requiring major adaptation. More generally, PHEV adopters exhibit profiles that are closer to the early adopters and early majority as defined by Rogers, and their income is closer to the national average if compared to BEV adopters. Therefore, in principle PHEVs are a better fit for mainstream passenger car users, including single-vehicle households.

The case of the Netherlands, which has until recently been the largest PHEV market in the world, is particularly interesting. Due to the structure of the incentives provided, PEVs and particularly PHEVs were predominantly adopted by company car users with profiles that are in many ways similar to those of the mainstream users of ICEVs, although with a bias towards higher income and education. This is because the strong government incentives made the adoption of PHEVs a rational decision motivated by cost savings. This also confirms that PHEVs are favoured by mainstream car users, who find them easier to live with than BEVs. By contrast, the few BEV adopters in the Netherlands had a rather different profile, being mainly technology enthusiast, high income private car buyers like in many other countries. Unlike most other north European countries, many of the early adopters in the Netherlands did not have access to private off-street parking. Instead they relied on public chargers in public parking bays in urban areas, which were built on demand by turning a public on-street parking bay into one reserved to PEVs only. At low levels of adoption this practically meant giving the PEV users their own private parking bay, which provided a particularly strong incentive to adoption especially in city centres where substantial pressure exists on parking availability. Those who adopted PHEVs before the incentives were discontinued are typically on 4–5 year long lease contracts, so over the next few years it will be interesting to see what kind of vehicle they will choose next, whether they will adopt a PEV again or will go back to ICEVs.

Sweden is in some ways similar to the Netherlands in that most PEVs are PHEVs and their adopters company car drivers. This can be explained by the particular incentive scheme, the need to drive over longer distances where BEVs would not be practical, and the availability of suitable models in the popular station wagon and SUV segments where BEVs are unavailable. Interestingly, PHEVs in Sweden are also predominantly adopted by multi-vehicle households with access to off-street parking. In Germany around one third of PEVs are privately owned and the remaining are either company cars or part of corporate fleets. The company car segment in Germany is not yet well understood due to the lack of data, however thanks to recently introduced incentives we can expect that the uptake of PEVs, and particularly PHEVs, will progress rapidly.

What Is Keeping Others From Adopting PEVs?

We have just discussed some key insights into the profiles of current adopters. Let us now look into the barriers that are preventing more consumers from adopting PEVs. Although this book is mainly about the study of the actual PEV owners and users, not those who have yet to adopt them, understanding what keeps others from adopting is just as important for policymakers and other stakeholders. Moreover, the study of the former can certainly help gain insight into the latter as well. In particular, understanding the extent to which adopters and non-adopters differ, and what these differences mean in terms of policy support required, is crucial. As we have seen earlier, in many countries adoption is still largely limited to so-called innovators and early adopters, and they have rather different awareness, attitudes and preferences compared to mainstream car buyers. For this reason, mainstream car buyers are directly tackled in several chapters of this book and the main insights that were generated are here briefly discussed in turn. It is worth adding though that, as explained by Stephen Skippon and George Beard in their Chapter “Beyond Consumer Innovators: Adoption of Plug-in Vehicles by Mainstream Consumers in the U.K.”, the study of the possible response to PEVs by mainstream ICEV buyers poses particular difficulties. This is mainly due to their lack of familiarity with the technology which results in biases. The impact of such biases can be mitigated through appropriate experimental designs, which include providing information and allowing the interviewee the time to reflect on it, or even providing the interviewee with the experience of driving a PEV over multiple days. Such strategies are also adopted in the studies presented in Chapters "The Present and Future Market for PEVs in Canada: Evidence from a Mixed-Method Research Program" and "What Can Driving Patterns Reveal About the Suitability of PEVs in Sweden? Analysis and Policy Implications".

Let us begin from the first barrier that can hinder the adoption of PEVs, i.e.: lack of awareness. As also postulated in Rogers' theory, the first step of the adoption process is acquiring the necessary awareness and knowledge of the new technology. Without this, passenger car buyers would not even consider PEVs as one of the possible options to choose from. In Chapter "What Conversations Between PEV Owners and Owners of Non-PEVs in California Tell Us About Sustaining a Transition", Ken Kurani shows how lack of awareness remains a powerful barrier to adoption of PEVs even in a leading market such as California. Moreover, Ken observes that between 2014 and 2017 the overall level of awareness and knowledge of PEVs among mainstream passenger car buyers in California has remained very low, despite the fact that the PEV market has grown substantially over those years. Similar results have been found for Canada by Jonn Axsen and co-workers. Ken's research identifies the limited direct interaction between PEV adopters and mainstream car buyers as one of the main reasons for the latter group's lack of awareness. The importance of the information flow between PEV drivers and drivers of ICEVs as an enabler of adoption is stressed in several other chapters of this book.

Erik Figenbaum's studies of the Norwegian PEV market however paint a different picture. Consumer to consumer communication on PEVs and their advantages have worked well in Norway and have been an important driver behind sustained sales growth. This can, at least in part, be explained by the long history of activities and stable support measures that have allowed consumers to gradually become aware and form a positive view of PEVs. A survey of the population in Norway showed that a 5% fleet share was sufficient to lead to a situation where the majority of the population knows someone who owns a BEV, more than half of the population has been passenger in one, and more than a third has driven one. In Sweden too communication between PEV owners and owners of ICEVs seems to be working well, even at a penetration level that is substantially lower than Norway's. The mechanisms that enable or hinder communication among these two groups in society are not yet well understood and deserve further study.

Once the necessary level of awareness has been reached, the next barrier to the adoption of PEVs is constituted by the possible negative perception of the technology and its attributes. Several studies presented in this book discuss this particular aspect. For example the study of the UK market by Stephen Skippon and George Beard (Chapter "Beyond Consumer Innovators: Adoption of Plug-in Vehicles by Mainstream Consumers in the U.K.") shows how early PEVs with limited performance suffered from a negative image and how this has improved as new, better models were introduced into the market. In California, the study by Ken Kurani shows that, just like awareness and knowledge, perceptions of PEVs and infrastructure coverage have not improved either over the years, due to untested assumptions regarding PEVs' "problems". It also shows that, somewhat counterintuitively, marketing PEVs as an environmentally friendly technology does not help promoting their adoption by mainstream car users and can even hinder it. This is because mainstream car users in California don't have particularly strong pro-environmental attitudes and their self-image is one of rational consumers that take financially sound decisions. Something similar has been observed in Canada and the UK too. However Ken's work has shown that the mainstream car users' perception of PEVs can become more positive if they get the chance to hear directly from the PEV owners and find out that these vehicles can make economic sense too.

Once consumers have overcome the barriers constituted by lack of knowledge and negative perceptions, they are ready to include PEVs among the possible options that they will consider when purchasing a vehicle. At this stage consumers will look more closely at the utility of the PEVs as opposed to their symbolic attributes, and decide whether to adopt or reject. Put simply, this means that prospective adopters will look for PEV models that can potentially suit their needs and tastes, and they will assess their purchase and running costs, their fit with their mobility needs, their quality and environmental attributes, and on this basis they will judge whether overall benefits outweigh costs.

All the main aspects that enter the adoption decision process have been addressed by the research presented in this book. In particular, the importance of the availability of attractive PEV models in the right segments for the target market cannot be overstated and the effect that the lack of it can have on adoption is discussed in detail in the Chapter "The Present and Future Market for PEVs in Canada: Evidence from a Mixed-Method Research Program". Moreover, in the case of BEVs, a crucial aspect is the extent to which their adoption requires adaptation in terms of travel behaviour. In other words, the potential adopters have to consider the range offered by the BEVs, the presence of charging infrastructure, and how both fit with their typical travel patterns. There are important policy implications to these aspects, because the way government incentives are designed tends to have an impact on the range of BEVs that are commercialised as well as on the type and extent of charging infrastructure that is rolled out. As for PHEVs, which do not require major adaptation from the point of view of travel patterns, an important question is around their environmental attributes and particularly the extent to which they can actually reduce emissions compared to conventional ICEVs. Considerations of quality are particularly relevant for the Chinese market, where many low-end BEVs are offered by the local OEMs which don't necessarily meet the expectations of the prospective buyers. Last but not least, the economic case for PEVs from the consumer perspective today is largely based on the presence of strong incentives, both monetary and non-monetary. The question is how long the incentives will remain in place and whether this will be sufficient for adoption to continue at the desired rate. These questions are briefly addressed in turn in the following sections.

How Long Does the Range of a BEV Need to Be?

Range is a particularly important attribute for BEVs and one that passenger car users considering adopting a BEV pay particular attention to. Yet, there are several complexities when discussing what an adequate range is from a consumer perspective. Firstly, the very definition of range from a consumer perspective is not entirely straightforward. Stephen Skippon and George Beard raise the issue of what exactly consumers participating in studies mean when they indicate a minimum range that they consider acceptable. Stephen and George hypothesise that many consumers regard range as sufficient if it allows them to travel to and from a distant place. This however may require cruising at highway speed, under which conditions the actual range of the BEV is much shorter than its nominal range. Moreover, range expectations among PEV users don't seem to be static but rather increase as the technology improves. In particular, the study of the German market by Patrick Plötz and Elisabeth Dütschke (Chapter "Electric Vehicle Adoption in Germany: Current Knowledge and Future Research") shows how the overall rating of the PEV experience by its users is steadily improving, which reflects the substantial increase in quality and performance of the PEV models on the market over the last few years. However, when it comes to BEV range, users' ratings don't seem to reflect the increase that has occurred over time, which suggests that BEV customers are becoming increasingly demanding.

While expectations on range are overall increasing across the markets, we need to make an important distinction between two broad categories that we have already mentioned: the single vehicle and the multi vehicle households. Based on the study of the driving patterns of ICEV and BEV users, it appears that in the European and North American markets relatively short range BEVs that are recharged at home overnight broadly meet the needs of multi vehicle households without requiring major adaptation, and yet there is strong demand for increased range and public charging infrastructure. The situation is obviously more extreme in the case of single vehicle households, where only those consumers who attach a particularly strong symbolic value to BEVs are prepared to adopt BEVs that are not comparable to ICEVs in terms of functionality. On the contrary, multi vehicle households don't show that symbolic value of BEVs has a strong weight, which can be explained by the fact that they can accept reduced functionality much better than a single vehicle household. However for long range BEVs the observed difference between the single vehicle and the multi vehicle households reduces. Therefore an offer of substantially longer range BEVs, also covering the large sedan and SUV segments, and an extensive network of rapid chargers would be required in order to unlock the rest of the market.

In particular, evidence from Norway shows that BEVs with nominal ranges between 160–250 km are adequate for most current users, who are predominantly multi-vehicle households, This holds true even considering that in real traffic the range reduces by around 25% and in cold weather by up to 50% (and the two are cumulative). Despite this, BEV users in Norway do state that they would like longer range, and ICEV owners say they would like even longer range if they were to adopt a BEV. Erik Figenbaum finds that a real driving range of 300 km could open up 80% of the market, and 200 km up to half of that, but only if combined with the availability of suitable models in the popular segments where BEVs are currently unavailable.

Frances Sprei's study of Swedish two vehicle households (Chapter "What Can Driving Patterns Reveal About the Suitability of PEVs in Sweden? Analysis and Policy Implications") provides similar results. Frances finds that BEVs with a real-driving range of 130 km (which in cold weather could go well below 100 km) if adopted as second car in the household do not require major adaptation. Yet, those two vehicle households that have trialled the BEVs found the range to be too short. Many participants stated that after the trial their perception of BEVs is more positive and that they enjoyed the experience, and yet they would not buy one. Some said they would consider leasing one while others said they would buy a PHEV instead. Expanding the analysis to the whole passenger car population in Sweden, GPS measurements of conventional vehicles show that, given a full recharge at night and no adaptation, only 20% of the vehicles would manage all their driving if they had a range of 130 km, as opposed to around 50% in Germany. If instead we look at ranges closer to 300 km, 65% of the vehicles would manage all their driving on a fully charged battery in the morning, which is lower than the figure discussed above for Norway. This is because for geographic reasons in Sweden vehicles travel particularly long distances, which could also partly explain why PHEVs so far have been more popular than BEVs.

Despite the peculiarities of the Chinese BEV market that we have previously discussed, the study by Shawn Ou and Zhenhong Lin (Chapter "China's Booming Plug-in Electric Vehicle Market—How Will It Continue?") highlights that range anxiety is cited as the main concern by consumers considering adopting a BEV. Shawn and Zhenhong also estimate the cost of adaptation to short range BEVs in China using GPS travel data and find that this could be quite substantial. However, it is apparent that the combination of low purchase price and extensive public charging infrastructure means that many Chinese consumers with limited disposable income are willing to live with short-range, low-speed BEVs.

How Important Is Public Charging Infrastructure?

Closely related to the issue of range limitations is that of the lack, actual or perceived, of public charging infrastructure. The latter has been clearly identified in many previous studies as a major barrier to adoption of PEVs and particularly BEVs. The studies however were largely based on surveys of owners of non-PEVs. Now that a substantial number of PEV users can also be studied, we can gain valuable insight from their actual charging behaviour, which may help design public infrastructure that is sufficiently utilised to be economically viable without being overly congested.

The study of PEV users across the markets analysed in this book shows that, where available, home charging is by far the preferred option and it is where most of the charging occurs. This is the case in most markets, where the majority of current PEV adopters tend to have access to off-street or private parking places equipped with a charging point. For this reason existing public charging infrastructure is still substantially underutilised in most markets and yet PEV users, and BEV users particularly, demand more of it and especially more rapid charging stations. A notable exception documented in this book is that of early adopters, or pioneers, in Canada who report to be making extensive use of the public charging infrastructure even though they have access to home charging and don't really need it. Utilisation of public charging infrastructure is obviously higher, but not necessarily high enough for the economics of the infrastructure to be favourable, in those urban areas where PEV users don't have access to private parking and therefore use public chargers as their primary chargers. We will discuss the cases of the Netherlands and China later.

As for non-PEV users, the evidence presented in this book reinforces the finding from previous research that the availability of an extensive network of public chargers is one of the pre-requisites for mainstream car users to consider adopting BEVs. In particular, the workshops conducted by Ken Kurani in California (Chapter "What Conversations Between PEV Owners and Owners of Non-PEVs in California Tell Us About Sustaining a Transition") show how the presence of public infrastructure, while not entirely dissipating non-PEVs users' doubts about the practicalities of PEVs, carries important positive meanings: (a) the presence of infrastructure signals the presence of PEVs; (b) any additional charger signifies that the charging infrastructure is becoming denser; and (c) it shows commitment to the transition from various actors.

Underutilisation of public charging infrastructure may not be surprising in those markets where PEV penetration is not particularly high, because infrastructure has to be built first in order to encourage vehicle uptake. However, even if we look at the case of Norway with its large population of PEVs, the picture is not radically different: Norwegians predominantly charge at home and only occasionally make use of public charging infrastructure. It is interesting to note though that the presence of fast chargers on motorways, combined with the longer range of newer models, has allowed Norwegian BEV owners to start travelling on longer routes where they would have otherwise used a different vehicle or means of transport. And yet fast chargers are used on average only once a month by typical BEV users and twice a month by Tesla users who tend to travel more regularly over long distances. The infrequent use of rapid fast charging on motorways even in Norway confirms that their economics overall remains problematic.

As mentioned earlier, the case of the Netherlands (Chapter "Plug-in (Hybrid) Electric Vehicle Adoption in the Netherlands: Lessons Learned") is quite different from that of Norway and several other countries discussed in this book in that high uptake of PEVs has been achieved in urban areas where dwellers typically don't have access to private parking and park overnight in public bays. This was made possible thanks to the build-up of a very extensive network of public charging points, the majority of which were installed on demand and used as primary chargers. Some chargers were also built as secondary chargers and were intended to support PEV drivers visiting the areas. This particular arrangement has allowed PEV adopters to recharge their vehicles overnight while being guaranteed a free parking place. Even then, Dutch city dwellers who adopted PEVs overwhelmingly chose PHEVs over

BEVs, at least until incentives on PHEVs where discontinued, because the ability of PHEVs to also run on gasoline removed the risk associated with not finding a free parking bay or the charging point being out of order.

The Dutch experience provides particularly valuable insights into the successful development of a network of public chargers for inner city residents who don't have access to private parking. It also clearly shows the several challenges that this poses, which comprise overall long connection times and consequent low utilisation, poor business case and pressure on the already limited availability of public parking bays. One possible approach to mitigating these challenges is a time-base charging fee. Survey results discussed in Chapter "Plug-in (Hybrid) Electric Vehicle Adoption in the Netherlands: Lessons Learned" suggest that PEV users would be in favour of it despite the increase in overall charging costs that the fee would bring. A radically different approach is to resort to the deployment of rapid charging stations in urban areas, which is a much more similar model to the conventional petrol station for ICEVs. Both have their advantages and disadvantages and more research is undoubtedly needed.

High levels of adoption of PEVs in urban areas where private parking availability is extremely limited have been achieved in China too. This has been supported by the development of a vast network of public chargers, both slow and fast, for primary as well as secondary use. Interestingly though the majority of PEVs in Chinese cities are BEVs, including short-range models, and not PHEVs as in the Netherlands. Even more remarkable is that many of these BEVs have been adopted by single-vehicle households that are often also first-time car buyers. These households generally report that the combination of the BEV range and the extensive public charging infrastructure provided in Chinese cities works well for them. However this observation can to a large extent be explained by the particular conditions of the Chinese market, and particularly cities like Beijing and Shanghai, where low-income households with no access to private chargers are prepared to accept the relative inconvenience of using low-range BEVs that completely rely on public charging infrastructure in order to satisfy their latent demand for private car transport. For this reason we cannot expect that similar levels of BEV range and availability of public chargers would be sufficient to meet the requirements of those single-vehicle households with no access to private chargers in North America and Europe as well.

While Chinese BEV owners are able to cope with using public infrastructure as their primary chargers, the study of six Chinese cities presented in Chapter "Plug-in Electric Vehicle Market Research on Consumer in Six Cities in China" shows that in China too home charging, where available, is the most important charging method, followed by workplace and public fast charging points. The study also shows that charging frequency is closely correlated with charging infrastructure availability: in cities with denser charging infrastructure, PEV owners charge more frequently than owners in the less infrastructure-dense cities. Similarly, in these cities it has been observed that shorter BEV ranges are correlated to more frequent charging behaviour and that the users of PHEVs with short ranges charge their vehicles very frequently. A related result is found by Gil Tal and co-workers in their recent study of charging behaviour of Californian PEV users (Chapter "Electric Cars in California:

Policy and Behavior Perspectives"). Gil and co-workers observe that different user segments show different charging patterns and that factors such as access to home charging, rate and cost of charging, PEV range and type (i.e.: BEV or PHEV) all play a role in determining where and how often PEV users charge their vehicles.

The wealth of knowledge that is starting to emerge from the study of charging behaviour of real PEV users can potentially be very valuable to design more effective charging infrastructures. It is apparent that past attempts to provide widespread public infrastructure coverage with limited understanding of the needs of prospective PEV drivers have led to very low rates of utilisation. This is particularly the case for the early build-up of slow destination charging infrastructure, which BEV users ended up not using at all. Fast charging is valued more and yet, as we have seen, even that is not used much. Building charging infrastructures that can be utilised sufficiently to make economic sense while also enabling adoption of BEVs by those households that do not have two cars and a garage may require more sophisticated approaches. These will need to be informed by a sound understanding of the driving patters and charging behaviour of the different segments of passenger car users, and how these may change as a result of the increasing range of new PEV models and of the different types of infrastructure offered.

Are PHEVs Important for Achieving High Levels of PEV Penetration?

Several studies presented in this book show how mainstream consumers are more interested in PHEVs than BEVs. As we have already seen, in general potential adopters are more interested in considering BEVs as second cars rather than first cars, and only if the range is sufficiently high. For single vehicle households PHEVs are a much more attractive proposition, thanks to the absence of range limitations which makes their functionality comparable to that of the ICEVs they replace. However, just like BEVs, actual adoption of PHEVs is driven by government incentives and these have been reducing in recent years, especially for PHEVs. Such reduction in incentives for PHEVs, which in some countries have caused their sales to drop, is at least in part due to the perception that these vehicles are not driven on electricity much. The extent to which PHEVs are actually driven on electricity is discussed in the following section. Let us first focus on the role that PHEVs can play in achieving high levels of PEV penetration.

As the markets that we have studied in this book show, the relative strength of support measures for BEVs and PHEVs plays a major role in determining which prevails. Typically in those markets where both BEVs and PHEVs are equally incentivised PHEVs dominate. BEVs on the other hand outsell PHEVs in those markets where the former are strongly supported and the latter are not. This is the case of Norway for example, where BEVs account for the majority of the PEV market thanks to the strong support they receive. PHEVs do not benefit from the same treatment and the structure of the incentives made available to them means that these vehicles are only at an advantage in the compact vehicle segment. Despite this, in Norway PHEVs still have a market in the larger segments too, which suggests that there is substantial latent demand for these vehicles and their uptake could be much higher if they were better supported. Crucially, Erik Figenbaum's study of the Norwegian market shows that, even in a market so successful for BEVs, more support to PHEVs would lead to overall substantially higher penetration of PEVs because the two types appeal to different segments of the population.

The study of the Canadian market by Jonn Axsen and Zoe Long shows similar results: mainstream consumers generally prefer PHEVs and if these are not sufficiently supported the uptake of PEVs overall will be slower. Jonn and Zoe also clearly demonstrate the importance of the availability of the right models of PEVs for their adoption. In particular, the lack of PEV offer in the most popular segments in North America such as SUVs and pickup trucks is a barrier to their mainstream adoption. The fact that the PHEV technology is a better fit than BEVs for these segments is another reason why the former are important in order to achieve high levels of PEV penetration rapidly.

Similarly, in Sweden one strong contributor to the rapid uptake of PEVs has been the availability of attractive PHEV models in the most popular segment, i.e.: station wagons. Another segment where PHEVs are already important and will remain so in future is company cars, because their users tend to drive above the annual average of the population and therefore cannot easily accommodate BEVs. Company cars are a substantial segment in many countries and, if adequately incentivised, can see a very rapid uptake of PHEVs as the cases of the Netherland and Sweden have shown. Lastly, we have already mentioned the limited uptake of BEVs in single vehicle households. Once the multi-vehicle household segment saturates, further uptake of BEVs may become harder and PHEVs may have an important role to play, until at least the range and recharging time issues of BEVs are fully resolved and their functionality approaches that of conventional ICEVs.

To What Extent Are PHEVs Really Driven on Electricity?

As already mentioned, several governments strongly decreased their support to PHEVs relative to BEVs due to the perception that PHEVs were not being driven on electricity much. This was based on anecdotal evidence and on one study of company car users in the Netherlands according to which PHEVs were not being charged regularly by their owners, and some were not being charged at all. However the evidence presented in this book, based on the actual use and charging habits of a large sample of PHEV users in the Netherlands and elsewhere, paints a rather different picture.

The evidence gathered by the scientists who have contributed to this book shows that PHEV charging behaviour can differ much between company car and private car users. Both in the US and in Europe, privately owned PHEVs have been shown to be driven mostly on electricity, while PHEVs purchased as company cars not so much.

Especially those company car drivers who have company fuel cards or can claim fuel costs as business expenses have little incentive to use electricity because the inconvenience of plugging in is not offset by personal savings. Therefore company car users are overall less likely to take full advantage of the electric driving capabilities of their vehicles.

The difference between company car and private car users can be clearly observed in the Dutch PEV market, which has been dominated by PHEVs until the recent decision by the government to discontinue the incentives for this type of PEVs. Rick Wolbertus and Robert van den Hoed (Chapter "Plug-in (Hybrid) Electric Vehicle Adoption in the Netherlands: Lessons Learned") have studied the behaviour of Dutch PHEV users over 4 years and have observed that the users of company vehicles tend to be indifferent to fuel savings. This is corroborated by the fact that they hardly show signs of the adaptation in behaviour over time that is typically observed with other segments of PEV owners. This lack of adaptation is particularly apparent in the case of short-all electric (AER) range PHEVs, while the longer-AER models tend to show utility factors (i.e.: percentage of electric driving out of total mileage driven) that are in line with those observed in other countries. Rick and Robert also found that those PHEVs that are privately owned have rather high utility factor, although marginally lower than the figure assumed in the European test driving cycle (NEDC). The study of the German market by Patrick Plötz and Elisabeth Dütschke confirms this finding and also shows how those PHEVs that are characterised by a longer AER have higher utility factors, close to the NEDC values. This result is further reinforced by the study of charging behaviour of PEV drivers in California by Gil Tal and co-workers, which clearly shows how longer AER PHEVs tend to be charged more frequently than their shorter range counterparts. As for Norway, Erik Figenbaum reports that the average utility factor of PHEVs was already reasonably high, at 50%, in 2016 and expects it to have gone up further since, thanks to the longer AER of newer PHEV models.

Another important finding of the analysis of German PHEV users by Patrick and Elisabeth is that long-AER PHEVs drive roughly the same annual distance on electricity than long range BEVs. In particular, a 60 km real driving AER appears to be sufficient for PHEVs to maximise their utility factor and hence the annual miles driven on electricity. These results have important policy implications: providing sufficient incentives for PHEVs, particularly those with longer AER and in larger vehicle segments, can substantially increase not only the overall adoption of PEVs but also the use of electricity in passenger car transport. The author of this chapter had arrived at a similar conclusion by studying different mixes of BEVs and PHEVs from a techno-economic perspective. [1] Lastly, it is worth mentioning that, while in the last few years several governments reduced support to PHEVs, the Canadian government has very recently moved in the opposite direction and increased federal incentives for long-AER PHEVs to match those provided for BEVs. This may perhaps be a sign that evidence such as that presented here is starting to reach policymakers.

[1] Contestabile, M., Alajaji, M., & Almubarak, B. (2017). Will current electric vehicle policy lead to cost-effective electrification of passenger car transport? *Energy Policy, 110,* 20–30.

How Easily Can Leading Markets Be Replicated?

When looking at the rapid uptake of PEVs achieved in countries like Norway, Sweden and China, governments may be tempted to set similar targets for their countries as well. However, success stories such as those of the leading PEV markets are difficult to replicate: they are the result of a combination of unique conditions and of policies that cannot be directly applied elsewhere. Therefore governments that want to follow in the footsteps of the leading countries and achieve rapid PEV market growth and high fleet shares are advised to set themselves realistic targets based on their own individual approach, learning from other countries where possible but also taking into account the specificities of their markets and the particular barriers that they face. To illustrate this point, we will briefly discuss some of the findings presented earlier in this book for two leading markets: Norway and China.

At over 50% market share, Norway is by far the most developed PEV market in the world. However, as the chapter by Erik Figenbaum clearly illustrates, this success can largely be attributed to the extremely favourable policy environment created by the Norwegian government, with nearly three decades of very stable and strong policy support to PEVs. Although some of the policy drivers may have changed over the years, the strong interest for environmental protection has not, and with a very clean electricity grid, thanks to the abundance of hydropower available in the country, the transport sector has received all the attention. Moreover, the exceptionally high taxation imposed on imported passenger cars, initially motivated by an attempt to support the development of a domestic car industry and only made possible by the very high disposable income of the Norwegians, has created the opportunity to strongly incentivise PEVs by simply waiving taxes rather than providing grants. This approach, which constitutes foregone government income rather than a budgetary expenditure, is inherently more stable because the waiving of taxes and other fees (such as road use charges) is inscribed in national laws and does not need to be approved on an annual basis. With a level of taxation so high, tax exemptions alone have proven sufficient to make PEVs cost-competitive with conventional ICE cars.

In addition to the strong and stable support policies, the PEV market in Norway benefits from the most favourable conditions possible, with the only exception of the cold winter weather which affects BEV range. Electricity is almost entirely renewable and is taxed much less than petrol and diesel, which are on the contrary very heavily taxed, and the result is that PEVs in Norway yield the highest fuel cost savings of all of Europe. Moreover Norwegian residential dwellings rely on electricity for heating as well, which means that the distribution grids are strong and that most homes can recharge PEVs at 3.6 kW, some even at 11 kW. In addition, with the exception of dense urban areas, many households have access to home charging and own multiple vehicles, which means that they can easily substitute BEVs with ICEVs for longer trips. BEV range therefore is generally not a major issue because, despite the negative impact of the cold weather, the stringent speed limits and limited motorway network means that average driving speeds are very low. All this combined with the long-standing and stable policy environment has allowed plenty of time for Norwegian car buyers to become aware of the technology, appreciate its unambiguous environmental credentials when powered by renewable electricity, understand its cost-competitiveness, and overall develop confidence in the technology and in the future stability of its market.

China's case is rather different but equally interesting and unique. China is currently the largest PEV market in the world and, although in terms of PEV market share it is in line with several other advanced markets, its rapid growth has been remarkable. Again, this can be attributed to the very favourable policy environment created by the government, mainly driven by the aim of developing a domestic PEV manufacturing industry for which the presence of a strong domestic market is a pre-requisite. The rapid expansion of the Chinese market can at least in part be attributed to the very substantial public procurement of PEVs in government and shared mobility fleets, which in China has happened on a much larger scale than elsewhere. The adoption of PEVs by private car buyers has however also been very substantial, particularly in the Chinese megacities where restrictions are in place on the registration of new ICEVs and from which PEVs are exempt. It is this unique combination of strong public procurement and restrictions on new licence plates for ICEVs, in addition to the substantial grants and other non-monetary incentives for PEVs that are common to other countries as well, that has made it possible for the Chinese market to grow faster than any other in the world in the last few years.

It is therefore apparent from the above discussion that it would be unwise for fast-following countries to set themselves targets similar to those already achieved by the Norway and China, at least not without being prepared to put in place similarly extreme incentives. A mixed-method research approach that addresses the interplay among technology, consumers and policy, such as that developed by Jonn Axsen and co-workers and discussed in Chapter "The Present and Future Market for PEVs in Canada: Evidence from a Mixed-Method Research Program", can be very valuable to support governments in setting realistic targets that are consistent with the type and intensity of the policy measures deployed.

How Long Do Governments Need to Incentivise PEVs?

Last but not least, the study of actual PEV adopters and users can shed some light on the very important question of how long government support is likely to be needed in order for the PEV market to become self-sustained. As discussed in several chapters of this book, PEV markets have been strongly driven by government support since their inception and they continue to be. As PEV sale volumes grow and battery price reduces, governments are becoming more ambitious and are setting themselves increasingly aggressive targets. At the same time though, they are starting to rein in the monetary incentives that they have been offering and some of the non-monetary incentives too, and in doing so are running the risk of slowing down the growth of the market and possibly even stalling it.

Striking the right balance between on the one hand offering support that is both effective and fiscally efficient, and on the other knowing when and how to start reining in incentives is very difficult. Those governments that in the past have ended up over-subsidising renewable power generation may now be particularly keen to reduce support to PEVs as early as they possibly can, and are encouraged in doing so by the falling cost of batteries and the expectation that PEVs will approach cost parity with ICEVs in the next few years. However policymakers should consider that building consumer awareness and confidence in the technology is also essential and it requires time. It is on this latter point that the evidence presented in this book is relevant.

The technology adoption modelling work discussed by Jonn Axsen and Zoe Long in Chapter "The Present and Future Market for PEVs in Canada: Evidence from a Mixed-Method Research Program" shows how coupling strong demand-side policy measures with supply-side measures and sustaining both for long enough can have a dramatic impact on PEV penetration. Their results also suggest that current support measures in various countries may be insufficient to achieve the desired PEV penetration targets. In particular, Jonn and Zoe's work clearly shows how the lack of familiarity, access to home charging and vehicle model availability, among others, can be powerful constraints to demand for PEVs. Jonn and Zoe also show how incentives that are either too weak or do not remove all barriers can lead to a stagnation in the market, while the study of the Netherlands by Rick and Robert clearly proves how discontinuing incentives can make the market collapse. As Rick and Robert argue, the support to PHEVs in the Netherlands is hardly a case of balanced, stable support and has been scrutinised for both over-subsidisation (2013–2015) and under-subsidisation (2017–2018).

The case of China is also interesting: as Shawn Ou and Zhenhong Lin discuss in Chapter "China's Booming Plug-in Electric Vehicle Market—How Will It Continue?", the very incentives that have made the record growth of the Chinese market possible are now either being phased out or are becoming less effective. In particular, the government is reining in the grants to PEV manufacturers while the registration privileges for PEVs in megacities are approaching the cap posed on this measure. At the same time the emphasis is shifting towards demand side measures and particularly the dual credit system, which combines fuel economy standards with a PEV mandate. The question though is whether the combination of the dual credit system and substantially weaker demand side measures will be sufficient to drive the sustained market growth that the Chinese government is targeting. The results of the survey-based study conducted by Yan Xing, Gil Tal and Yunshi Wang (Chapter "Plug-in Electric Vehicle Market Research on Consumer in Six Cities in China") suggests that under a scenario where the PEV registration privilege is no longer available, some segments of current BEV buyers would go back to ICEVs.

What Next?

In this book we have discussed the substantial amount of evidence that is emerging from the study of the behaviour of PEV adopters as well as of mainstream car users across several leading markets worldwide. The studies presented paint a complex picture of consumers and their PEV adoption and use behaviour. The book also clearly shows how consumer behaviour continues to evolve, as a result of individual learning, interpersonal communication and the evolution of technology and policy. In order to set attainable penetration targets and design support policies that are both effective and efficient, a thorough understanding of the heterogeneity of the market and the complex nature of the PEV diffusion process is needed. The book has shed light on many of the difficult questions currently faced by policymakers and other stakeholders. It has also raised new questions and it has pointed at several areas for future research. The importance of continuing to study the adoption and use of PEVs, and of feeding the new knowledge so generated into the policy process, is clear. As Jonn Axsen and Zoe Long put it in their chapter, researching PEV adoption is itself a reflexive process where the researchers learn as they go along, just like the consumers who interact with PEVs do.

Future research will undoubtedly benefit from the rapidly growing volume of data that can be gathered from actual PEV adopters and users. As adoption progresses, time series data will become increasingly available which will allow to study the evolution of the market using quantitative methods, thus adding to the insights already generated on adoption decision-making using qualitative methods and quantitative analysis of cross-sectional data. More consumer data on real charging behaviour will help design more efficient and effective charging infrastructures. Future research will also increasingly resort to mixed-methods approaches, examples of which are provided in several chapters of this book. Mixed-method research requires different expertise and therefore more collaboration across research teams will be essential. Similarly, more international collaboration on multi-disciplinary projects will be very useful. At the same time, the study on non-adopters and rejecters should also continue, using the necessary strategies to control for possible biases as discussed by Stephen Skippon and George Beard in their chapter.

Just as it is important that in future we continue to research the same questions addressed in this book using better data and methods, this book has pointed at several other questions that would be worth tackling too. In particular, it would be important to develop a better understanding of how consumers define and perceive BEV range. Similarly, their perceptions of PHEVs, both positive and negative, would need to be better understood. Consumer response to non-monetary incentives is not yet well understood and more work would be needed there too. Another aspect that deserves further investigation is how PEV users adapt their driving style and patterns over time. More research should also be done seeking to better understand how communication between PEV users and users of non-PEVs works and how it can be facilitated. These are just some of the questions that follow from the research presented in this book. Not explicitly considered in this book but very important too is how owning a

PEVs may impact modal choices, particularly in relation to public transport. More work can also be done exploring how PEV charging behaviour may evolve in future, particularly as a result of the interaction of PEV users with new technologies such as managed charging, wireless charging and vehicle-to-grid (V2G).

Lastly, it is paramount that experts devote substantial efforts to making this field of research, its methods, insights and future directions more accessible to policymakers. This is necessary if policymakers are to design and implement the kind of adaptive policies that will deliver high levels of electrification of passenger car transport rapidly and efficiently.

The manufacturer's authorised representative in the EU is Springer Nature Customer Service Centre GmbH, Europaplatz 3, 69115 Heidelberg, Germany. If you have any concerns regarding our products, please contact ProductSafety@springernature.com

Printed and bound by CPI Group (UK) Ltd, Croydon, CR0 4YY
30/09/2026
02225647-0009